# Genetics and Breeding of Edible Mushroom

# Genetics and Breeding of Edible Mushroom

Guest Editor

**Xiangli Wu**

MDPI

Basel • Beijing • Wuhan • Barcelona • Belgrade • Novi Sad • Cluj • Manchester

*Guest Editor*
Xiangli Wu
Institute of Agricultural
Resources and Regional
Planning
Chinese Academy of
Agricultural Sciences
Beijing
China

*Editorial Office*
MDPI AG
Grosspeteranlage 5
4052 Basel, Switzerland

This is a reprint of the Special Issue, published open access by the journal *Agriculture* (ISSN 2077-0472), freely accessible at: https://www.mdpi.com/journal/agriculture/special_issues/71C78891Z3.

For citation purposes, cite each article independently as indicated on the article page online and as indicated below:

| Lastname, A.A.; Lastname, B.B. Article Title. *Journal Name* **Year**, *Volume Number*, Page Range. |
|---|

**ISBN 978-3-7258-5325-0 (Hbk)**
**ISBN 978-3-7258-5326-7 (PDF)**
**https://doi.org/10.3390/books978-3-7258-5326-7**

# Contents

agriculture

MDPI

Article

# Construction of an SNP Fingerprinting Database and Population Genetic Analysis of *Auricularia heimuer*

**Kaisheng Shao [1], Qiuyu Feng [1], Fangjie Yao [1,2,*], Lixin Lu [2], Ming Fang [2], Xiaoxu Ma [2] and Xu Sun [1]**

1 Engineering Research Centre of Chinese Ministry of Education for Edible and Medicinal Fungi, Jilin Agricultural University, Changchun 130118, China; shaoks217@163.com (K.S.); 17643420266@163.com (Q.F.); sunxu0512@163.com (X.S.)
2 Laboratory of the Genetic Breeding of Edible Mushroom, College of Horticulture, Jilin Agricultural University, Changchun 130118, China; lixinl@jlau.edu.cn (L.L.); fangming@jlau.edu.cn (M.F.); maxiaoxu@jlau.edu.cn (X.M.)
* Correspondence: yaofj@jlau.edu.cn

**Abstract:** *Auricularia heimuer* is the second most widely cultivated edible fungus in China, with significant food and medicinal value, and is highly popular throughout Asia and globally. However, the differentiation of *A. heimuer* is simple, as its morphology is characterized by a small "black disc", making it difficult to distinguish among germplasms with highly similar agronomic traits, thus posing challenges for germplasm identification. To address this issue, this study conducted whole-genome resequencing analysis on 150 *A. heimuer* germplasms. Through filtering 9,589,911 SNPs obtained from 280 G resequencing data, a total of 1,202,947 high-quality SNP sites were identified. Based on these high-quality SNPs, population structure analysis, principal component analysis (PCA), and phylogenetic tree analysis revealed that the 150 *A. heimuer* germplasms could be divided into five groups, with wild strains from the same geographical origin exhibiting significant geographical clustering patterns. This finding underscores the relationship between the genetic diversity of wild *A. heimuer* and its geographical distribution in China. A further selection of 71 SNP sites was made, and 61 KASP markers were successfully developed using kompetitive allele-specific PCR (KASP) technology, with 54 of them demonstrating good polymorphism. The average values for the polymorphism information content (PIC), minor allele frequency (MAF), gene diversity, and heterozygosity of these core KASP markers were 0.34, 0.35, 0.34, and 0.43, respectively. Based on the 54 core KASP markers, a DNA fingerprinting map of the 150 *A. heimuer* germplasms was constructed in this study. The findings provide important molecular marker resources and theoretical support for the identification of *A. heimuer* germplasm, molecular marker-assisted breeding, and the selection of superior varieties.

**Keywords:** *Auricularia heimuer*; SNP; KASP; DNA fingerprinting; population genetics

## 1. Introduction

*Auricularia heimuer* is an important edible and medicinal fungus worldwide [1]. Due to its good flavor [2], high nutritional value [3,4], and strong anticancer effects [5], it has become increasingly popular in daily diets. The production of *A. heimuer* has been continuously increasing over recent decades, reaching 7.0643 million tons in 2021, making it the second-largest edible fungus in China [6]. However, due to the simple differentiation of *A. heimuer*, characterized only by a black ear-like disk, it is difficult to distinguish among germplasm resources with similar agronomic traits and other phenotypic characteristics. This poses significant challenges for germplasm identification and breeding. Therefore, conducting investigations and collections of *A. heimuer* germplasm resources, subsequent genetic diversity assessment and DNA fingerprinting are essential for accurately characterizing germplasm resources and facilitating the breeding of high-performing varieties.

With the development of molecular biology techniques, significant progress has been made in molecular marker research. Due to their resistance to environmental influences, abundant polymorphism, and suitability for high-throughput detection, molecular markers have been widely applied in various research areas such as germplasm identification and variety improvement [7]. Single nucleotide polymorphisms (SNPs), as third-generation molecular markers, have become widely adopted in agricultural genetic studies due to their low cost, high throughput, strong reproducibility, even genomic distribution, and ease of automated data collection [8–10]. Currently, the International Union for the Protection of New Varieties of Plants (UPOV) uses SNP as the molecular detection guideline [11].

Kompetitive allele-specific PCR (KASP), a competitive amplification technique based on allele-specific PCR, allows the design of specific composite markers for a given population, or the development of functional molecular markers based on specific functional loci. This provides great flexibility in the detection loci and sample types [12,13]. Compared with other SNP genotyping approaches, KASP technology stands out for its high precision, cost-effectiveness, flexibility across diverse experimental conditions, and suitability for large-scale SNP screening [14]. These characteristics make KASP technology suitable for crop gene identification, genetic diversity analysis, and the construction of fingerprint maps. The construction of DNA fingerprint maps is crucial for ensuring the varietal specificity and species authenticity of edible fungi [15,16]. Currently, KASP technology has been widely used in the construction of fingerprint maps for crops such as wheat [17], rice [18], cucumber [19], and cigar tobacco [16]. In edible fungi, 12 core SNP markers were selected from 60 *Grifola frondosa* germplasms using KASP genotyping technology, and a fingerprint map of *G. frondosa* germplasm resources was constructed [20]. However, there have been no reports on the construction of fingerprint maps using SNP markers in *A. heimuer* to date.

In this study, whole-genome resequencing data from 150 *A. heimuer* germplasm accessions were employed to identify high-confidence SNP loci. Utilizing these SNPs, we conducted comprehensive analyses of population structure and genetic diversity within the *A. heimuer* collection. Polymorphic SNPs with uniform genomic distribution were selected and subsequently converted into KASP markers. These markers were used to genotype all 150 accessions, enabling the identification of core loci for constructing a germplasm DNA fingerprinting system. The resulting fingerprint map provides a robust molecular tool for variety identification and offers a valuable reference for the conservation and genetic improvement of *A. heimuer* germplasm in future breeding efforts.

## 2. Materials and Methods

### *2.1. Materials*

Test Strain

A total of 150 *A. heimuer* germplasms were collected for this experiment, sourced from 14 provinces across China, as well as two wild germplasms collected from Russia and Korea. All wild germplasms were collected by our team for many years. Strains are preserved at the College of Horticulture, Jilin Agricultural University (Table S1). The mycelia were inoculated onto potato dextrose agar (PDA) medium and placed in a constant-temperature incubator at 25 °C for dark cultivation until the mycelia covered the entire Petri dish. The mycelia were then collected with a spatula into centrifuge tubes for further use. This experiment was conducted in March 2024.

### 2.2. Methods

#### 2.2.1. DNA Extraction and Library Construction

Genomic DNA was extracted from *A. heimuer* mycelia using the CTAB methods [21]. The extracted DNA was subjected to quality control tests and DNA samples that met the quality criteria (>3 μg; concentration >30 ng/μL; OD260/OD280 = 1.80–2.00) were used for further analysis. The extracted DNA was randomly fragmented using ultrasonic treatment, followed by end repair. The End Repair Mix2 from the kit was used to remove the protruding bases at the 5′ end of the DNA sequence, add a phosphate group, and fill in any missing bases at the 3′ end. To prevent self-legation of DNA fragments and facilitate the ligation with sequencing adapters (which carry a protruding T base at their 3′ end), an A nucleotide was added to the 3′ ends of the DNA inserts. Sequencing adapters containing unique index sequences were then ligated to the 5′ ends to enable anchoring of DNA molecules onto the flow cell during sequencing. Following adapter ligation, magnetic bead-based selection using BECKMAN AMPure XP (Indianapolis, IN, USA). Beads was carried out to eliminate adapter dimers and purify the library. The adapter-tagged DNA fragments were subsequently amplified via PCR to enrich the sequencing library. A second round of purification using the same bead system was performed to remove PCR byproducts. Finally, target fragments were size-selected and further purified by 2% agarose gel electrophoresis to obtain the final library for sequencing.

#### 2.2.2. Variant Detection and Annotation

The reference genome data for *A. heimuer* were obtained from the NCBI database (https://www.ncbi.nlm.nih.gov). High-quality sequencing reads (clean reads) were aligned to the reference genome sequence using BWA software (v.0.7.17) [22], generating a sequence alignment map (SAM format file). The SAM file was then sorted and merged using SAMtools, and duplicate reads were removed using Picard. Based on the alignment results, sequencing depth and coverage were calculated using a custom Perl script. The valid BAM files underwent SNP detection through the HaplotypeCaller module of GATK [23], generating Variant Call Format (VCF) files, which were then quality filtered using the VCFtools Variant Filtration function (parameters: QD < 2.0 | | FS > 60.0 | | MQ < 40.0 | | SOR > 10.0). Detected variants were annotated using ANNOVAR (version released on 8 June 2020), including SNPs (synonymous/non-synonymous mutations), while structural variants (SVs) were identified using Break Dancer. High-quality SNPs were selected from the entire database based on the following filtering criteria: average coverage depth > 5×, minor allele frequency (MAF) > 0.05, average quality value (AverageQ) > 30, and minimum integrity > 0.9. The filtering tools used were BCFtools and VCFtools [24].

#### 2.2.3. Genetic Diversity and Population Structure Analysis

Cluster analysis was performed using the neighbor-joining method in MEGA7.0.26 software [25], and a cluster tree was constructed. Principal component analysis (PCA) of the *A. heimuer* germplasms was conducted using NTSYS-PC 2.10 software. The population genetic structure was analyzed using the admixture software (v1.3.0), with the range for the optimal number of population clusters (K) set from 2 to 9. Each K value was run five times, and the appropriate K value was selected based on the principle of the lowest error rate, which determined the number of populations.

#### 2.2.4. Design of KASP Markers and Genotyping

To develop KASP markers, candidate SNP sites were extracted along with 100 bp of flanking sequence on both sides. These sequences were aligned to the reference genome using BLASTN to exclude non-specific regions, ensuring marker specificity. For each selected SNP locus, two allele-specific forward primers and one common reverse primer were designed. Fluorescent tags were incorporated at the 5′ ends of the primers: F1 (FAM): GAAGGTGACCAAGTTCATGCT and F2 (VIC): GAAGGTCGGAGTCAACGGATT. The primers were synthesized and diluted to a final concentration of 10 μM using TE buffer (pH 8.0). Prior to genotyping, they were mixed in a 1:1:3 ratio of the two allele-specific primers to the common primer and then used for PCR amplification on the KASP platform. For each 5 μL reaction system, 1.25 μL of the primer mixture was added. DNA samples were diluted in bulk to a concentration corresponding to the lowest sample concentration, with 1.25 μL of the diluted DNA sample added to each 5 μL reaction system. The PCR reaction system consisted of 5 μL, including 2.5 μL of 2 × KASP Master Mix, 1.25 μL of the KASP primer mix, and 1.25 μL of template (30 ng genomic DNA). The 96-well PCR reaction plate was sealed, shaken, and centrifuged to ensure proper mixing of the reaction system. After centrifugation, PCR was performed with the following conditions: 95 °C for 10 min for pre-denaturation; 95 °C for 20 s for denaturation, followed by 61–55 °C for 60 s for annealing and extension in 10 Touchdown cycles (each cycle reducing by 0.6 °C); the second round of PCR was performed with 95 °C for 20 s for denaturation and 55 °C for 60 s for annealing and extension, for 30 cycles. After the reaction, fluorescence detection was conducted using the BMG POLARstar Omega 5.10 R2 reader, and the data were analyzed using KlusterCaller software (LGC Biosearch Technologies, Hoddesdon, UK).

#### 2.2.5. DNA Fingerprinting Map Construction

Based on the obtained high-quality SNPs, DNA fingerprinting was constructed using a Perl script [26]. The aim was to identify as many varieties as possible using the fewest markers, achieving simplicity, effectiveness, and cost-efficiency [27]. Core markers were selected based on their high detection efficiency, pronounced polymorphism, and capacity to differentiate all tested varieties. Selection criteria included PIC values and allele frequency distribution. The finalized set of core markers was subsequently employed to establish the DNA fingerprinting profile for the *A. heimuer* germplasm.

## 3. Results

### 3.1. *Quality Control and Alignment Analysis of Resequencing Data*

In this study, resequencing analysis was conducted on 150 *A. heimuer* strains, yielding a total of 280 G of data. The data for each sample were statistically analyzed, with an average GC content of 54.53%, an average Q20 value of 98.40%, and an average Q30 value of 95.52%. Detailed results can be found in Table S2. The sequencing data included some adapter-containing and low-quality reads, which required further filtering. The basic information on data filtering is presented in Table S3. The high-quality data obtained after filtering were aligned to the reference genome, with an average alignment rate of 90.09% and an average sequencing depth of 34.47 X. Sequencing depth (Figure 1A) and cumulative sequencing depth (Figure 1B), as well as sequence alignment results and sequencing depth distribution, are shown in Table S4 and Figure S1. The sequencing quality is high, allowing for subsequent analysis.

### 3.2. *SNP Detection and Annotation*

SNP detection was performed using GATK software (v.4.6.0.0) [28]. To ensure the reliability of the SNP sites, further filtering was applied to the obtained SNPs, resulting in 1,202,947 high-quality SNPs for subsequent analysis. Among these, 630,506 were transitions and 571,991 were transversions, with a transition-to-transversion ratio of 1:1.1. These high-quality SNPs are ideal for constructing the *A. heimuer* fingerprint map and conducting genetic diversity analysis. Furthermore, a distribution analysis of SNPs in the coding regions and the entire genome revealed that 71.99% of SNPs in the coding regions were synonymous mutations, and 27.32% were non-synonymous mutations (Figure 2A). In the genome, SNPs were distributed as follows: 12.04% in intergenic regions, 34.86% in introns, 22.17% in exons, 12.95% in the upstream 1 kb region of the transcription start site, and 11.65% in the downstream 1 kb region of the transcription termination site. Additionally, 6.23% of SNPs were found in both the upstream 1 kb region of one gene and the downstream 1 kb region of another gene (Figure 2B). Finally, an SNP distribution map was constructed based on the number and density of SNPs (Figure 2C). The darker the color in the map, the higher the concentration of SNPs.

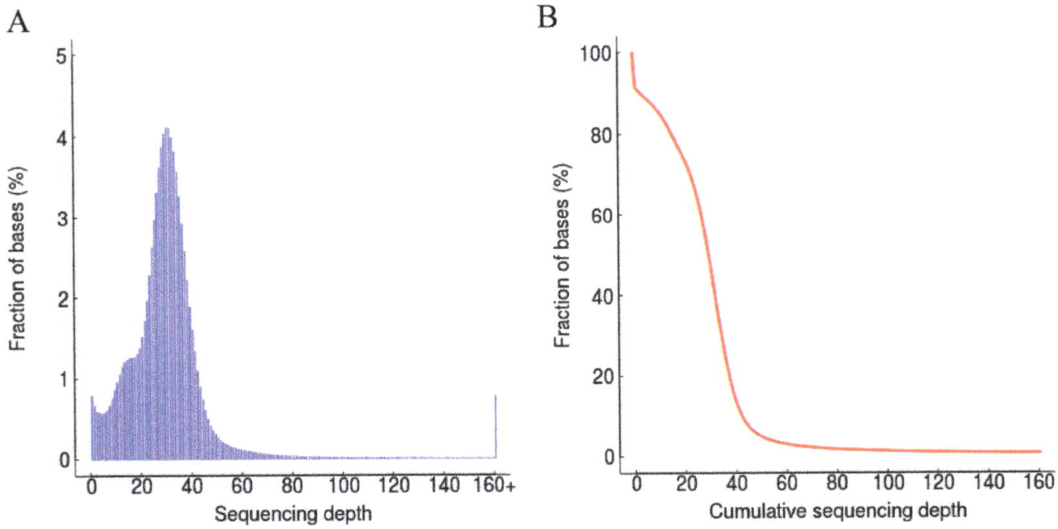


**Figure 1.** Sequencing depth and coverage of 150 *A. heimuer* strains. (**A**) Distribution of sequencing depths of the strains. (**B**) Proportion of accumulated bases at different sequencing depths.

### *3.3. Genetic Diversity and Population Structure Analysis*

Based on the high-quality SNP data obtained from resequencing, a phylogenetic tree of 150 *A. heimuer* germplasms was constructed using the neighbor-joining method (Figure 3A). The 150 germplasms were divided into five major groups, named pop-1, pop-2, pop-3, pop-4, and pop-5. Wild strains from Gansu Province were clustered in the pop-2 group, wild strains from Jilin were mainly clustered in the pop-3 group, the pop-4 group contained most of the wild strains from Yunnan, and wild strains from Inner Mongolia were predominantly grouped in the pop-5 group. This suggests that the genetic diversity of wild *A. heimuer* in China may be associated with its geographical distribution. The population structure of *A. heimuer* was analyzed using the admixture software, and the optimal number of clusters (K) was determined based on the cross-validation (CV) error value (Figure 3B). Line plots were drawn for each K value, and the CV error was repeated 10 times. The optimal number of clusters was determined by the minimum cross-validation error rate. When K = 5, the minimum cross-validation error rate was 0.27, indicating that the 150 strains could be divided into five groups. The different colors in the plot suggest that the genetic background of the tested *A. heimuer* strains is complex, with different strains being both interconnected and independent (Figure 3C,D). Based on the genotypic data, we further performed principal component analysis (PCA), which revealed that the 150 *A. heimuer* strains were divided into four groups. Specifically, pop-1 and pop-4 were grouped together, and wild strains from Gansu were clustered in the pop-2 group.

This suggests a correlation between the PCA results and the geographic origin of the germplasm. The first and second principal components explained 19.69% and 9.02% of the data variation. The results show that the five groups interact closely with each other, and wild *A. heimuer* strains from the same geographic origin are clustered in the same group. The 150 strains are genetically closely related and exhibit a certain correlation with the geographic origin of the germplasm.

**Figure 2.** *A. heimuer* genome SNP analysis. (**A**) The number of different types of SNPs in the coding regions. (**B**) The number of SNPs in different genomic regions. (**C**) SNP density distribution across each chromosome. The *x*-axis represents chromosome length, while the *y*-axis represents the number of chromosomes. Different colors represent the number of SNPs in different regions.

### 3.4. Conversion and Selection of KASP Markers

Based on the resequencing results, the total SNP sites obtained were filtered with the following criteria: removal of SNPs with a missing rate greater than 0.1, a missing value less than 0.05, minor allele frequency (MAF) > 0.05, and heterozygosity (het) < 0.1. A total of 141,573 SNP sites were selected, which were evenly distributed across the *A. heimuer* genome. The polymorphism information content (PIC) can be used to measure the polymorphism of DNA molecular markers in a population. Further filtering was performed with PIC > 0.2 and Pi > 0.4, resulting in 167 high-quality SNP sites. Additional screening for SNPs with no missing genotypic data, predominantly homozygous variation, and detected in as many individuals as possible led to the selection of 71 SNP sites. KASP primers were designed for 71 SNP sites, and 61 (85.9%) successfully converted into KASP markers. The PIC values of the 61 KASP markers ranged from 0.221 to 0.405, with an average of 0.342. Among the 61 core markers, four had PIC values lower than 0.3. The average MAF value of the 61 markers was 0.353, ranging from 0.271 to 0.489. The average observed heterozygosity was 0.428, and the average gene diversity was 0.344, with a range of 0.201 to 0.492. These results indicate that the 61 KASP markers exhibit sufficient polymorphism (Figure 4). The 61 KASP markers are highly reliable for genetic diversity analysis of *A. heimuer* germplasm resources.

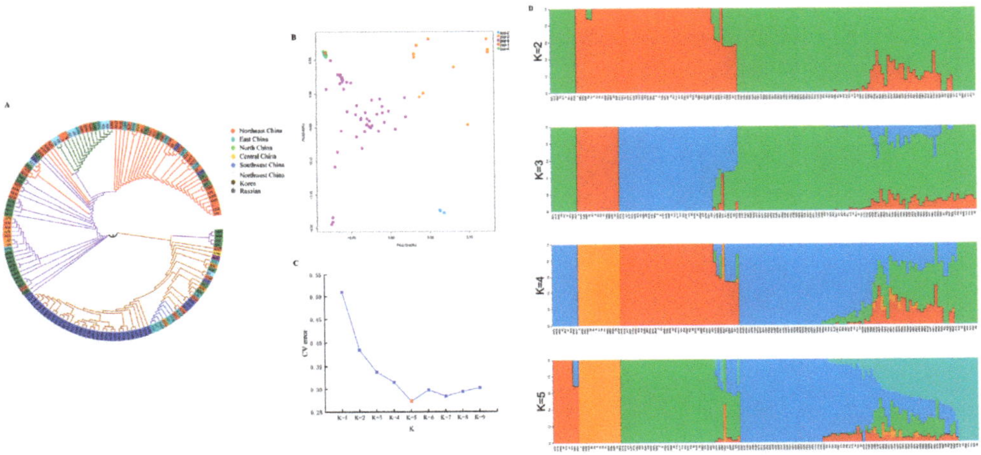


**Figure 3.** Genetic analysis of 150 *A. heimuer* germplasm based on SNP polymorphic sites. (**A**) Phylogenetic tree of 150 *A. heimuer* strains. (**B**) Principal component analysis (PCA). (**C**) Cross-validation error rates corresponding to different K values. (**D**) Population structure analysis. Note: The populations are represented with consistent colors in (**A**,**D**), facilitating comparison of the performance of each population across different analyses.

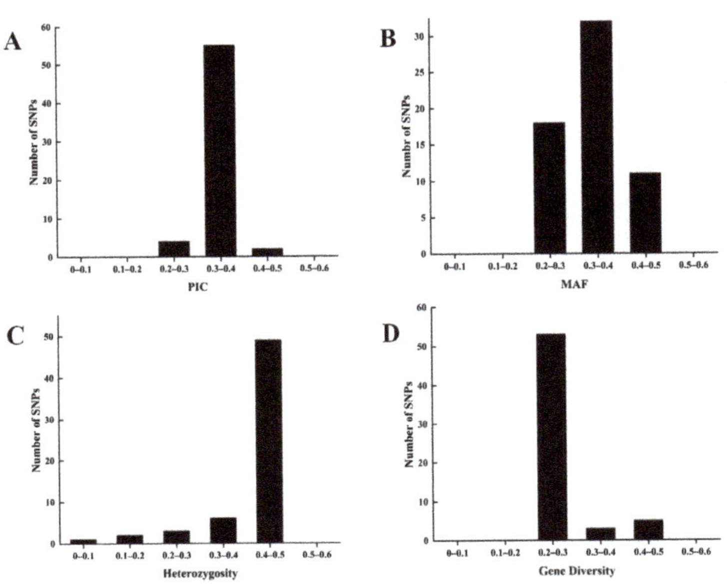


**Figure 4.** Genetic information content based on 61 KASP markers. (**A**) PIC. (**B**) MAF. (**C**) Heterozygosity. (**D**) Genetic diversity.

### 3.5. Genotyping Validation of KASP Markers

Genotyping validation of the 61 KASP markers was conducted by genotyping 150 *A. heimuer* germplasms using KASP primers. Based on the genotyping results, 54 high-accuracy KASP markers were selected. Figure 5 shows the genotyping results for some KASP markers, where red and blue dots represent two different homozygous genotypes, and green dots represent heterozygous genotypes. Based on the genotyping results, one marker failed to amplify, and six markers were discarded due to monomorphism or missing rates exceeding 10%. Ultimately, 54 high-quality KASP markers were retained as the core set of KASP markers (Table 1). Table S5 provides detailed information on the 54 KASP markers, including marker names, locations, mutation types, and primer sequences.

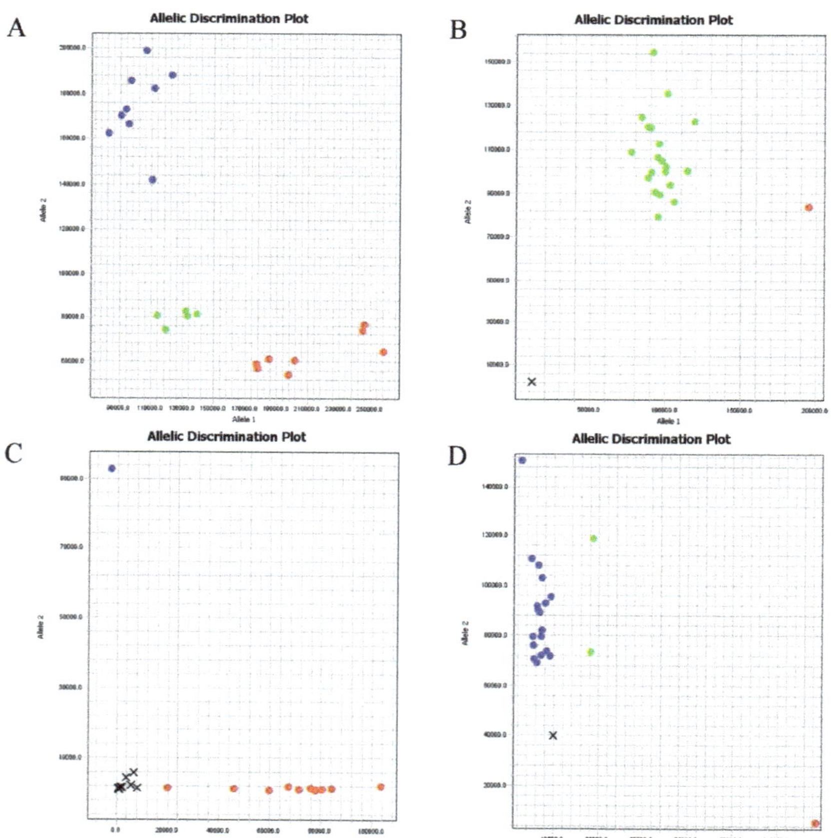


**Figure 5.** Fluorescence determination results of representative KASP markers. (**A**) KASP marker with good genotyping. (**B**) KASP marker with heterozygous monomorphic pattern. (**C**) KASP marker with monomorphic pattern and high missing rate. (**D**) KASP marker with homozygous monomorphic pattern. Note: Red dots (homozygous type 1), blue dots (homozygous type 2), green dots (heterozygous type) and black "×" (negative control).

**Table 1.** Information of 54 pairs of core KASP primers. Note: Ref = reference allele; Alt = alternative allele; PIC = polymorphism information content; MAF = minor allele frequency.

| Marker ID | Position | Ref | Alt | PIC | MAF | Heterozygosity | Gene Diversity |
|---|---|---|---|---|---|---|---|
| rs1 | 545,620 | C | T | 0.320 | 0.276 | 0.399 | 0.200 |
| rs2 | 566,712 | G | A | 0.321 | 0.279 | 0.402 | 0.201 |
| rs3 | 1,351,884 | G | T | 0.363 | 0.392 | 0.477 | 0.238 |
| rs4 | 2,235,372 | T | C | 0.363 | 0.390 | 0.476 | 0.338 |
| rs5 | 1,542,598 | T | C | 0.364 | 0.395 | 0.478 | 0.239 |
| rs6 | 1,543,068 | C | G | 0.363 | 0.392 | 0.477 | 0.238 |
| rs7 | 1,061,751 | A | G | 0.362 | 0.388 | 0.475 | 0.237 |
| rs8 | 545,005 | A | G | 0.365 | 0.403 | 0.481 | 0.241 |
| rs9 | 591,235 | G | T | 0.372 | 0.450 | 0.495 | 0.447 |
| rs10 | 220,084 | T | C | 0.360 | 0.381 | 0.472 | 0.236 |
| rs11 | 964,656 | T | C | 0.375 | 0.479 | 0.499 | 0.450 |
| rs12 | 1,145,038 | G | A | 0.354 | 0.359 | 0.460 | 0.230 |
| rs13 | 83,293 | T | G | 0.368 | 0.420 | 0.487 | 0.244 |
| rs14 | 268,204 | T | A | 0.355 | 0.362 | 0.462 | 0.231 |
| rs15 | 375,892 | G | T | 0.365 | 0.403 | 0.481 | 0.241 |
| rs16 | 822,184 | G | T | 0.331 | 0.299 | 0.419 | 0.309 |
| rs17 | 60,862 | G | A | 0.342 | 0.323 | 0.437 | 0.219 |
| rs18 | 418,050 | G | A | 0.324 | 0.284 | 0.407 | 0.203 |
| rs19 | 888,504 | T | C | 0.367 | 0.414 | 0.485 | 0.243 |
| rs20 | 652,372 | A | G | 0.321 | 0.279 | 0.402 | 0.301 |
| rs21 | 638,390 | G | A | 0.373 | 0.455 | 0.496 | 0.448 |
| rs22 | 243,519 | G | A | 0.338 | 0.315 | 0.432 | 0.216 |
| rs23 | 317,039 | G | A | 0.361 | 0.385 | 0.473 | 0.237 |
| rs24 | 318,102 | G | A | 0.364 | 0.396 | 0.478 | 0.239 |
| rs25 | 319,778 | C | T | 0.363 | 0.392 | 0.477 | 0.238 |
| rs26 | 502,236 | T | C | 0.357 | 0.370 | 0.466 | 0.233 |
| rs27 | 168,451 | T | C | 0.360 | 0.380 | 0.471 | 0.236 |
| rs28 | 63,914 | T | C | 0.353 | 0.354 | 0.457 | 0.229 |
| rs29 | 413,082 | A | G | 0.324 | 0.283 | 0.406 | 0.203 |
| rs30 | 330,071 | G | A | 0.373 | 0.458 | 0.497 | 0.248 |
| rs31 | 80,160 | A | G | 0.329 | 0.294 | 0.415 | 0.207 |
| rs32 | 150,530 | C | A | 0.361 | 0.385 | 0.473 | 0.237 |
| rs33 | 152,642 | G | A | 0.363 | 0.392 | 0.477 | 0.238 |
| rs34 | 226,439 | C | T | 0.329 | 0.293 | 0.415 | 0.207 |
| rs35 | 19,582 | C | A | 0.332 | 0.300 | 0.420 | 0.210 |
| rs36 | 40,366 | A | T | 0.358 | 0.373 | 0.468 | 0.234 |
| rs37 | 68,568 | A | T | 0.343 | 0.327 | 0.440 | 0.220 |
| rs38 | 83,938 | A | T | 0.329 | 0.293 | 0.415 | 0.207 |
| rs39 | 146,075 | T | G | 0.362 | 0.387 | 0.474 | 0.437 |
| rs40 | 150,233 | C | A | 0.322 | 0.280 | 0.403 | 0.202 |
| rs41 | 151,011 | A | T | 0.325 | 0.287 | 0.409 | 0.204 |
| rs42 | 149,869 | A | G | 0.370 | 0.430 | 0.490 | 0.245 |
| rs43 | 33,848 | T | C | 0.359 | 0.375 | 0.469 | 0.234 |
| rs44 | 62,732 | T | C | 0.351 | 0.347 | 0.453 | 0.227 |
| rs45 | 63,398 | G | A | 0.351 | 0.348 | 0.454 | 0.227 |
| rs46 | 64,460 | A | G | 0.348 | 0.339 | 0.448 | 0.224 |
| rs47 | 84,293 | C | G | 0.335 | 0.307 | 0.425 | 0.213 |
| rs48 | 20,830 | C | T | 0.339 | 0.317 | 0.433 | 0.217 |
| rs49 | 173,979 | C | T | 0.249 | 0.277 | 0.292 | 0.246 |
| rs50 | 864,211 | A | C | 0.128 | 0.422 | 0.137 | 0.269 |
| rs51 | 2,022,315 | G | A | 0.375 | 0.489 | 0.500 | 0.250 |
| rs52 | 227,420 | T | C | 0.251 | 0.393 | 0.295 | 0.247 |
| rs53 | 101,264 | A | C | 0.124 | 0.277 | 0.133 | 0.266 |
| rs54 | 364,281 | T | G | 0.093 | 0.293 | 0.098 | 0.249 |

### *3.6. Construction of DNA Fingerprinting*

By integrating a custom-developed set of 41 core KASP SNP markers with 150 strains of *A. heimuer* germplasm, a high-throughput molecular fingerprint database covering genetic diversity was constructed. Additionally, a genotype-based accurate identification system for *A. heimuer* varieties was established (Figure 6). In this system, each row represents an SNP genotype, and each column represents a sample. Yellow, green, blue, and purple represent nucleotide genotypes C/C, A/A, T/T, and G/G, respectively. Missing data and heterozygous loci are displayed in gray. These 41 KASP markers exhibit high polymorphism, strong identification capability, and can be directly applied for variety genotyping and identification.

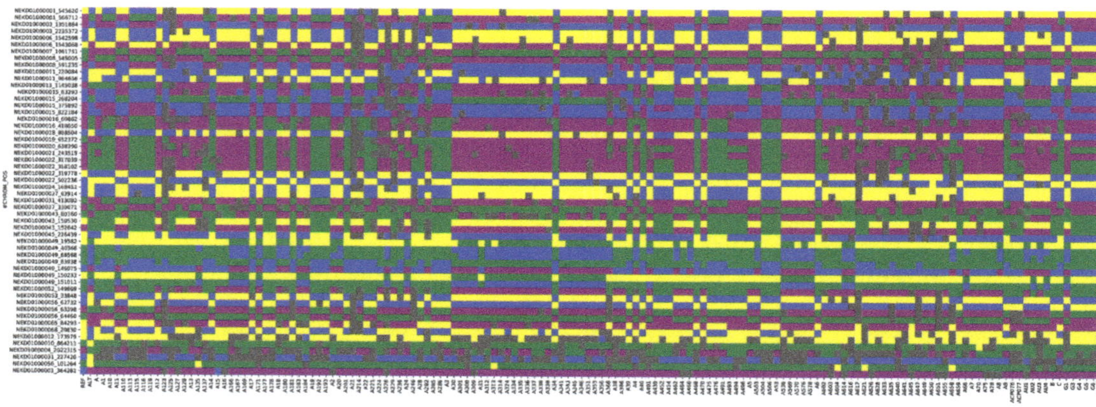

**Figure 6.** Fingerprint profile of 150 strains of *A. heimuer* germplasm. Each row represents an SNP locus, and each column represents a sample. Yellow, green, blue, and purple represent nucleotide genotypes C/C, A/A, T/T, and G/G, respectively. Missing and heterozygous data are displayed in gray.

## 4. Discussion

*A. heimuer*, an important edible and medicinal mushroom, holds significant nutritional value. However, the phenomenon of "nomenclature confusion" in *A. heimuer* and the lack of accurate identification methods for variety protection have, to some extent, hindered breeding efforts. To better utilize germplasm resources and protect the rights of *A. heimuer* varieties, it is essential to understand the phylogenetic relationships and population genetic structure of different varieties at the genomic level. With the development of DNA molecular marker technologies, various types of markers, such as RAPD, AFLP, SSR, and SCAR, have been applied to the identification and genetic diversity studies of *A. heimuer* germplasm resources [29–32]. However, these markers have gradually been phased out due to their instability and complexity in operation. With the advancement of high-throughput sequencing, single nucleotide polymorphisms (SNPs), due to their stability and high-throughput reproducibility, have become ideal genetic analysis markers. With the application of KASP technology, SNP markers have been widely used in genetic analysis, variety identification, and fingerprint profiling in crops such as potato [33], tomato [34], and cowpea [35]. However, a SNP fingerprinting system for *A. heimuer* has yet to be established, and there is an urgent need to develop an efficient and accurate fingerprinting platform for the authenticity and population genetic analysis of *A. heimuer* varieties.

PIC, MAF, gene diversity, and heterozygosity averages of the KASP markers developed in this study were 0.34, 0.35, 0.34, and 0.43, respectively, indicating a high level of genetic information and stability. Compared to traditional SSR markers, SNP markers offer higher coverage and resolution, allowing for a more accurate reflection of the genetic background of germplasm. The successful application of KASP technology has further enhanced the detection efficiency and practicality of SNP markers, providing a powerful tool for the rapid identification of *A. heimuer* germplasm resources and molecular marker-assisted breeding.

*A. heimuer* germplasm resources are a crucial material foundation for the breeding of new *A. heimuer* varieties and genetic theoretical research. China possesses a rich diversity of *A. heimuer* germplasm resources. Accurate identification of these resources is beneficial for their collection and conservation, as well as for the full development and study of *A. heimuer*. Genetic analysis of population structure can enhance the integration and utilization efficiency of germplasm resources. Yin et al. [36] analyzed 72 wild *A. heimuer* strains using 30 pairs of SSR primers, dividing them into six groups and selecting nine core SSR primers to construct a fingerprint profile for *A. heimuer*, providing important reference information for genetic breeding, germplasm resource conservation, and intelligent management. Furthermore, Jiao et al. [29] studied the genetic diversity and population structure of 52 wild *A. heimuer* germplasm resources using 13 pairs of EST-SSR primers, dividing them into three groups and discovering that the genetic diversity of *A. heimuer* from Northeast China was particularly rich. These studies have laid an important foundation for the scientific management and efficient utilization of *A. heimuer* germplasm resources. In this study, through population structure analysis, principal component analysis, and genetic phylogenetic tree analysis, 150 *A. heimuer* germplasm samples were divided into five groups. The strains in pop-1 and pop-4 are geographically proximate, which may facilitate gene flow and consequently result in similar genetic structures. This inference is also supported by the population structure plot. Wild strains from the same geographic region exhibited significant geographic clustering characteristics, a result consistent with the findings of Meng [37], suggesting that geographic isolation and environmental adaptability may be the primary factors driving genetic differentiation in *A. heimuer*. The correlation between geographic distribution and genetic diversity may stem from *A. heimuer*'s adaptive selection to different ecological environments during long-term evolution, such as variations in temperature, humidity, and substrate types. In addition, the limited natural dispersal ability of wild *A. heimuer* and geographic isolation may reduce gene flow, thereby intensifying genetic differentiation between populations. This finding provides scientific evidence for the regional protection and utilization of *A. heimuer* germplasm resources.

In this study, a set of 54 core KASP markers was employed to establish a DNA fingerprinting profile for 150 *A. heimuer* germplasm accessions, enabling the precise differentiation of all tested samples. These results offer a providing crucial support for the identification of new varieties and the application and protection of plant variety rights. Moreover, by identifying molecular markers associated with target traits, early selection of superior individuals becomes possible, thereby improving breeding efficiency. The genetic information obtained in this study can serve as a reference for future screening of elite germplasm and the design of breeding parent combinations. In the future, as the new *A. heimuer* germplasm resources in China continue to increase, more SNP markers will need to be integrated, and molecular markers tightly linked to key agronomic traits should be developed. Additionally, combining GWAS and phenotypic data will allow for further identification of genes associated with important agronomic traits in *A. heimuer*, as well as genes that regulate phenotypic trait variation, to promote the fine mapping of candidate functional regions.

Although this study has made significant progress, there are still some limitations. Firstly, while the sample size covers the major *A. heimuer* production areas in China, there is still a need to further expand the sample size, especially by increasing the collection of germplasm from peripheral distribution areas and foreign sources, to fully reveal the genetic diversity of *A. heimuer*. As germplasm resources are more widely collected, the germplasm bank is also continuously expanding, which greatly increases the management costs of the germplasm bank and the difficulty of selecting specific germplasm. At the same time, breeders are unable to effectively evaluate and identify large amounts of genetic material, leading to the loss of germplasm and a reduction in genetic diversity over time [38]. The future development of variety identification technology not only requires accuracy and speed but also demands simplicity and automation. Therefore, using SNP markers to identify varieties of *A. heimuer* in China's existing germplasm resources will help screen out redundant germplasm, ensuring the standardization and authenticity of the *A. heimuer* germplasm bank, while laying a foundation for future genetic breeding efforts. Currently, high-throughput SNP detection for large sample sizes has demonstrated clear advantages. In the future, as the number of varieties increases and SNP markers are further enriched, SNP detection technology will have a broad application prospect in the authentication and specificity identification of new varieties.

## 5. Conclusions

In this study, a whole-genome resequencing analysis of 150 *A. heimuer* germplasm samples was conducted, resulting in the identification of 1,202,947 high-quality SNP loci. Furthermore, 54 core KASP markers with good polymorphism were developed. Population structure analysis indicated that the *A. heimuer* germplasm could be divided into five groups, with genetic diversity significantly correlated with geographic distribution. The DNA fingerprinting constructed based on core KASP markers provides an efficient and accurate technical method for the identification of *A. heimuer* germplasm resources. The research findings lay an important theoretical foundation for the conservation of *A. heimuer* germplasm resources, molecular marker-assisted breeding, and the selection of superior varieties, and have significant scientific and practical value.

**Supplementary Materials:** The following supporting information can be downloaded at https://www.mdpi.com/article/10.3390/agriculture15080884/s1, Figure S1: Sequencing depth distribution map; Table S1: Information of 150 *A. heimuer* stains; Table S2: Sequencing data statistics; Table S3: Statistical results of sequence comparison; Table S4: SNP number and type of each sample statistics; Table S5: Fifty-four pairs of core marker information.

**Author Contributions:** F.Y. designed the experiments; L.L. and M.F. revised the manuscript; X.M. guided the experiment; X.S. and Q.F. conducted the formal analysis; K.S. prepared the materials for the experiments, analyzed the data, and wrote the manuscript. All authors have read and agreed to the published version of the manuscript.

**Funding:** This study was supported by the National Key Research and Development Program of China (No. 2023YFD1201604-4).

**Institutional Review Board Statement:** Not applicable.

**Data Availability Statement:** The data presented in this study are available in the article and Supplementary Materials.

**Acknowledgments:** The authors thank the reviewers for their valuable suggestions.

**Conflicts of Interest:** The authors declare no conflicts of interest.

## References

1. Wu, F.; Yuan, Y.; Malysheva, V.F.; Du, P.; Dai, Y.-C. Species clarification of the most important and cultivated Auricularia mushroom "Heimuer": Evidence from morphological and molecular data. *Phytotaxa* **2014**, *186*, 241–253. [CrossRef]
2. Islam, T.; Yao, F.; Kang, W.; Lu, L.; Xu, B. A systematic study on mycochemical profiles, antioxidant, and anti-inflammatory activities of 30 varieties of Jew's ear (*Auricularia auricula-judae*). *Food Sci. Hum. Wellness* **2022**, *11*, 781–794. [CrossRef]
3. Xu, S.; Zhang, Y.; Jiang, K. Antioxidant activity in vitro and in vivo of the polysaccharides from different varieties of *Auricularia auricula*. *Food Funct.* **2016**, *7*, 3868–3879. [CrossRef] [PubMed]

4. Islam, T.; Yu, X.; Xu, B. Phenolic profiles, antioxidant capacities and metal chelating ability of edible mushrooms commonly consumed in China. *LWT-Food Sci. Technol.* **2016**, *72*, 423–431. [CrossRef]
5. Yoon, S.-J.; Yu, M.-A.; Pyun, Y.-R.; Hwang, J.-K.; Chu, D.-C.; Juneja, L.R.; Mourão, P.A.S. The nontoxic mushroom *Auricularia auricula* contains a polysaccharide with anticoagulant activity mediated by antithrombin. *Thromb. Res.* **2003**, *112*, 151–158. [CrossRef]
6. Yao, F.; Zhang, Y.; Lu, L.; Fang, M. Research progress on genetics and breeding of *Auricularia auricula-judae*. *J. Fungal Res.* **2015**, *13*, 125–128+122. [CrossRef]
7. Oliveira, M.; Azevedo, L. Molecular Markers: An Overview of Data Published for Fungi over the Last Ten Years. *J. Fungi* **2022**, *8*, 803. [CrossRef] [PubMed]
8. Kumar, R.; Das, S.P.; Choudhury, B.U.; Kumar, A.; Prakash, N.R.; Verma, R.; Chakraborti, M.; Devi, A.G.; Bhattacharjee, B.; Das, R.; et al. Advances in genomic tools for plant breeding: Harnessing DNA molecular markers, genomic selection, and genome editing. *Biol. Res.* **2024**, *57*, 80. [CrossRef]
9. Tsykun, T.; Rellstab, C.; Dutech, C.; Sipos, G.; Prospero, S. Comparative assessment of SSR and SNP markers for inferring the population genetic structure of the common fungus *Armillaria cepistipes*. *Heredity* **2017**, *119*, 371–380. [CrossRef]
10. Li, H.; Shi, L.; Tang, W.; Xia, W.; Zhong, Y.; Xu, X.; Xie, B.; Tao, Y. Comprehensive Genetic Analysis of Monokaryon and Dikaryon Populations Provides Insight Into Cross-Breeding of *Flammulina filiformis*. *Front. Microbiol.* **2022**, *13*, 887259. [CrossRef]
11. Jördens, R. Progress of plant variety protection based on the International Convention for the Protection of New Varieties of Plants (UPOV Convention). *World Pat. Inf.* **2005**, *27*, 232–243. [CrossRef]
12. Dipta, B.; Sood, S.; Mangal, V.; Bhardwaj, V.; Thakur, A.K.; Kumar, V.; Singh, B. KASP: A high-throughput genotyping system and its applications in major crop plants for biotic and abiotic stress tolerance. *Mol. Biol. Rep.* **2024**, *51*, 508. [CrossRef] [PubMed]
13. Semagn, K.; Babu, R.; Hearne, S.; Olsen, M. Single nucleotide polymorphism genotyping using Kompetitive Allele Specific PCR (KASP): Overview of the technology and its application in crop improvement. *Mol. Breed.* **2014**, *33*, 1–14. [CrossRef]
14. Wang, F.-q.; Fan, X.-c.; Zhang, Y.; Sun, L.; Liu, C.-h.; Jiang, J.-f. Establishment and application of an SNP molecular identification system for grape cultivars. *J. Integr. Agric.* **2022**, *21*, 1044–1057. [CrossRef]
15. Yang, Y.; Lyu, M.; Liu, J.; Wu, J.; Wang, Q.; Xie, T.; Li, H.; Chen, R.; Sun, D.; Yang, Y.; et al. Construction of an SNP fingerprinting database and population genetic analysis of 329 cauliflower cultivars. *BMC Plant Biol.* **2022**, *22*, 522. [CrossRef]
16. Wang, Y.; Lv, H.; Xiang, X.; Yang, A.; Feng, Q.; Dai, P.; Li, Y.; Jiang, X.; Liu, G.; Zhang, X. Construction of a SNP Fingerprinting Database and Population Genetic Analysis of Cigar Tobacco Germplasm Resources in China. *Front. Plant Sci.* **2021**, *12*, 618133. [CrossRef]
17. Grewal, S.; Hubbart-Edwards, S.; Yang, C.; Devi, U.; Baker, L.; Heath, J.; Ashling, S.; Scholefield, D.; Howells, C.; Yarde, J.; et al. Rapid identification of homozygosity and site of wild relative introgressions in wheat through chromosome-specific KASP genotyping assays. *Plant Biotechnol. J.* **2020**, *18*, 743–755. [CrossRef]
18. Tang, W.; Lin, J.; Wang, Y.; An, H.; Chen, H.; Pan, G.; Zhang, S.; Guo, B.; Yu, K.; Li, H.; et al. Selection and Validation of 48 KASP Markers for Variety Identification and Breeding Guidance in Conventional and Hybrid Rice (*Oryza sativa* L.). *Rice* **2022**, *15*, 48. [CrossRef]
19. Zhang, J.; Yang, J.; Zhang, L.; Luo, J.; Zhao, H.; Zhang, J.; Wen, C. A new SNP genotyping technology Target SNP-seq and its application in genetic analysis of cucumber varieties. *Sci. Rep.* **2020**, *10*, 5623; Erratum in *Sci. Rep.* **2021**, *11*, 8010. https://doi.org/10.1038/s41598-021-86981-x. [CrossRef]
20. Diao, B.; Xu, Z.; Liu, M.; Zhang, G.; Wang, G.; Zhang, Y.; Tian, X. Establishment and application of a SNP molecular identification system in *Grifola frondosa*. *Front. Microbiol.* **2024**, *15*, 1417014. [CrossRef] [PubMed]
21. Zhou, X.; Li, Q.; Zhao, J.; Tang, K.; Lin, J.; Yin, Y. Comparison of rapid DNA extraction methods applied to PCR identification of medicinal mushroom *Ganoderma* spp. *Prep. Biochem. Biotechnol.* **2007**, *37*, 369–380. [CrossRef] [PubMed]
22. Li, H.; Durbin, R. Fast and accurate short read alignment with Burrows–Wheeler transform. *Bioinformatics* **2009**, *25*, 1754–1760. [CrossRef]
23. McKenna, A.; Hanna, M.; Banks, E.; Sivachenko, A.; Cibulskis, K.; Kernytsky, A.; Garimella, K.; Altshuler, D.; Gabriel, S.; Daly, M.; et al. The Genome Analysis Toolkit: A MapReduce framework for analyzing next-generation DNA sequencing data. *Genome Res.* **2010**, *20*, 1297–1303. [CrossRef] [PubMed]
24. Danecek, P.; Auton, A.; Abecasis, G.; Albers, C.A.; Banks, E.; DePristo, M.A.; Handsaker, R.E.; Lunter, G.; Marth, G.T.; Sherry, S.T.; et al. The variant call format and VCFtools. *Bioinformatics* **2011**, *27*, 2156–2158. [CrossRef]
25. Kumar, S.; Stecher, G.; Tamura, K. MEGA7: Molecular Evolutionary Genetics Analysis Version 7.0 for Bigger Datasets. *Mol. Biol. Evol.* **2016**, *33*, 1870–1874. [CrossRef]
26. Xu, C.; Ren, Y.; Jian, Y.; Guo, Z.; Zhang, Y.; Xie, C.; Fu, J.; Wang, H.; Wang, G.; Xu, Y.; et al. Development of a maize 55 K SNP array with improved genome coverage for molecular breeding. *Mol. Breed.* **2017**, *37*, 20. [CrossRef] [PubMed]

27. Wang, M.L.; Zhu, C.; Barkley, N.A.; Chen, Z.; Erpelding, J.E.; Murray, S.C.; Tuinstra, M.R.; Tesso, T.; Pederson, G.A.; Yu, J. Genetic diversity and population structure analysis of accessions in the US historic sweet sorghum collection. *Theor. Appl. Genet.* **2009**, *120*, 13–23. [CrossRef]
28. Cibulskis, K.; McKenna, A.; Fennell, T.; Banks, E.; DePristo, M.; Getz, G. ContEst: Estimating cross-contamination of human samples in next-generation sequencing data. *Bioinformatics* **2011**, *27*, 2601–2602. [CrossRef]
29. Jiao, L.; Han, C.; Zhu, J.; Zhang, P.; Ma, Y.; Dai, X.; Zhang, Y. Transcriptome analysis and development of EST-SSR markers in the mushroom *Auricularia heimuer*. *Sci. Rep.* **2024**, *14*, 12340. [CrossRef]
30. Sun, X.; Yang, C.; Ma, Y.; Zhang, J.; Wang, L. Research progress of *Auricularia heimuer* on cultivation physiology and molecular biology. *Front. Microbiol.* **2022**, *13*, 1048249. [CrossRef]
31. Du, P.; Cui, B.-K.; Zhang, C.-F.; Dai, Y.-C. Genetic diversity of wild *Auricularia auricula-judae* revealed by ISSR analysis. *Biochem. Syst. Ecol.* **2013**, *48*, 199–205. [CrossRef]
32. Yao, F.-J.; Lu, L.-X.; Wang, P.; Fang, M.; Zhang, Y.-M.; Chen, Y.; Zhang, W.-T.; Kong, X.-H.; Lu, J.; Honda, Y. Development of a Molecular Marker for Fruiting Body Pattern in *Auricularia auricula-judae*. *Mycobiology* **2018**, *46*, 72–78. [CrossRef]
33. Gazendam, I.; Mojapelo, P.; Bairu, M.W. Potato Cultivar Identification in South Africa Using a Custom SNP Panel. *Plants* **2022**, *11*, 1546. [CrossRef] [PubMed]
34. Xu, W.; Gong, C.; Mai, P.; Li, Z.; Sun, B.; Li, T. Genetic diversity and population structure analysis of 418 tomato cultivars based on single nucleotide polymorphism markers. *Front. Plant Sci.* **2024**, *15*, 1445734. [CrossRef]
35. Ongom, P.O.; Fatokun, C.; Togola, A.; Salvo, S.; Oyebode, O.G.; Ahmad, M.S.; Jockson, I.D.; Bala, G.; Boukar, O. Molecular Fingerprinting and Hybridity Authentication in Cowpea Using Single Nucleotide Polymorphism Based Kompetitive Allele-Specific PCR Assay. *Front. Plant Sci.* **2021**, *12*, 734117. [CrossRef]
36. Yin, L.; Yao, F.; Shi, C.; Lu, L.; Wang, T.; Liu, W. Genetic diversity of wild *Auricularia heimuer* germplasm resources based on SSR markers. *Acta Edulis Fungi* **2022**, *29*, 1–9. [CrossRef]
37. Meng, Q.; Xie, Z.; Jiang, Z.; Xu, H.; Guo, J.; Li, Y.; Wang, C.; Hu, Z.; Yan, S.; Liu, H. Population genetic structure analysis of the wild esculenta clade (yellow morels) of *Morchella* spp. in Qinghai. *Mycosystema* **2021**, *40*, 1328–1342+1370–1372. [CrossRef]
38. Lee, Y.-J.; Mun, J.-H.; Jeong, Y.-M.; Joo, S.-H.; Yu, H.-J. Assembly of a radish core collection for evaluation and preservation of genetic diversity. *Hortic. Environ. Biotechnol.* **2018**, *59*, 711–721. [CrossRef]

agriculture

Article

# Genetic Diversity and Genome-Wide Association Study of *Pleurotus pulmonarius* Germplasm

**Qian Li [1,2,†], Xuebing Ying [3,†], Yashu Yang [1,2] and Wei Gao [1,2,*]**

1 Institute of Agricultural Resources and Regional Planning, Chinese Academy of Agricultural Sciences, Beijing 100081, China; dt1999liqian22@163.com (Q.L.); yangyashusdau@163.com (Y.Y.)
2 State Key Laboratory of Efficient Utilization of Arid and Semi-Arid Arable Land in Northern China, Beijing 100081, China
3 Lin'an District Agricultural and Forestry Technology Extension Center, Hangzhou 311300, China; yingxuebing2024@163.com
* Correspondence: gaowei01@caas.cn
† These authors contributed equally to this work.

**Abstract:** *Pleurotus pulmonarius* is prized by consumers for its distinct flavor, strong aroma, and dense, crispy texture. Although China has extensive germplasm resources for *P. pulmonarius*, only a limited number of cultivars are commercially available. A comprehensive evaluation and detailed analysis of *P. pulmonarius* germplasm, alongside the exploration of superior germplasm resources, are essential for developing new varieties. In this study, we resequenced the genomes of 47 *P. pulmonarius* strains collected nationwide, identifying a total of 4,430,948 single nucleotide polymorphism (SNP) loci. After filtering based on minor allele frequency and data integrity, 181,731 high-quality SNP markers were retained. Phylogenetic analysis grouped the strains into six clusters, with strains from similar geographical regions clustering together. Most CBS strains formed a single cluster; cultivated varieties exhibited higher genetic similarity, whereas wild strains displayed greater diversity. Principal component analysis (PCA) and population structure analyses, using the same SNP markers, corroborated the phylogenetic findings. DNA fingerprinting, derived from 369 core SNPs, further underscored the genetic diversity among strains. Significant morphological variation was observed, with strains in groups ZP, CBS, and WHLJ exhibiting notably higher yields and cap widths compared to other groups. Correlation analysis revealed associations among various phenotypes, while genome-wide association study (GWAS) identified multiple SNP markers within candidate genes linked to agronomic traits, most of which were controlled by multiple genes. This research offers a molecular-level characterization and evaluation of *P. pulmonarius* germplasm resources, providing a scientific basis for enriching available germplasm and advancing breeding materials.

**Keywords:** *Pleurotus pulmonarius*; genetic diversity; genome-wide association study

## 1. Introduction

*Pleurotus pulmonarius*, a member of the class Agaricomycetes and subclass Agaricomycetidae, is one of the main species within the oyster mushroom group. According to the China Edible Fungi Association, *P. pulmonarius* production in China reached 636,700 tons in 2022. China has abundant *P. pulmonarius* germplasm resources, with wild populations found in regions such as Jiangsu, Zhejiang, Guangdong, Fujian, Henan, and Guangxi. The successful cultivation of *P. pulmonarius* was first achieved in Luoyuan County, Fuzhou City, in 1998, leading to its rapid and widespread adoption thereafter [1]. Known for its high nutritional value, *P. pulmonarius* is rich in protein, amino acids, sugars, unsaturated fatty acids, vitamins, folic acid, and essential trace elements. Its distinct aroma, firm yet tender texture, and mild flavor make it highly desirable to consumers.

Studying and assessing the genetic diversity of *P. pulmonarius* germplasm is essential for breeding applications, providing a foundation for identifying superior germplasm and selecting parental lines with optimal genetic backgrounds. Mating type analysis of 30 *P. pulmonarius* cultivars revealed only four A factors and three B factors, indicating high genetic similarity among cultivars. The A factor controls nuclear binding and binucleation, while the B factor regulates hyphal fusion and nuclear migration [2]. Clonal propagation through tissue culture is commonly practiced, occasionally leading to the misidentification of distinct strains under the same name. Therefore, accurate variety identification and genetic relationship analysis between cultivated and wild strains are crucial for ensuring high-quality production and advancing scientific research. With the rapid development of molecular biology, molecular markers have become invaluable for analyzing interspecies genetic differences. In addition to traditional morphological identification and antagonistic reactions, molecular markers such as RFLP, RAPD, ISSR, AFLP, and SNPs are widely used to analyze the genetic diversity and phylogenetic relationships of *P. pulmonarius* germplasm [3]. However, these methods often face limitations in marker quantity, reproducibility, and stability. In contrast, single-nucleotide polymorphisms (SNPs) are abundant, stable, and suitable for automated, high-throughput applications, making them the most widely used genetic markers today. As a primary genotyping marker, SNPs are extensively applied in identifying edible fungi and analyzing genetic diversity across strains, including species like *Lentinus edodes* [4–6], *Flammulina filiformis* [7,8], and *Agaricus bisporus* [9]. SNPs are also critical in analyzing *P. pulmonarius* germplasm diversity, identifying hybrids, and constructing genetic linkage maps. Genome resequencing enables the acquisition of SNP markers evenly distributed across the genome [10]. For example, SNP markers have been used to identify laccase genes in *Lentinus edodes* [6]. Okuda et al. constructed a genetic linkage map for *P. pulmonarius* using 150 single-spore hybrid populations and 300 AFLP markers along with two mating-type factors and sporeless traits. This map comprises 12 linkage groups, covering a total length of 971 cM [11]. Vidal-Diez et al. recently assembled and annotated high-quality genomes of *P. pulmonarius* strains ss2 and ss5. Each genome consists of 23 scaffolds, with sizes ranging from 5.55 Mb to 11 kb (ss2) and 5.06 Mb to 21 kb (ss5), and N50 values of 3.2 Mb and 3.4 Mb, respectively, providing valuable reference genomic data for developing genome-wide SNP markers [12].

Genome-wide association study (GWAS) is commonly employed to identify genes associated with agronomic traits across the entire genome [13–16]. However, applications of GWASs in edible fungi remain limited. Li et al. used 297 genome-wide markers to genotype 89 cultivated varieties of *L. edodes*. Among these, 43 markers were significantly associated with four phenotypic traits, leading to the identification of 97 nearby genes and establishing a foundation for molecular breeding in *L. edodes* [17]. Corner et al. classified the genus *Pleurotus* into three categories based on hyphal types [18]. Morphological characteristics in edible fungi are highly susceptible to environmental influences [19]. Using a GWAS, Yu identified that DNA damage repair and protein translation processes may play roles in *L. edodes* responses to cadmium (Cd) stress, providing valuable genomic and population data for functional genomics and artificial breeding research [20]. Zhang applied bulked-segregant analysis (BSA) and an extreme-phenotype GWAS (XP-GWAS) to map cap color traits in *Pleurotus cornucopiae* and identified the tyrosinase-encoding gene *PcTYR* through comparative transcriptome analysis [21]. Traditional QTL analysis primarily relies on genetic diversity among parent populations, and its detection efficiency varies between populations. As it often involves multiple genes, accurately pinpointing candidate genes is challenging. In contrast, GWAS enhances allele diversity and improves mapping accuracy. Advances in sequencing technology and mutation analysis tools have significantly increased marker density, while the incorporation of mixed linear models with random and fixed effects has substantially reduced false-positive results.

In this study, the genomes of 47 *P. pulmonarius* strains collected from various regions across the country were resequenced to detect genome-wide genetic polymorphisms. SNP markers were screened to construct an SNP fingerprint for the strains. Using these genome-wide SNP markers, a phylogenetic tree was developed to analyze the genetic relationships among the strains. Additional population genetic analyses, including population structure, principal component analysis, and linkage disequilibrium, were performed. Fruiting body-related traits of the strains were measured, and the phenotypic data were statistically analyzed. Finally, GWAS between SNP markers and phenotypic traits was conducted to identify SNP markers with significant influence on the traits. This study provides a theoretical foundation for future screening of superior germplasms, gene mining for trait variation, and targeted breeding efforts.

## 2. Materials and Methods

### 2.1. Experimental Materials

The tested strains were all stored in the China Center for Mushroom Spawn Standards and Control (CCMSSC) (Table 1).

**Table 1.** Tested strains and their origins.

| Sample Name | Group | Sample Name for Analysis | Name Interpretation |
|---|---|---|---|
| CCMSSC00313 | ZP [1] | ZP1 | Cultivation |
| CCMSSC00493 | ZP | ZP2 | Cultivation |
| CCMSSC00494 | ZP | ZP3 | Cultivation |
| CCMSSC00498 | ZP | ZP4 | Cultivation |
| CCMSSC00499 | ZP | ZP5 | Cultivation |
| CCMSSC03815 | ZP | ZP6 | Cultivation |
| CCMSSC03836 | ZP | ZP7 | Cultivation |
| CCMSSC03886 | ZP | ZP8 | Cultivation |
| CCMSSC03897 | ZP | ZP9 | Cultivation |
| CCMSSC04423 | ZP | ZP10 | Cultivation |
| CCMSSC04537 | ZP | ZP11 | Cultivation |
| R08026 | ZP | ZP12 | Cultivation |
| R08027 | ZP | ZP13 | Cultivation |
| CCMSSC01106 | WSC | WSC1 | Sichuan wild |
| CCMSSC01108 | WSC | WSC2 | Sichuan wild |
| CCMSSC01109 | WSC | WSC3 | Sichuan wild |
| CCMSSC01110 | WSC | WSC4 | Sichuan wild |
| CCMSSC01111 | WSC | WSC5 | Sichuan wild |
| CCMSSC01114 | WSC | WSC6 | Sichuan wild |
| CCMSSC01116 | WSC | WSC7 | Sichuan wild |
| CCMSSC01118 | WSC | WSC8 | Sichuan wild |
| CCMSSC01119 | WSC | WSC9 | Sichuan wild |
| CCMSSC01120 | WSC | WSC10 | Sichuan wild |
| CCMSSC01123 | WSC | WSC11 | Sichuan wild |
| CCMSSC04199 | WHLJ | WHLJ | Heilongjiang wild |
| CCMSSC04458 | CBS | CBS1 | CBS |
| CCMSSC04459 | CBS | CBS2 | CBS |
| CCMSSC04460 | CBS | CBS3 | CBS |
| CCMSSC04461 | CBS | CBS4 | CBS |
| CCMSSC04462 | CBS | CBS5 | CBS |
| CCMSSC01301 | WYN | WYN1 | Yunnan wild |
| CCMSSC04584 | WYN | WYN2 | Yunnan wild |
| CCMSSC04585 | WYN | WYN3 | Yunnan wild |
| CCMSSC04586 | WYN | WYN4 | Yunnan wild |
| CCMSSC04587 | WYN | WYN5 | Yunnan wild |
| CCMSSC04588 | WYN | WYN6 | Yunnan wild |
| CCMSSC04592 | WYN | WYN7 | Yunnan wild |
| CCMSSC04593 | WYN | WYN8 | Yunnan wild |
| CCMSSC04591 | WSX | WSX | Shaanxi wild |
| CCMSSC04594 | WLN | WLN1 | Liaoning wild |
| CCMSSC04595 | WLN | WLN2 | Liaoning wild |

**Table 1.** *Cont.*

| Sample Name | Group | Sample Name for Analysis | Name Interpretation |
|---|---|---|---|
| CCMSSC04597 | WJL | WJL1 | Jilin wild |
| CCMSSC04598 | WJL | WJL2 | Jilin wild |
| CCMSSC04599 | WHB | WHB1 | Hubei wild |
| CCMSSC04600 | WHB | WHB2 | Hubei wild |
| CCMSSC04601 | WHB | WHB3 | Hubei wild |
| CCMSSC04602 | WHB | WHB4 | Hubei wild |

[1] The cultivated strains are labeled with the format 'ZP' followed by a numerical code, while the wild strains are labeled with 'W' followed by the abbreviation of the collection region and a numerical code.

### 2.2. DNA Extraction and Sequencing

Genomic DNA from the tested strains was extracted from mycelia grown on PDA (Difco™ Potato Dextrose Agar) plates using the cetyltrimethylammonium bromide (CTAB) method [22]. DNA concentration and purity were assessed with a NanoDrop 2000 spectrophotometer (Thermo Fisher Scientific, Waltham, MA, USA). Sequencing libraries of 350 bp fragments were constructed and sequenced on the Illumina/BGI platform (Beijing BioMarker Technology Co., Ltd., Beijing, China), generating 150 bp paired-end reads.

### 2.3. Identification and Screening of SNP

Single nucleotide polymorphisms (SNPs) were primarily detected using the GATK software (v4.0) toolkit [23]. Redundant reads were filtered with samtools (v1.9) based on localization results of clean reads aligned to the reference genome to ensure accuracy [24]. GATK's HaplotypeCaller algorithm was employed to detect SNP and InDel mutations, generating gVCF files for each sample, followed by joint genotyping across groups. Rigorous filtering was applied to ensure reliability. The main filtering parameters included using vcfutils.pl in bcftools to exclude SNPs within 5 bp of an InDel and adjacent InDels within 10 bp; limiting variations to a maximum of two in any 5 bp window; and excluding variants with a Phred quality score (QUAL) below 30, a quality-to-depth ratio (QD) below 2.0, a mapping quality (MQ) below 40, and a Fisher Strand value (FS) above 60. Default GATK parameters were used for additional filtering. These steps ensured the accurate and reliable detection of SNPs and InDels.

### 2.4. Population Genetic Analysis and Construction of Fingerprints

In the analysis, all selected high-quality SNP markers were used for population genetic analysis. MEGA X software [25] was applied to construct phylogenetic trees for each sample using the neighbor-joining method with the p-distance calculation model and 1000 bootstrap replications. Population genetic indicators were calculated using VCFtools (v0.1.15) [26]. Admixture software assessed population structure [27], while EIGENSOFT [28] performed principal component analysis to cluster the samples. Genetic diversity analysis was conducted using PLINK software [29]. A DNA fingerprint was created to identify markers suitable for genetic typing based on specific criteria: markers were evenly distributed across the genome, exhibited 100% completeness (no missing data), had a minor allele frequency (MAF) above 20%, a polymorphic information content (PIC) above 0.35, conformed to Hardy–Weinberg equilibrium ($p$-value > 0.01), and showed no nearby mutations within 100 bp of the marker [30].

### 2.5. Cultivation Test and Trait Screening

The 47 tested strains were cultivated in polycarbonate plastic bottles (diameter: 70 mm, height: 90 mm, volume: 280 mL) filled with 180 g (wet weight) of substrate. The substrate consisted of cottonseed shell (94%), wheat bran (5%), and gypsum (1%) with approximately 65% water content. The bottles were sterilized for 2 h at 0.15 MPa and 126 °C. Six replicate bottles were inoculated for each strain. Once fully colonized, the bottles were transferred to a fruiting room maintained at 16 °C to 18 °C, with 80% to 90% relative humidity and a 12 h photoperiod (300 to 350 lx). $CO_2$ concentration was kept below 1000 ppm through ventilation.

Phenotypic data were collected for various agronomic traits, including the period from inoculation to primordia generation (PPG) and the period from inoculation to harvesting (PIH) for each bottle. Measurements of mature fruiting bodies included cap length (CL), cap width (CW), stipe length (SL), and stipe diameter (SD). Yield of fruiting bodies per bottle (YFB) was also recorded. All strains were cultivated under the same climate-controlled conditions. Broad-sense heritability ($H^2$) was calculated across tester lines as $H^2 = \sigma_G^2 / \left[\sigma_G^2 + \left(\frac{\sigma_e^2}{n}\right)\right]$, where $\sigma_G^2$ represents the genotypic variance, $\sigma_e^2$ represents the residual variance, and n is the number of testers (indicate heterokaryon set for which this was the case) ($n$ = 42). The phenotypic data were recorded individually and analyzed statistically. One-way analysis of variance and correlation analysis were performed using SPSS software (v27.0) [31]. The best linear unbiased prediction (BLUP) [32] method was used to estimate trait values for association analysis. Descriptive statistics were calculated for each phenotype, including the number of samples after removing missing values (SamNum), mean (Mean), standard deviation (SD), median (Median), minimum (Min), maximum (Max), range (Range), and coefficient of variation (CV). This comprehensive approach was designed to assess the growth and phenotypic characteristics of *P. pulmonarius* strains under controlled cultivation conditions, enabling a detailed analysis of mushroom production traits.

### 2.6. Genome-Wide Association Study

SNP data were filtered based on a minor allele frequency (MAF $\geq$ 0.05) and data integrity (INT $\geq$ 0.8). Phenotypic data were used to construct frequency distribution histograms. GEMMA software [33] employing the linear mixed model (LMM) was used for GWAS, represented by the model:

$$y = W\alpha + x\beta + \mu + e$$

The analysis generated $p$-values, which were used to create Manhattan and QQ plots. A threshold was set by dividing the number of effective markers by quality control criteria, with regions meeting $-\log_{10}(p) \geq 5$ designated as candidate intervals. Functional annotation was applied to genes identified within these candidate regions to clarify their potential biological roles. All experimental steps were completed in 2023.

## 3. Results

### 3.1. Quality Statistics of Sequencing Data and Discovery of SNPs

Whole-genome resequencing of *P. pulmonarius* strains yielded clean reads ranging from 6,119,174 to 11,448,795, with an average of 7,998,114 reads per strain. The Q30 quality score ranged from 88.4% to 95.81%, with an average of 92.66%. GC content varied from 43.58% to 50.35%, averaging 47.33%. In total, 112.15 Gbp of clean data was generated across the 47 strains, with an alignment rate to the reference genome of 75.02%, an average coverage depth of 41×, and genome coverage of 89.84% (at least one base covered). These metrics satisfied the requirements for sequence analysis.

Using the GATK toolkit, 4,430,948 SNPs were detected between the samples and the reference genome (NCBI: ASM1298053v1). The largest number of SNPs was observed between cultivated varieties ZP1 and ZP13 (234,642 SNPs), while the smallest number was observed between ZP2 and ZP12 (2151 SNPs). Analysis of SNP distribution across the genome showed that 65.51% of SNPs were located within gene-coding regions, 13.97% were within 5 kb upstream of genes, and 5.95% were in intronic regions. Synonymous coding mutations accounted for 87.13% of the total, whereas nonsynonymous coding mutations constituted 12.69%. Other types of mutations, such as SNPs within 5 kb downstream of genes, accounted for 10.59% (Figure 1).

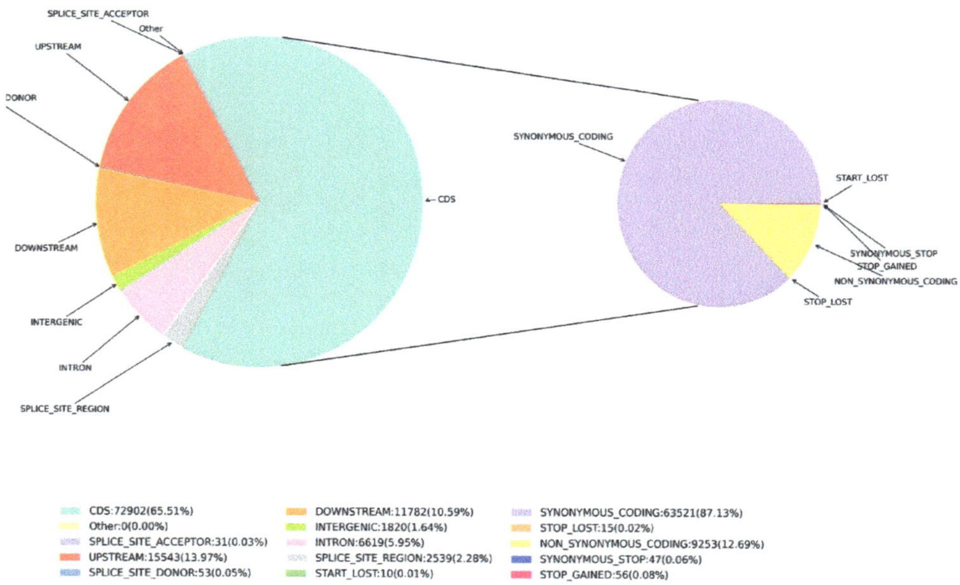


**Figure 1.** SNP proportions at different genomic locations: INTERGENIC represents intergenic regions, INTRON represents intronic regions, CDS represents coding sequence regions, UPSTREAM/DOWNSTREAM represent SNP sites located within 5 kb upstream/downstream of genes, UTR_5_PRIME/UTR_3_PRIME represent SNP sites in the 5′ UTR and 3′ UTR of genes, SPLICE_SITE_ACCEPTOR/SPLICE_SITE_DONOR represent splice site mutations (within the first 2 bp of an exon), SPLICE_SITE_REGION represents splice site region mutations (1−3 bp variation in exons or 3–8 bp variation in introns), START_GAINED represents gained start codons (in non-coding regions). SNP mutation types within CDS regions annotation: START_LOST represents lost start codons, SYNONYMOUS_START/NON_SYNONYMOUS_START represent synonymous/non-synonymous start codon mutations, SYNONYMOUS_CODING represents synonymous coding mutations, NON_SYNONYMOUS_CODING represents non-synonymous coding mutations, STOP_GAINED/STOP_LOST represent gained/lost stop codons.

### *3.2. Population Genetic Analysis Based on Whole-Genome Wide SNP Markers*

#### 3.2.1. Phylogenetic Relationship

After filtering, a total of 181,731 high-quality SNP markers were obtained. Phylogenetic analysis revealed distinct clusters among the tested *P. pulmonarius* strains. Cultivated varieties (ZP8, ZP11, ZP10, ZP7, ZP3, ZP6, ZP4, ZP12, ZP2, ZP9, and CBS3) grouped closely within Cluster I, indicating high genetic similarity among cultivars (Figure 2). In contrast, two other cultivated varieties, ZP5 and ZP13, clustered with a wild strain from Yunnan in a separate branch within Cluster I, suggesting that these cultivars may have been domesticated from wild strains originating in Yunnan. Most strains collected from CBS clustered in Cluster II, except for wild strain WSC5 from Sichuan and cultivated strain ZP1, which also fell into this cluster but on different branches.

The wild strains exhibited relatively high genetic diversity, with strains from similar geographical regions clustering together. For instance, wild strains from Liaoning, Jilin, and Heilongjiang grouped in Cluster III, while those from Sichuan were predominantly found in Clusters IV and V, and those from Hubei in Cluster VI.

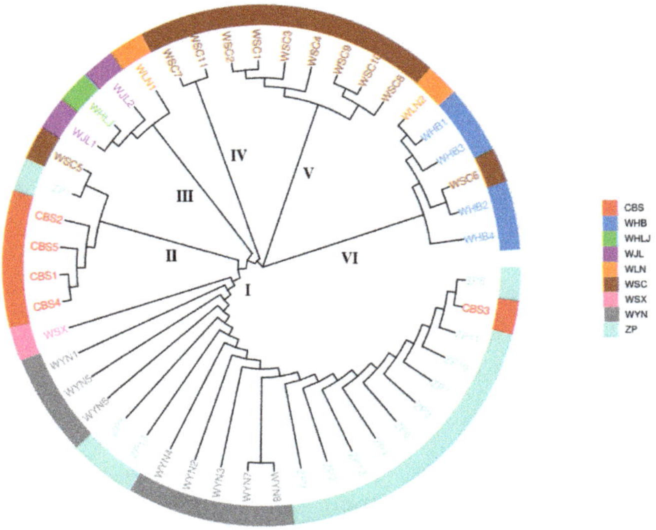


**Figure 2.** Illustration of 47 *P. pulmonarius* phylogenetic trees based on SNP. The Roman numerals (I–VI) indicate different clusters. Strain names of each group are listed in Table 1.

PCA analysis illustrated the clustering pattern of the samples, confirming the results of the phylogenetic analysis (Figure 3, Table S1). The two-dimensional plot of PC1 and PC3, as well as the three-dimensional plot of PC1, PC2, and PC3, depicted the spatial distribution of each strain based on genetic variation. PC1 accounted for 36.58% of the variation, PC2 explained 12%, and PC3 accounted for 6.52%. The analysis revealed four distinct clusters among the strains. With the exception of ZP1 and ZP5, most strains and CBS3 formed a cohesive group characterized by PC1 values ranging from 0.2125 to 0.2122, closely matching the branches observed in the phylogenetic tree. Strains such as WYN8, WYN7, WYN3, WYN4, and WYN2 exhibited PC1 values between 0.1507 and 0.1368, while ZP5, ZP13, WYN6, ZP1, WSC5, WYN5, WYN1, and WSX ranged from 0.0505 to 0.0491, aligning closely with branches in Clusters I and II in the phylogenetic tree.

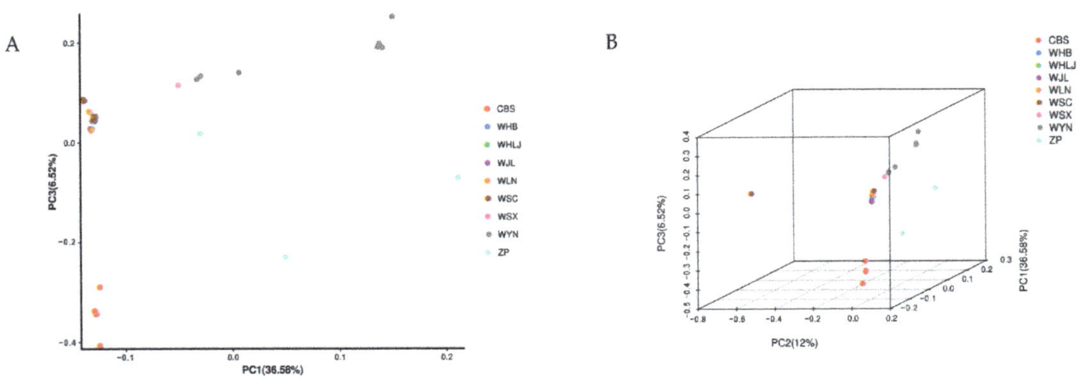


**Figure 3.** Principal component analysis of 47 *P. pulmonarius* based on SNP. The samples are aggregated into two-dimensional (**A**) and three-dimensional (**B**) by PCA analysis: PC1 represents the first principal component; PC2 represents the second principal component; PC3 represents the third principal component. A point represents a sample, and a color represents a grouping.

Wild strains, represented by Clusters III, IV, V, and VI from the phylogenetic tree, showed a broader distribution range of PC1, spanning from −0.1236 to −0.1385. The PCA results closely aligned with the phylogenetic tree, reflecting high genetic similarity among cultivated varieties and significant variability and divergence among wild strains.

3.2.2. Genetic Diversity Analysis

Polymorphic markers among strains within each group exhibited average polymorphic information content (PIC) ranging from 0.226 to 0.327 (Table 2). Among domestic wild strains, the average minor allele frequency (MAF) of Sichuan wild strains was notably low at 0.202, with predominant alleles more widely represented. The average Nei diversity index for this group was only 0.303. In contrast, wild strains from Liaoning and Jilin exhibited higher average MAF values of 0.328 and 0.326, respectively, with Nei diversity index averages of 0.552 and 0.550, respectively. Further comparison showed that average heterozygosity values were notably high in Jilin wild strains (0.606) and Liaoning wild strains (0.598), along with their respective PIC values of 0.326 and 0.327, indicating significantly greater genetic diversity in Northeast China compared to Sichuan. Wild strains from Yunnan and Hubei displayed lower average PIC values of 0.260 and 0.281, respectively. The genetic characteristics of CBS strains were similar to those of wild strains from Yunnan, with relatively lower diversity among the tested strains. Cultivated strains exhibited an average MAF of 0.207, an average Nei diversity index of 0.287, observed heterozygosity of 0.435, and an average PIC of 0.226. These values indicated lower genetic diversity in cultivated strains compared to wild strains.

**Table 2.** Analysis table of population genetic diversity of *P. pulmonarius*.

| Group | Average MAF | Nei Diversity Index | Observed Heterozygous Number | Polymorphism Information Content |
|---|---|---|---|---|
| CBS | 0.229 | 0.200–0.667 (0.361) [1] | 0.200–1.000 (0.243) | 0.164–0.375 (0.265) |
| WHB | 0.255 | 0.250–0.667 (0.399) | 0.250–1.000 (0.409) | 0.195–0.375 (0.281) |
| WJL | 0.326 | 0.500–0.667 (0.550) | 0.500–1.000 (0.606) | 0.305–0.375 (0.326) |
| WLN | 0.328 | 0.500–0.667 (0.552) | 0.500–1.000 (0.598) | 0.305–0.375 (0.327) |
| WSC | 0.202 | 0.091–0.556 (0.303) | 0.091–1.000 (0.293) | 0.083–0.375 (0.238) |
| WYN | 0.231 | 0.125–0.667 (0.343) | 0.125–1.000 (0.296) | 0.110–0.375 (0.260) |
| ZP | 0.207 | 0.077–0.533 (0.287) | 0.077–1.000 (0.435) | 0.071–0.375 (0.226) |

[1] The number in parentheses represents the average value, and the number range outside parentheses represents the minimum and maximum.

3.2.3. Population Structure

Based on the 181,731 high-quality SNP markers, Admixture software was used to analyze the population structure of the samples. The analysis considered K values ranging from 1 to 10 to determine optimal clustering. At K = 2, the cross-validation error rate was minimized at 0.514, categorizing the 47 tested strains into two main clusters, with 17 strains showing ambiguous classification (Figure 4, Table S2). The first cluster included 11 cultivated strains previously identified along with CBS3, while the second cluster predominantly consisted of 10 wild strains from Sichuan, 5 from Northeast China, and 4 from Hubei. This analysis underscored the genetic differentiation between cultivated varieties and specific wild strains, highlighting substantial genetic divergence.

Additionally, at K = 4, the cross-validation error rate was lower than at K = 3, with a rate of 0.548, and the strains were distinctly divided into four clusters. Compared to clustering at K = 2, strains CBS1, CBS2, CBS5, and CBS4 formed a separate cluster, while ZP1 and WSC were also categorized into their own cluster.

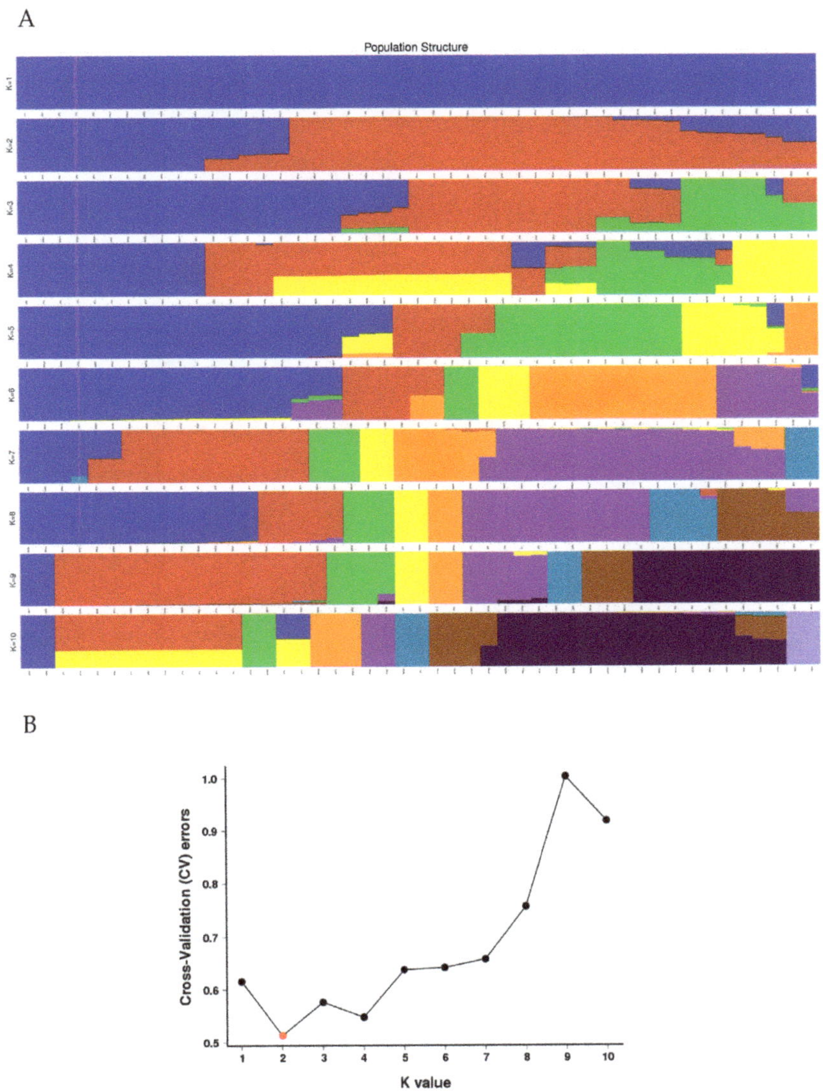


**Figure 4.** Population structure at different K values. (**A**) shows the clustering results for K values ranging from 1 to 10. (**B**) illustrates the clustering of the research population with the number of subgroups (K value) preset from 1 to 10, followed by cross-verification of the clustering results. The optimal number of clusters is determined by identifying the valley in the cross-validation error rate.

A total of 369 candidate core SNP markers were identified, facilitating the construction of a DNA fingerprint map to differentiate the 47 *P. pulmonarius* strains (Figure 5, Table S3). The fingerprint map clearly depicted the close genetic relationships and divergence among the strains. The minor allele frequency was reported as 0.416 (Table S4). These core SNPs exhibited high quality, even distribution across the genome, and strong discriminatory power among strains. The map illustrated that cultivated strains had high genetic similarity with minimal SNP variation, whereas wild strains exhibited greater diversity with numerous variations, indicating substantial genetic differences among wild strains.

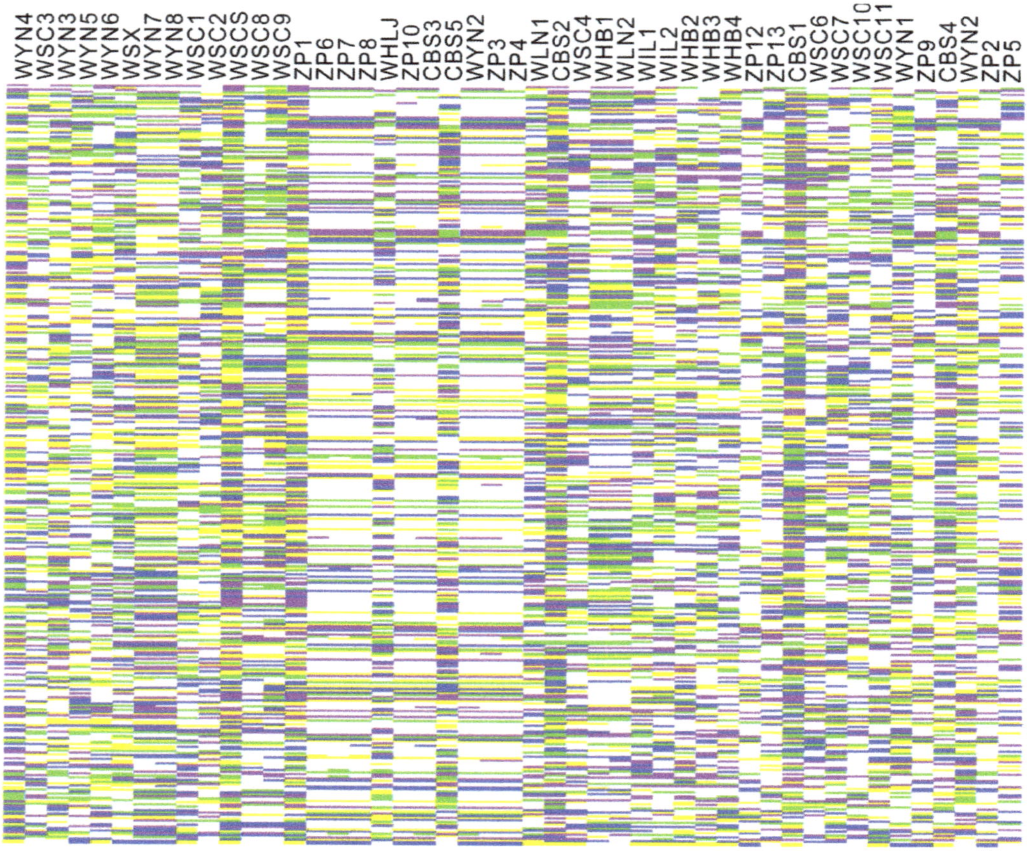


**Figure 5.** Screening of core tags and construction of SNP fingerprint atlas. SNP fingerprint atlas for 47 strains: Each row in the fingerprint atlas corresponds to a specific SNP, and the columns represent different strains. Yellow indicates C/C genotype, green indicates A/A, blue indicates T/T, and purple indicates G/G. The missing data are represented by gray, while the heterozygosity check points between strains are represented by white.

### 3.3. Evaluation of Agronomic Traits in *P. pulmonarius*

A total of 42 *P. pulmonarius* strains were capable of producing fruiting bodies (Table S5). The frequency distribution indicated continuous variation for nearly all seven agronomic traits, suggesting that these traits are quantitative and controlled by multiple genes (Figure 6). Since the period from inoculation to primordia generation (PPG) represents the earliest time point for primordia formation, it was excluded from the ANOVAs. The ANOVA results revealed significant variation among strains for the remaining six traits ($\alpha = 0.05$) (Table S6).

The boxplot clearly illustrates the variation for each trait (Figure 7). The most pronounced differences among strains were observed for the trait YFB, while PPG exhibited the least variation. However, it is worth noting that there were four significant outliers for PPG, indicating that although most strains showed a similar period from inoculation to primordia formation, a few strains had notably different times.

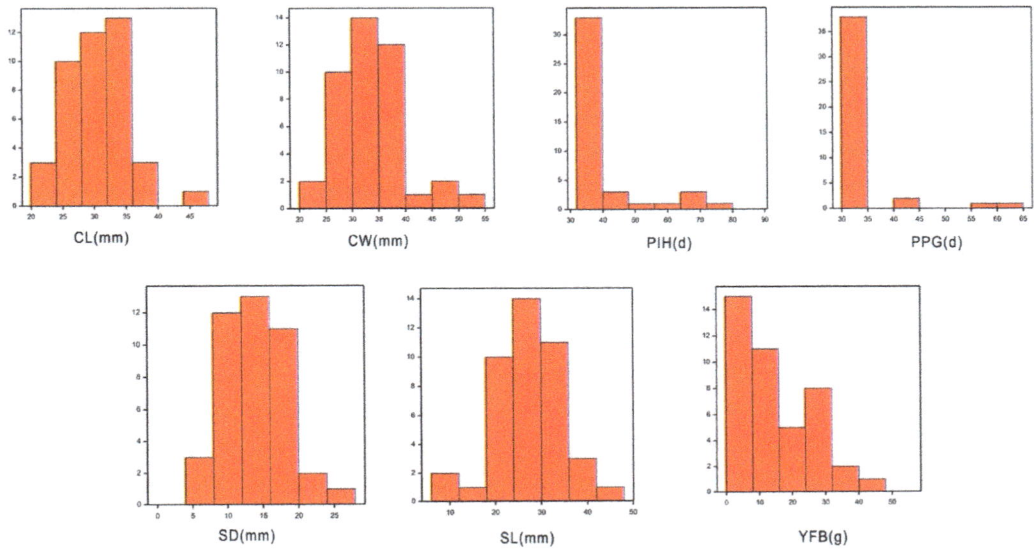


**Figure 6.** Histograms showing the frequency distribution of 7 agronomic traits. The *x*-axis represents the value range of the traits, while the *y*-axis indicates the frequency. The top left to the bottom right are the frequency distribution histograms of cap length (CL), cap width (CW), the period from inoculation to harvesting (PIH), the period from inoculation to primordia generation (PPG), stipe diameter (SD), stipe length (SL), and yield of fruiting body per bottle (YFB).

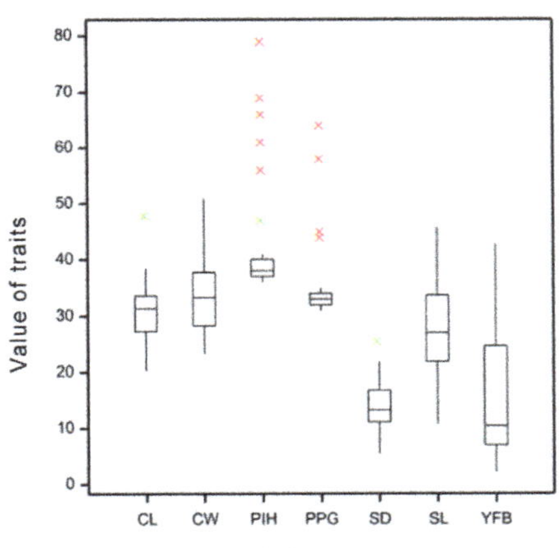


**Figure 7.** Boxplots describing the variation in each trait. The line inside the box represents the median, the whiskers extend to the minimum and maximum values. Red Crosses indicate outliers. The *y*-axis indicates the value range of agronomic traits, while the *x*-axis represents traits, from left to right are cap length (CL), cap width (CW), the period from inoculation to harvesting (PIH), the period from inoculation to primordia generation (PPG), stipe diameter (SD), stipe length (SL), and yield of fruiting body per bottle yield per bottle (YFB).

The average period from inoculation to primordium generation (PPG) was approximately 35 days (Table 3). Notably, strains ZP3, ZP4, WHLJ, and CBS5 exhibited faster primordium formation, occurring in as few as 31 days, while ZP2 required a longer period of 64 days. The broad-sense heritability for stipe diameter (SD) and yield of fruiting bodies (YFB) was relatively high, indicating that the variations observed for these traits were primarily influenced by genetic factors. The average period from inoculation to harvest (PIH) was around 42 days; however, strains ZP4, WHLJ, and CBS3 required only 36 days, while WSC5 took 79 days to reach harvest.

**Table 3.** *P. pulmonarius* phenotypic descriptive statistical table.

| Trait | SamNum [1] | Mean | Sd [2] | Median | Min [3] | Max [4] | Range | CV [5] |
|---|---|---|---|---|---|---|---|---|
| CL | 42 | 30.75 | 5.06 | 31.36 | 20.32 | 47.85 | 27.53 | 16.45% |
| CW | 42 | 33.81 | 6.33 | 33.29 | 23.29 | 50.99 | 27.70 | 18.72% |
| SD | 42 | 13.80 | 4.01 | 13.33 | 5.50 | 25.56 | 20.07 | 29.05% |
| SL | 42 | 27.47 | 7.47 | 27.03 | 10.73 | 45.77 | 35.04 | 27.18% |
| PPG | 42 | 34.86 | 6.54 | 33 | 31 | 64 | 33 | 18.76% |
| PIH | 42 | 42.40 | 10.37 | 38 | 36 | 79 | 43 | 24.45% |
| YFB | 42 | 15.30 | 11.06 | 10.41 | 2.15 | 42.76 | 40.61 | 72.32% |

[1] SamNum: number of samples after removing missing values; [2] Sd: standard deviation; [3] Min: minimum; [4] Max: maximum; [5] CV: coefficient of variation.

Cap size varied significantly among the 42 strains. For example, strain CBS3 had a cap length (CL) of 47.85 mm and a cap width (CW) of 50.99 mm, while strain WJL1 had a CL of only 20.32 mm and a CW of 23.29 mm. Stipe length (SL) also varied greatly, with WSC7 having the shortest length at 10.73 mm and CBS5 the longest at 45.77 mm. Stipe diameter (SD) showed notable variation as well, with WSC5 having the smallest diameter at 5.5 mm, while ZP9 had the largest diameter at 25.56 mm. In terms of YFB, strain ZP3 achieved the highest fruiting body weight at 42.76 g per bottle. Cultivated strains generally produced higher fruiting body weights with more considerable variation, as indicated by a coefficient of variation of 72.32%.

Significant differences were detected for CW, SD, and YFB among the groups. Strains in groups ZP, CBS, and WHLJ exhibited significantly larger yields and cap widths, while strains in groups WHB, WJL, and WLN showed significantly larger stipe diameters compared to other groups (Figure 8). No significant differences were observed for the other four traits among the groups.

A

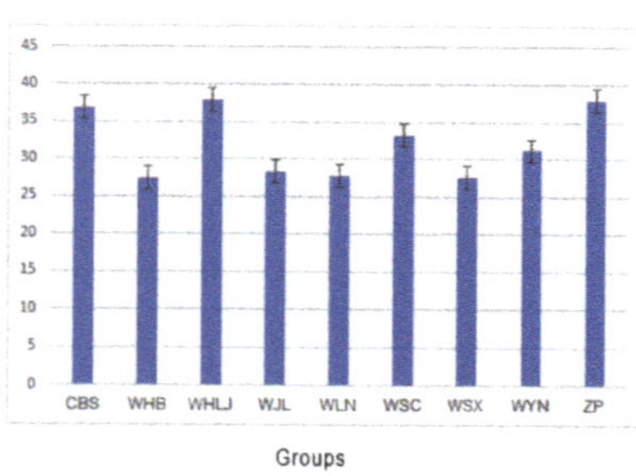


B

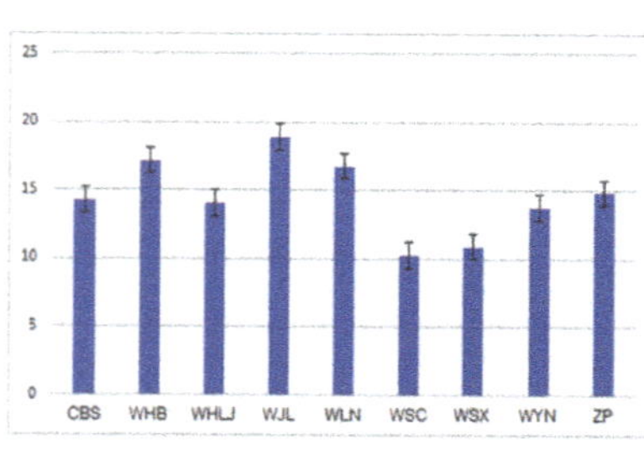


**Figure 8.** *Cont.*

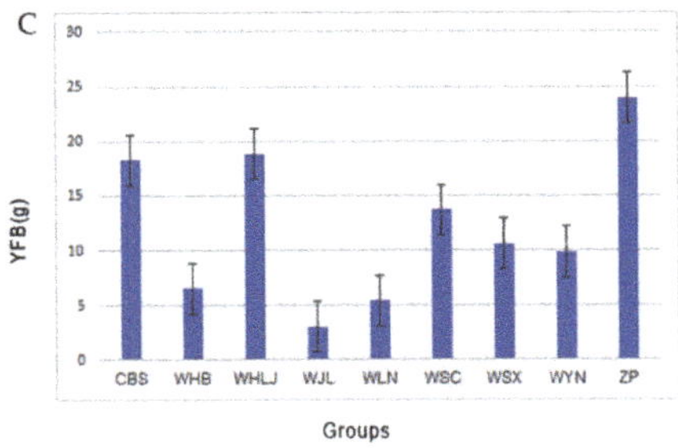


**Figure 8.** Variations among groups in agronomic traits: (**A**) variations among groups in cap width (CW); (**B**) variations among groups in stipe diameter (SD); (**C**) variations among groups in the yield of fruiting body per bottle (YFB). The *x*-axis represents different groups, and the *y*-axis represents the range of trait values.

Correlation analysis revealed a significant positive correlation between PPG and PIH (correlation coefficient = 0.77). Cap length and width were also significantly positively correlated (correlation coefficient = 0.91). Additionally, YFB showed a positive correlation with CL, CW, and SL, decreasing in strength in that order. Conversely, YFB exhibited a negative correlation with SD (Figure 9, Table S7). The negative correlation between YFB and SD suggests that a thicker stipe may adversely affect overall quality. Furthermore, harvesting time was negatively correlated with cap length and width, indicating that later fruiting resulted in smaller cap sizes.

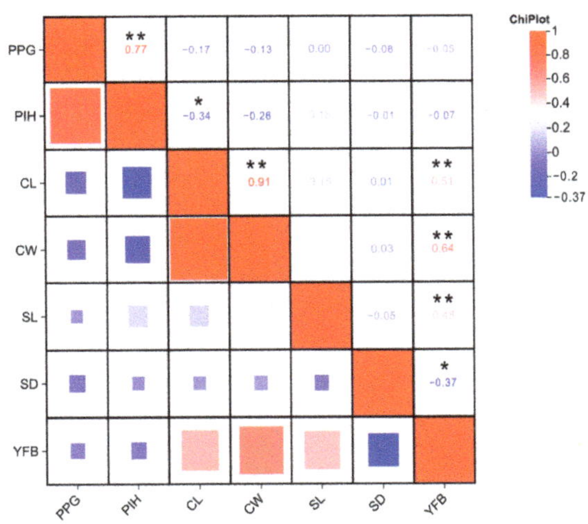


**Figure 9.** Phenotypic correlation analysis heat map. From top to bottom and left to right, are in the same order, corresponding to the period from inoculation to primordia generation (PPG), period from inoculation to harvesting (PIH), cap length (CL), cap width (CW), stipe length (SL), stipe diameter (SD), and yield of fruiting body per bottle yield per bottle (YFB). Sample size = 42; * and ** indicate the significance at 0.05 and 0.01 level, respectively.

### 3.4. GWAS Analysis Results

In this GWAS, significant SNPs across various scaffolds were identified for key traits (Figure 10, Table 4). On scaffold SJKF01000008.1, SNP rs_474319 explained only 0.02% of the variance in cap length (CL), indicating a minimal effect. Conversely, SNP rs_338157 on scaffold SJKF01000016.1 had a strong influence on cap width (CW) with a proportion of variance explained (PVE) of 40.45%. Scaffolds SJKF01000006.1 and SJKF01000003.1 also showed high PVEs for CL with SNPs rs_3071735 (47.30%) and rs_3643242 (48.76%).

For stipe diameter (SD), SNP rs_3145491 on scaffold SJKF01000005.1 showed a smaller effect (PVE 4.29%). In stipe length (SL), SNPs rs_499031 and rs_2285326 on scaffolds SJKF01000007.1 and SJKF01000005.1, respectively, made significant contributions with PVEs of 28.29% and 32.72%. The highest impact was observed on scaffold SJKF01000002.1, where SNP rs_4651725 influenced yield with a PVE of 65.72%.

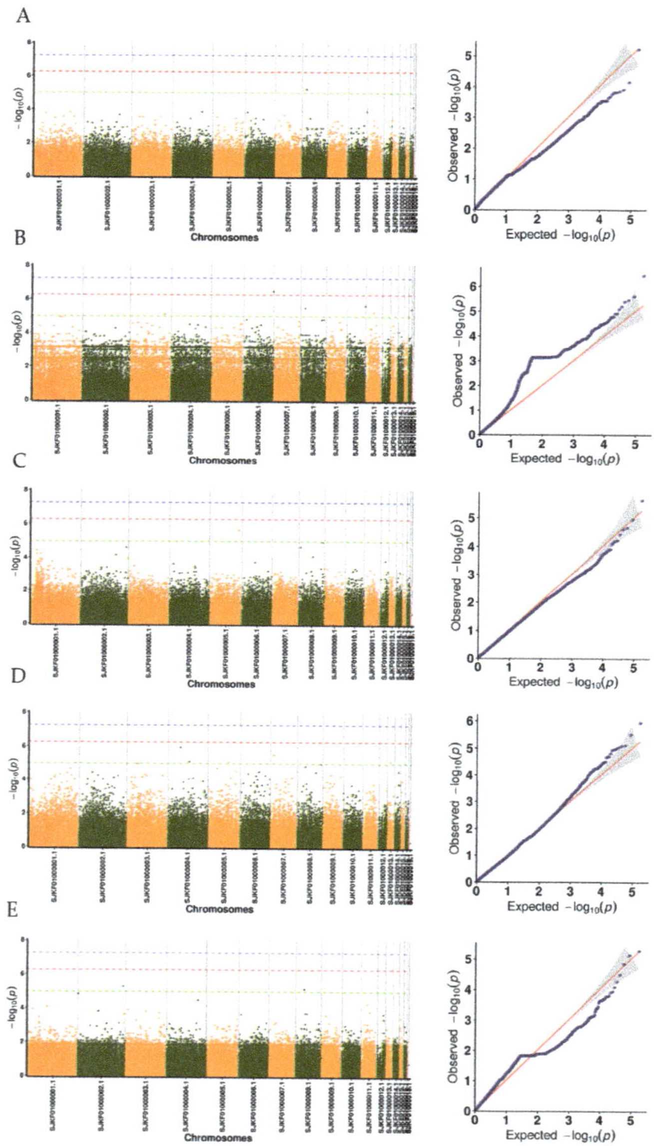


**Figure 10.** Manhattan plot (**left**) and quantile–quantile (Q-Q) plot (**right**) of SNP *p*-values in the whole population. In the Manhattan plot, *y*-axis presents the observed SNP $-\log_{10}(p)$ and *x*-axis the SNP positions across the 20 scaffolds. Horizontal lines show the genome-wide significant threshold; the green dashed line represents the value of $-\log_{10}(p) = 5$. In the Q-Q plot, the *y*-axis and *x*-axis represent the observed SNP $-\log_{10}(p)$ and expected $-\log_{10}(p)$, respectively. From top to bottom: (**A**) cap length (CL), (**B**) cap width (CW), (**C**) stipe diameter (SD), (**D**) stipe length (SL), and (**E**) yield of fruiting body per bottle yield per bottle (YFB).

**Table 4.** Proportion of variance explained (PVE%) and minor allele frequency (MAF) per significant SNP detected in whole population sampling.

| Trait | Scaffold | $p$-Value | $-\log_{10}p$ | Allele | MAF [1] | SNP ID | PVE [2] |
|---|---|---|---|---|---|---|---|
| CL [3] | SJKF01000008.1 | $6.03 \times 10^{-6}$ | 5.22 | A/G | 0.39 | rs_474319 | 0.02% |
| | SJKF01000016.1 | $3.92 \times 10^{-6}$ | 5.41 | A/C | 0.41 | rs_338157 | 40.45% |
| | SJKF01000008.1 | $3.36 \times 10^{-6}$ | 5.47 | A/G | 0.39 | rs_474319 | 0.38% |
| CW | SJKF01000011.1 | $6.11 \times 10^{-6}$ | 5.21 | T/C | 0.22 | rs_949638 | 0.30% |
| | SJKF01000010.1 | $2.40 \times 10^{-6}$ | 5.62 | C/T | 0.41 | rs_1934108 | 0.43% |
| | SJKF01000006.1 | $3.44 \times 10^{-7}$ | 6.46 | T/A | 0.17 | rs_3071735 | 47.30% |
| | SJKF01000003.1 | $8.07 \times 10^{-6}$ | 5.09 | T/C | 0.2 | rs_3643242 | 48.76% |
| SD | SJKF01000005.1 | $2.34 \times 10^{-6}$ | 5.63 | A/G | 0.21 | rs_3145491 | 4.29% |
| | SJKF01000013.1 | $7.60 \times 10^{-6}$ | 5.12 | G/T | 0.28 | rs_459476 | 19.54% |
| | SJKF01000007.1 | $3.04 \times 10^{-6}$ | 5.52 | G/A | 0.13 | rs_499031 | 28.29% |
| SL | SJKF01000004.1 | $1.14 \times 10^{-6}$ | 5.94 | C/T | 0.12 | rs_1310326 | 20.78% |
| | SJKF01000004.1 | $7.99 \times 10^{-6}$ | 5.1 | T/C | 0.43 | rs_2224699 | 4.55% |
| | SJKF01000005.1 | $9.00 \times 10^{-6}$ | 5.05 | G/A | 0.27 | rs_2285326 | 32.72% |
| YFB | SJKF01000008.1 | $7.24 \times 10^{-6}$ | 5.14 | T/G | 0.17 | rs_885485 | 0.64% |
| | SJKF01000002.1 | $5.17 \times 10^{-6}$ | 5.29 | A/C | 0.27 | rs_4651725 | 65.72% |

[1] MAF: minor allele frequency; [2] PVE: phenotypic variance explained; [3] cap length (CL); cap width (CW); stipe diameter (SD); stipe length (SL); yield of fruiting body per bottle yield per bottle (YFB).

To avoid false positives, the long PPG periods of four strains (ZP1, ZP2, WCS5, and WYN1) and the PIH periods of five strains (WSC5, ZP2, WYN1, WJL1, WLN1) were excluded as outliers in the GWASs. After their exclusion, no significant SNP loci were identified for traits PPG and PIH.

## 4. Discussion

### 4.1. Genetic Diversity Analysis of *P. pulmonarius* Based on SNP

Traditional germplasm identification methods are characterized by long identification cycles, high costs, and susceptibility to environmental factors. Therefore, DNA molecular markers are crucial for germplasm identification at the molecular level. Compared to other molecular markers, SNP-based diversity analysis provides a more accurate and detailed interpretation of phenotypic differences at a lower cost, making it the most widely applied molecular marker [34,35]. SNPs, which represent the most common types of nucleotide sequence polymorphisms—including insertions and deletions—are abundant throughout the genome. These co-dominant markers are highly reproducible and suitable for genetic diversity and population structure mapping [36–40]. Previous studies have shown that SNPs are closely associated with individual traits, where different SNP positions may affect gene expression and phenotype through various mechanisms. The functional impact depends on whether the nucleotide variation affects protein structure or function.

Despite its relatively short cultivation history in China, *P. pulmonarius* holds promising economic potential and market prospects. Assessing genetic diversity in germplasm resources is essential for resource identification, new gene discovery, and breeding of new varieties. Current challenges in *P. pulmonarius* breeding include issues such as misidentification of varieties, insufficient research on breeding practices, and limited depth in genetic studies. Additionally, some *P. pulmonarius* varieties suffer from degeneration and poor disease resistance after several years of introduction. Understanding the genetic relationships and population structure of *P. pulmonarius* strains from different regions of China at the genomic level is essential for addressing these challenges.

To advance the breeding and application of *P. pulmonarius* germplasm resources, this study conducted whole-genome resequencing of 47 strains collected nationwide. A total of 181,731 high-quality SNP loci were identified after filtering, indicating substantial genetic variation among the strains. The significant number and proportion of SNPs make these strains well suited for subsequent genetic evolution andGWAS. These SNPs were used for DNA identity verification, phylogenetic relationship analysis, and genetic diversity assessment. Phylogenetic analysis revealed that most currently cultivated *P. pulmonarius* germplasm originates from the same source, exhibiting close genetic relationships and potential synonymy issues. In contrast, wild germplasm showed greater genetic diversity. Strains from the same geographic origin were grouped into the same cluster, a pattern also observed in other mushroom species [41]. The results of principal component analysis and population structure analysis were consistent with the phylogenetic tree, underscoring the reliability of SNP markers in assessing genetic relatedness in *P. pulmonarius* strains.

DNA fingerprinting, based on molecular markers or specific sequences, was used to assess genetic diversity and effectively differentiate strains. DNA fingerprinting has been widely applied in genetic diversity analyses of plants and microbes [42], providing valuable tools and technical support for germplasm identification and SNP-based fingerprinting methods [43]. In this study, a fingerprint atlas was constructed using 369 core SNP loci, offering unique genetic identification for each strain. These results provide a foundation for studies on genetic diversity, quality control, and molecular breeding.

### 4.2. Evaluation of Agronomic Traits

*P. pulmonarius*, one of the most widely cultivated edible fungi in China, faces challenges due to chaotic germplasm and limited genetic diversity in cultivars. These limitations stem from tissue separation, monosporal and polysporal hybridization, long-term artificial selection, and strain introductions across cultivation areas. These factors complicate the effective utilization and innovation of *P. pulmonarius* germplasm resources, restricting available breeding materials. Furthermore, unlike green plants, edible fungi exhibit lower morphological differentiation, and oyster mushroom fruiting body morphology is highly influenced by environmental conditions. Therefore, optimizing multiple traits is essential to maximize productivity when selecting superior varieties, necessitating the evaluation of germplasm based on multiple traits rather than a single trait [44].

In this study, we conducted comprehensive statistical analyses on seven key traits related to mushroom production in 47 *P. pulmonarius* strains. According to the ANOVA results, all seven traits were identified as quantitative traits primarily determined by genetic variation. The high broad-sense heritability values indicated that genetic factors are the primary contributors to the observed variation. Correlation analyses revealed significant associations, such as between the period of primordia generation and the period from inoculation to harvesting, as well as between cap length and cap width. Strains from Groups ZP, CBS, and WHLJ generally exhibited superior agronomic traits compared to others, suggesting their potential as breeding materials. The outliers observed for the traits PPG and PIH may be attributed to the low mycelial activity of certain strains, underscoring the need for improved strain preservation to ensure a robust analysis and evaluation system for *P. pulmonarius* germplasm. These findings provide valuable insights for optimizing breeding and cultivation strategies for *P. pulmonarius*.

### 4.3. Genome-Wide Association Study

GWASs rely on a comprehensive set of variants covering the entire genome. Initially, these studies used low-throughput markers like RFLP, AFLP, and SSR. However, with advancements in technology, the focus has shifted to higher-throughput and more stable SNPs, which now dominate the field. A GWAS identifies specific alleles contributing to desirable traits, facilitating genetic improvement efforts. While GWASs have been extensively applied in plants and animals to uncover genes associated with agronomic traits, their use in edible fungi remains limited. In this study, 47 *P. pulmonarius* strains from diverse geographic origins, showing significant phenotypic variation in agronomic traits, were used for a GWAS. All seven traits were identified as polygenic, controlled by multiple genes. Effective GWAS models, such as the linear mixed model (LMM) provided by GEMMA software, helped mitigate false positives and identified numerous key SNPs and candidate genes associated with these traits.

For instance, SNP rs_474319 on scaffold SJKF01000008.1, though statistically significant, had a low proportion of variance explained (PVE) at 0.02% for cap length, indicating a minimal impact on this trait. In contrast, SNPs like rs_338157 on SJKF01000016.1, rs_3071735 on SJKF01000006.1, and rs_3643242 on SJKF01000003.1 exhibited high PVE values (ranging from 40.45% to 48.76%), underscoring their significant roles in the genetic control of cap width and cap length. Similarly, SNPs associated with stipe length, such as rs_499031 on scaffold SJKF01000007.1 and rs_2285326 on SJKF01000005.1, showed high PVEs (28.29% and 32.72%, respectively), highlighting their substantial influence on this trait. These findings suggest that the genetic architecture of traits like cap length, cap width, and stipe length is shaped by SNPs distributed across multiple scaffolds, reflecting the complex genetic regulatory networks underlying these phenotypic characteristics.

Additionally, scaffold SJKF01000005.1 exhibited pleiotropic effects on both stipe diameter (SD) and stipe length (SL), while scaffold SJKF01000008.1 showed pleiotropy on cap length (CL), cap width (CW), and yield of fruiting bodies (YFB). The observed phenotypic correlations were effectively explained by genetic correlations, providing valuable insights for future breeding programs and molecular improvement strategies. Notably, SNP rs_4651725, associated with the YFB trait, exhibited a high PVE of 65.72% despite a relatively modest -log10P value of 5.29, making it less prominent in the Manhattan plot. This discrepancy may arise because the *p*-value primarily reflects statistical significance, which can be influenced by limited sample size, discontinuous phenotypic expression from the micro-cultivation system, and lack of strong natural selection, potentially causing the SNP to appear less significant. Additionally, the large PVE value might result from the high coefficient of variation of YFB among strains. The PVE quantifies the proportion of phenotypic variance explained by the SNP, highlighting its contribution to the traits. Thus, an SNP that marginally exceeds the significance threshold may still demonstrate a substantial PVE.

After removing outliers, no significant SNP associations were found for the period from inoculation to harvest or the primordium formation period. To detect significant SNPs associated with these traits, future experiments should increase sample size and conduct repeated trials to address outliers.

To date, GWASs have been used in various species, including staple crops like rice [45], model organisms such as *Arabidopsis thaliana* [46], and other plants like sorghum [47]. Their combination with genomics, transcriptomics, and metabolomics has provided a deeper understanding of species biology and genetics. This multidisciplinary approach helps clarify the genetic basis of complex traits. Looking ahead, GWAS technology is expected to continue evolving, with promising applications in the field of edible fungi. Future developments are likely to enhance its effectiveness in understanding genetic structures, optimizing breeding strategies, and contributing to the development of improved fungal varieties.

## 5. Conclusions

In this study, sequencing of 47 *P. pulmonarius* strains identified 4,430,948 SNPs. Analysis using 181,731 SNP markers revealed six distinct subpopulations, and a DNA fingerprint was developed using 369 core markers. Genetic diversity analysis showed minimal variation within cultivated strains from the same area but significant diversity among wild strains from different regions. GWASs identified key SNPs and candidate genes associated with five phenotypic traits, providing valuable insights for targeted breeding of improved edible fungus strains.

**Supplementary Materials:** The following supporting information can be downloaded at: https://www.mdpi.com/article/10.3390/agriculture14112023/s1, Table S1: PCA data; Table S2: K value; Table S3: Core SNP; Table S4: Core SNP filtering; Table S5: Phenotypic statistics of 47 *Pleurotus pulmonarius* strains; Table S6: ANOVA among different groups; Table S7: Correlation analysis.

**Author Contributions:** Formal analysis, Q.L. and X.Y.; Resources, W.G.; Visualization, Y.Y.; Writing—original draft, Q.L. and X.Y.; Writing—review and editing, W.G.; Supervision and project administration, W.G. All authors have read and agreed to the published version of the manuscript.

**Funding:** This research was funded by Lin'an District People's Government and Chinese Academy of Agricultural Sciences Collaborative Project (NH202244), the Jiangxi Facility Vegetable Project (JXNK202303-06), and the Beijing Innovation Consortium of Agriculture Research System (BAIC03).

**Institutional Review Board Statement:** Not applicable.

**Data Availability Statement:** Data are contained within the article or Supplementary Material.

**Conflicts of Interest:** The authors declare no conflicts of interest.

## References


1. Zhu, J.; Liu, X.; Xie, B.; Liu, F.; Deng, Y.; Xiong, F. Analysis of germplasm resources of *Pleurotus geesteranus* by ITS-RFLP. *J. Fujian Agric. For. Univ. (Nat. Sci. Ed.)* **2009**, *38*, 186–191.
2. Liu, X.; Ye, L.; Zhang, L.; Xie, B.; Wu, X. Mating type analyses of cultivated *Pleurotus pulmonarius* in China. *Mycosystema* **2021**, *40*, 3109–3117.
3. Chen, G.G.; Lin, Y.; Liu, X.R.; Zhao, G.H.; Chen, J. Recent advances in genetics and breeding of *Pleurotus pulmonarius*. *Chin. J. Trop. Crops* **2017**, *38*, 1377–1381.
4. Lee, H.Y.; Moon, S.; Shim, D.; Hong, C.P.; Lee, Y.; Koo, C.D.; Chung, J.W.; Ryu, H. Development of 44 Novel Polymorphic SSR Markers for Determination of Shiitake Mushroom (*Lentinula edodes*) Cultivars. *Genes* **2017**, *8*, 109. [CrossRef]
5. Saito, T.; Sakuta, G.; Kobayashi, H.; Ouchi, K.; Inatomi, S. Genetically Independent Tetranucleotide to Hexanucleotide Core Motif SSR Markers for Identifying *Lentinula edodes* Cultivars. *Mycobiology* **2019**, *47*, 466–472. [CrossRef]
6. Kim, K.H.; Ka, K.H.; Kang, J.H.; Kim, S.; Lee, J.W.; Jeon, B.K.; Yun, J.K.; Park, S.R.; Lee, H.J. Identification of Single Nucleotide Polymorphism Markers in the Laccase Gene of Shiitake Mushrooms (*Lentinula edodes*). *Mycobiology* **2015**, *43*, 75–80. [CrossRef]
7. Liu, X.B.; Feng, B.; Li, J.; Yan, C.; Yang, Z.L. Genetic diversity and breeding history of Winter Mushroom (*Flammulina velutipes*) in China uncovered by genomic SSR markers. *Gene* **2016**, *591*, 227–235. [CrossRef]
8. Liu, X.B.; Li, J.; Yang, Z.L. Genetic diversity and structure of core collection of winter mushroom (*Flammulina velutipes*) developed by genomic SSR markers. *Hereditas* **2018**, *155*, 3. [CrossRef]
9. Wang, L.; Gao, W.; Wang, Q.; Qu, J.; Zhang, J.; Huang, C. Identification of commercial cultivars of *Agaricus bisporus* in China using genome-wide microsatellite markers. *J. Integr. Agric.* **2019**, *18*, 580–589. [CrossRef]
10. Rostoks, N.; Ramsay, L.; Mackenzie, K.; Cardle, L.; Bhat, P.R.; Roose, M.L.; Svensson, J.T.; Stein, N.; Varshney, R.K.; Marshall, D.F.; et al. Recent history of artificial outcrossing facilitates whole-genome association mapping in elite inbred crop varieties. *Proc. Natl. Acad. Sci. USA* **2006**, *103*, 18656–18661. [CrossRef] [PubMed]
11. Okuda, Y.; Murakami, S.; Matsumoto, T. A genetic linkage map of *Pleurotus pulmonarius* based on AFLP markers, and localization of the gene region for the sporeless mutation. *Genome* **2009**, *52*, 438–446. [CrossRef]
12. Vidal-Diez, D.U.G.; Lee, Y.Y.; Stajich, J.E.; Schwarz, E.M.; Hsueh, Y.P. Genomic analyses of two Italian oyster mushroom *Pleurotus pulmonarius* strains. *G3* **2021**, *11*, jkaa007. [CrossRef]
13. Breseghello, F.; Sorrells, M.E. Association mapping of kernel size and milling quality in wheat (*Triticum aestivum* L.) cultivars. *Genetics* **2006**, *172*, 1165–1177. [CrossRef]
14. Huang, X.; Wei, X.; Sang, T.; Zhao, Q.; Feng, Q.; Zhao, Y.; Li, C.; Zhu, C.; Lu, T.; Zhang, Z.; et al. Genome-wide association studies of 14 agronomic traits in rice landraces. *Nat. Genet.* **2010**, *42*, 961–967. [CrossRef]
15. Olsen, K.M.; Halldorsdottir, S.S.; Stinchcombe, J.R.; Weinig, C.; Schmitt, J.; Purugganan, M.D. Linkage disequilibrium mapping of *Arabidopsis* CRY2 flowering time alleles. *Genetics* **2004**, *167*, 1361–1369. [CrossRef]
16. Risch, N.; Merikangas, K. The future of genetic studies of complex human diseases. *Science* **1996**, *273*, 1516–1517. [CrossRef]
17. Li, C.; Gong, W.; Zhang, L.; Yang, Z.; Nong, W.; Bian, Y.; Kwan, H.S.; Cheung, M.K.; Xiao, Y. Association Mapping Reveals Genetic Loci Associated with Important Agronomic Traits in *Lentinula edodes*, Shiitake Mushroom. *Front. Microbiol.* **2017**, *8*, 237. [CrossRef]
18. Cardoso, W.S.; Soares, F.; Queiroz, P.V.; Tavares, G.P.; Santos, F.A.; Sufiate, B.L.; Kasuya, M.; de Queiroz, J.H. Minimum cocktail of cellulolytic multi-enzyme complexes obtained from white rot fungi via solid-state fermentation. *3 Biotech* **2018**, *8*, 46. [CrossRef]
19. Corner, E.J.H. *The Agaric Genera Lentinus, Panus, and Pleurotus, with Particular Reference to Malaysian Species*; J. Cramer: Vaduz, Liechtenstein, 1981; p. 169. ISBN 85192445.
20. Yu, H.; Zhang, L.; Shang, X.; Peng, B.; Li, Y.; Xiao, S.; Tan, Q.; Fu, Y. Chromosomal genome and population genetic analyses to reveal genetic architecture, breeding history and genes related to cadmium accumulation in *Lentinula edodes*. *BMC Genom.* **2022**, *23*, 120. [CrossRef]
21. Zhang, Y.; Huang, C.; van Peer, A.F.; Sonnenberg, A.; Zhao, M.; Gao, W. Fine Mapping and Functional Analysis of the Gene *PcTYR*, Involved in Control of Cap Color of *Pleurotus cornucopiae*. *Appl. Environ. Microbiol.* **2022**, *88*, e0217321. [CrossRef]
22. Huang, X.; Duan, N.; Xu, H.; Xie, T.N.; Xue, Y.R.; Liu, C.H. CTAB-PEG DNA Extraction from Fungi with High Contents of Polysaccharides. *Mol. Biol.* **2018**, *52*, 718–726. [CrossRef]

23. Mckenna, A.; Hanna, M.; Banks, E.; Sivachenko, A.; Cibulskis, K.; Kernytsky, A.; Garimella, K.; Altshuler, D.; Gabriel, S.; Daly, M.; et al. The Genome Analysis Toolkit: A MapReduce framework for analyzing next-generation DNA sequencing data. *Genome Res.* **2010**, *20*, 1297–1303. [CrossRef]
24. Danecek, P.; Bonfield, J.K.; Liddle, J.; Marshall, J.; Ohan, V.; Pollard, M.O.; Whitwham, A.; Keane, T.; Mccarthy, S.A.; Davies, R.M.; et al. Twelve years of SAMtools and BCFtools. *Gigascience* **2021**, *10*, giab008. [CrossRef] [PubMed]
25. Kumar, S.; Stecher, G.; Li, M.; Knyaz, C.; Tamura, K. MEGA X: Molecular Evolutionary Genetics Analysis across Computing Platforms. *Mol. Biol. Evol.* **2018**, *35*, 1547–1549. [CrossRef]
26. Danecek, P.; Auton, A.; Abecasis, G.; Albers, C.A.; Banks, E.; Depristo, M.A.; Handsaker, R.E.; Lunter, G.; Marth, G.T.; Sherry, S.T.; et al. The variant call format and VCFtools. *Bioinformatics* **2011**, *27*, 2156–2158. [CrossRef]
27. Alexander, D.H.; Novembre, J.; Lange, K. Fast model-based estimation of ancestry in unrelated individuals. *Genome Res.* **2009**, *19*, 1655–1664. [CrossRef]
28. Price, A.L.; Patterson, N.J.; Plenge, R.M.; Weinblatt, M.E.; Shadick, N.A.; Reich, D. Principal components analysis corrects for stratification in genome-wide association studies. *Nat. Genet.* **2006**, *38*, 904–909. [CrossRef]
29. Slifer, S.H. PLINK: Key Functions for Data Analysis. *Curr. Protoc. Hum. Genet.* **2018**, *97*, e59. [CrossRef]
30. Wang, Y.; Lv, H.; Xiang, X.; Yang, A.; Feng, Q.; Dai, P.; Li, Y.; Jiang, X.; Liu, G.; Zhang, X. Construction of a SNP Fingerprinting Database and Population Genetic Analysis of *Cigar Tobacco* Germplasm Resources in China. *Front. Plant Sci.* **2021**, *12*, 618133. [CrossRef]
31. Valeri, L.; Vanderweele, T.J. Mediation analysis allowing for exposure-mediator interactions and causal interpretation: Theoretical assumptions and implementation with SAS and SPSS macros. *Psychol. Methods* **2013**, *18*, 137–150. [CrossRef]
32. Mucha, A.; Wierzbicki, H. Linear models for breeding values prediction in haplotype-assisted selection—An analysis of QTL-MAS Workshop 2011 Data. *BMC Proc.* **2012**, *6* (Suppl. S2), S11. [CrossRef] [PubMed]
33. Zhou, X.; Stephens, M. Genome-wide efficient mixed-model analysis for association studies. *Nat. Genet.* **2012**, *44*, 821–824. [CrossRef] [PubMed]
34. Trick, M.; Long, Y.; Meng, J.; Bancroft, I. Single nucleotide polymorphism (SNP) discovery in the polyploid Brassica napus using Solexa transcriptome sequencing. *Plant Biotechnol. J.* **2009**, *7*, 334–346. [CrossRef]
35. Schlotterer, C. The evolution of molecular markers--just a matter of fashion? *Nat. Rev. Genet.* **2004**, *5*, 63–69. [CrossRef]
36. Cabezas, J.A.; Ibanez, J.; Lijavetzky, D.; Velez, D.; Bravo, G.; Rodriguez, V.; Carreno, I.; Jermakow, A.M.; Carreno, J.; Ruiz-Garcia, L.; et al. A 48 SNP set for grapevine cultivar identification. *BMC Plant Biol.* **2011**, *11*, 153. [CrossRef] [PubMed]
37. Gao, W.; Qu, J.; Zhang, J.; Sonnenberg, A.; Chen, Q.; Zhang, Y.; Huang, C. A genetic linkage map of *Pleurotus tuoliensis* integrated with physical mapping of the de novo sequenced genome and the mating type loci. *BMC Genom.* **2018**, *19*, 18. [CrossRef]
38. An, H.; Lee, H.Y.; Shim, D.; Choi, S.H.; Cho, H.; Hyun, T.K.; Jo, I.H.; Chung, J.W. Development of CAPS Markers for Evaluation of Genetic Diversity and Population Structure in the Germplasm of Button Mushroom (*Agaricus bisporus*). *J. Fungi* **2021**, *7*, 375. [CrossRef]
39. Yang, Q.; Zhang, J.; Shi, X.; Chen, L.; Qin, J.; Zhang, M.; Yang, C.; Song, Q.; Yan, L. Development of SNP marker panels for genotyping by target sequencing (GBTS) and its application in soybean. *Mol. Breed.* **2023**, *43*, 26. [CrossRef] [PubMed]
40. Gao, W.; Weijn, A.; Baars, J.J.; Mes, J.J.; Visser, R.G.; Sonnenberg, A.S. Quantitative trait locus mapping for bruising sensitivity and cap color of *Agaricus bisporus* (button mushrooms). *Fungal Genet. Biol.* **2015**, *77*, 69–81. [CrossRef]
41. Gu, M.; Chen, Q.; Zhang, Y.; Zhao, Y.; Wang, L.; Wu, X.; Zhao, M.; Gao, W. Evaluation of Genetic Diversity and Agronomic Traits of Germplasm Resources of *Stropharia rugosoannulata*. *Horticulturae* **2024**, *10*, 213. [CrossRef]
42. Tian, H.L.; Wang, F.G.; Zhao, J.R.; Yi, H.M.; Wang, L.; Wang, R.; Yang, Y.; Song, W. Development of maizeSNP3072, a high-throughput compatible SNP array, for DNA fingerprinting identification of Chinese maize varieties. *Mol. Breed.* **2015**, *35*, 136. [CrossRef] [PubMed]
43. Yang, J.; Zhang, J.; Han, R.; Zhang, F.; Mao, A.; Luo, J.; Dong, B.; Liu, H.; Tang, H.; Zhang, J.; et al. Target SSR-Seq: A Novel SSR Genotyping Technology Associate with Perfect SSRs in Genetic Analysis of Cucumber Varieties. *Front. Plant Sci.* **2019**, *10*, 531. [CrossRef] [PubMed]
44. Gao, W.; Baars, J.; Maliepaard, C.; Visser, R.; Zhang, J.; Sonnenberg, A. Multi-trait QTL analysis for agronomic and quality characters of *Agaricus bisporus* (button mushrooms). *AMB Express* **2016**, *6*, 67. [CrossRef] [PubMed]
45. Huang, X.; Zhao, Y.; Wei, X.; Li, C.; Wang, A.; Zhao, Q.; Li, W.; Guo, Y.; Deng, L.; Zhu, C.; et al. Genome-wide association study of flowering time and grain yield traits in a worldwide collection of rice germplasm. *Nat. Genet.* **2011**, *44*, 32–39. [CrossRef]
46. Atwell, S.; Huang, Y.S.; Vilhjalmsson, B.J.; Willems, G.; Horton, M.; Li, Y.; Meng, D.; Platt, A.; Tarone, A.M.; Hu, T.T.; et al. Genome-wide association study of 107 phenotypes in *Arabidopsis* thaliana inbred lines. *Nature* **2010**, *465*, 627–631. [CrossRef]
47. Morris, G.P.; Ramu, P.; Deshpande, S.P.; Hash, C.T.; Shah, T.; Upadhyaya, H.D.; Riera-Lizarazu, O.; Brown, P.J.; Acharya, C.B.; Mitchell, S.E.; et al. Population genomic and genome-wide association studies of agroclimatic traits in sorghum. *Proc. Natl. Acad. Sci. USA* **2013**, *110*, 453–458. [CrossRef]

agriculture

Article

# Genomic Inference Unveils Population Bottlenecks and a North-to-South Migration Pattern of Wild *Cordyceps militaris* Across China

**Tianqiao Yong *,†, Yuanchao Liu †, Manjun Cai †, Lijun Zhuo, Xiaoxian Wu, Huiyang Guo, Huiping Hu *, Yichuang Gao, Shaodan Chen, Yizhen Xie and Wei Zhong**

National Health Commission Science and Technology Innovation Platform for Nutrition and Safety of Microbial Food, Guangdong Provincial Key Laboratory of Microbial Safety and Health and State Key Laboratory of Applied Microbiology Southern China, Institute of Microbiology, Guangdong Academy of Sciences, Guangzhou 510070, China; liuyc1020@163.com (Y.L.); caimanjun4439@webmail.hzau.edu.cn (M.C.); zhuolijun@gdim.cn (L.Z.); xiaoxian9911@163.com (X.W.); guohuiyang0309@163.com (H.G.); gaoyichuang@gdim.cn (Y.G.); chensd@gdim.cn (S.C.); xieyz@gdim.cn (Y.X.); zhongwei@gdim.cn (W.Z.)

* Correspondence: tianqiao@mail.ustc.edu.cn (T.Y.); huhp@gdim.cn (H.H.); Tel.: +86-20-8768-8132 (T.Y. & H.H.); Fax: +86-20-8768-0942 (T.Y. & H.H.)

† These authors contributed equally to this work.

**Abstract:** The Ascomycete genus *Cordyceps* affects plant crops significantly, filling an important ecological niche. *Cordyceps militaris* (L.) Fr. presents many health benefits for humans, but its population history has not been reported. The objective of this research was to report the collection, population structure, demographic history, diversity, and cytosine deaminases of 43 wild strains of *C. militaris* in China through resequencing using an Illumina HiseqTM platform. All strains were assigned to the warm, subtropical, and middle temperate zone populations, confirmed by ADMIXTURE-1.3.0, PCA, and phylogenic analysis. Their population sizes declined historically, suggesting that this species suffered from bottlenecks in the wild. LD decays ($r^2$) revealed a north-to-south migration pattern of wild *C. militaris*, consistent with the MSMC2-v2.1.4 analysis. The regions of high Pi were aggregating at the chromosomes CP023325.1 (51) and CP023323.1 (9), playing a key role in adaptation, especially for the sites on cytosine deaminase. Within the species, genetic differentiation was relatively high among the three populations ($F_{st}$ = 0.083, 0.092, and even 0.109). According to the artificial intelligence-assisted (RoseTTAFold) predicted structures of the cytosine deaminases, they were classified into eight clades with unique, distinct, and structurally conserved domains, offering a potential suite of single- and double-stranded deaminases of great promise as tunable base editors for therapeutic and agricultural breeding applications. These provided new insights for mining novel proteins from macrofungi, structurally and functionally.

**Keywords:** genomics; population genetics; distribution; diversity; functional genes

## 1. Introduction

The Ascomycete genus *Cordyceps* [1–3], which affects plant crops significantly, consists of 500 plus species [4,5] and is associated with fighting injurious insects, arthropods, nonarthropod microinvertebrates, and pathogens and parasites on food crops by causing infectious diseases against them and then filling an important ecological niche in the ecosystem [6–8]. Generally, these fungi always penetrate the cuticles of their host when competing with serious and complex interactions with the host, and a successful infection is only built when all barriers and reactions of the host are overcome or defeated [9]. Given their outstanding performance in infecting injurious insects and protecting agricultural crops, these fungi are usually developed as biocontrol agents, which demonstrate superiorities of biosafety, non-pollution, and degradation; since they come from nature and are utilized in nature to cause significant pest population reductions, they are important in many agricultural systems [10].

*Cordyceps militaris* (L.) Fr. occurs throughout much of the Northern Hemisphere and fills an important ecological niche as a pathogenic fungus against lepidopteran insect pupae, controlling the outbreaks of many harmful lepidopterans that harm agricultural production [11,12], such as *Pyrausta nubilalis* Hubern and *Dendrolimus punctatus* Walker, the pupae of which are dormant in the soil and emerge as adult moths that eat various plants and reduce production. Moreover, *C. militaris* follows the classic life cycle of entomopathogenic fungi, secreting multiple extracellular enzymes to invade its host efficiently during the pathogen–host interaction [13], including proteases [13], chitinases [14], and lipases [15], which degrade the structural components of the host. Among these, lipase (EC 3.1.1.3), which plays a crucial role in infecting hosts, is a water-soluble enzyme for catalyzing the hydrolysis reaction of apolar and water-insoluble ester substrates, such as long-chain triacylglycerols, and initiating the conidial adhesion on the epicuticle surface of its host in the early stage by releasing free fatty acids and alkenes and enhancing the hydrophobic interactions between the conidial surface and epicuticle through destroying the hydrophobic barrier of the cuticles of the insects and providing nutrients for fungal colonization and growth, which consist of lipoproteins, acylglycerol, and long-chain alkene esters [15]. These functional elements and mechanisms represent next-generation biotechnologies that may benefit people by converting various biomasses into major food sources.

For centuries, *C. militaris* has been used as a traditional herbal medicine in China, Japan, Korea, and other Asian countries [16] due to the large range of bioactive compounds it produces and its various benefits for protecting health [17], such as anti-hyperuricemia properties. A series of pharmacologically active components have been isolated and characterized, consisting of cordycepin, cordycepic acid, flavones, carotenes, macrolides, and polysaccharides, some of which play a role as valuable chemical markers for quality control, including cordycepin (3′-deoxyadenosine), which was the earliest and arguably most significant natural product first isolated in 1950 from *C. militaris* as the major bioactive component [18]. These components exhibit broad bioactivity as antimicrobial and antivirus agents and polyadenylation inhibitors, some of which are undergoing clinical trials. Their fungal biosynthesis was unveiled to be coupled with the production of pentostatin, which is an important drug against cancer, clinically [19]. Their biosynthesis involves HisG-type ATP phosphoribosyltransferase, xidoreductase (PenB), and phosphoribosylaminoimidazolesuccinocarboxamide synthetase. Also, the characteristic component cordycepin has served as an important skeleton for designing and synthesizing anti-virus drugs and prodrugs since it has a structure similar to that of a nucleotide, which may disturb the DNA replication of viruses [20]. Overall, it inhibits tumors [21] and is an immune or metabolic modulator for preventing the deterioration of inner organs such as the lungs, liver [22], and kidneys. With these benefits and so many useful genes, *C. militaris* was cultivated widely in China and other countries for commercial purposes.

*C. militaris* features with its sexual fruiting bodies growing on the mycosed pupae, earning it the common name "pupa grass" in China [3]. Presently, *C. militaris* is cultivated in laboratories and industrial factories on worms or grain culture such as rice, mung beans, corn, and sunflower seeds, or by using other protein sources that can replace worms and insects because of their high market value. The cereal culture method is exploited most frequently in artificial culturing [23] since it results in fruiting body formation and the formula may be adjusted to fit the desire for high yield and high quantity of cordycepin and adenosine. Also, on another aspect, *C. militaris* cultivated with silkworm pupae has been reported to induce allergic events after consumption by some people due to the cross reactivity and incompatibility with silkworm pupae. However, certain aspects remain unclear about the nature and evolution of the interaction of the fungi with its hosts, the synthesis of its polysaccharides, and the evolution and diversity of its nucleases [24], proteases [13,25,26], cellulase [27,28] and deaminase [29], which may be keys for developing next-generation biotechnologies.

Previously, we built the Scientific Database of Edible and Medicinal Fungi in Institute of Microbiology, Guangdong Academy of Sciences over a period of 10 years. Based on that database, we obtained a wild strain of *C. militaris* with the merits of high cordycepin content, reaching 3.72 mg/kg (dried weight), and high polysaccharide content, reaching 6.7 g/100 g (dried weight) in fruiting bodies [30]. Significantly, it was confirmed that pairing the opposite mating-type isolates of *C. militaris* is a perquisite for the induction of sexual fruiting body formation, where sex mating is key for chromatin and DNA recombination and the evolution of the species [31–33]. Importantly, it was discovered by our group that it was a significant modulator against the metabolism disorder hyperuricemia due to the uric acid-lowering effects of an extract of *C. militaris* [34] (Yong et al., 2016) and its component cordycepin [35,36] through down-regulating URAT1, which is a novel and significant target for hyperuricemia and gout [37,38]. Also, its geographical origins were included [39]. On the other hand, it was reported that the exploration of mitochondrial genomes presents novel opportunities for enhancing mushroom cultivation biotechnology and medicinal applications [40]. Overall, most of studies on *Cordyceps* were concentrated on its pharmacologic effects and cultivation [41].

However, the population structure and demographic history have not been included so far. Quantifying its population structure, demographic history, and diversity and then mining its functional genes based on these remain challenging. We hypothesized that combining resequencing and clustering of predicted protein structures may be an effective strategy [42]. In this paper, 43 wild strains of *C. militaris* were collected across China, and their DNA sequences were re-sequenced for the first time using an Illumina HiseqTM platform (Figure 1a) to quantify the population structure, demographic history, and diversity and then mining their functional genes. Their population structure was determined by the ADMIXTURE, PCA and phylogenic tree analysis. Demographic analysis was conducted with MSMC2 to reveal that historical dynamics of their population sizes and their divergence history. Linkage disequilibrium analysis was conducted to reveal the evolution and distribution history of *C. militaris* strains. Hardy–Weinberg Equilibrium analysis was performed to illustrate the proportion of sites that have undergone natural selection. The nucleotide diversity Pi was analyzed to unveil the genome features and the sites that played a key role in adaption to the climate and other alterations. On the other hand, the Tajima's D was examined to reveal the deviations from neutrality in order to identify evolutionary pressures and genomic variation. Genetic differentiation ($F_{st}$) analysis was carried out to reveal the genomic modules of divergences and differentiations between the three populations of *C. militaris*. Specifically, the cytosine deaminase of *C. militaris* was selected to carry out a combination of AI-assisted protein structure predictions, structural alignments, and clustering since it may be a key for next-generation biotechnology. This work demonstrates how genomic variation leads to functional diversity.

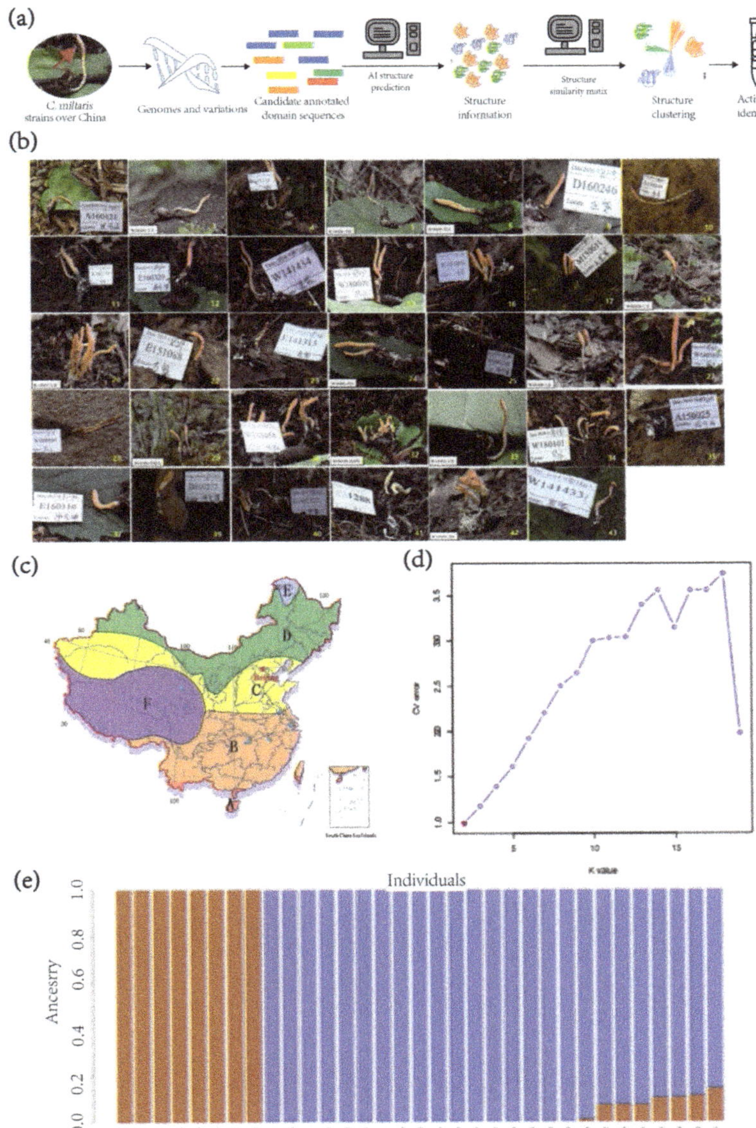


**Figure 1.** Collected *C. militaris* strains, locations, temperature zones, and population structure. (**a**) Workflow of resequencing of *C. militaris* and the key protein clustering of cytosine deaminase based on AI-predicted structures. The structures of annotated and polymorphic domain sequences were predicted by AI and subsequently clustered based on structural similarities. (**b**) *C. militaris* strains were collected from various researches for locations across China. These included the strains A160421, A150549, M150611, W141904, W180129, A200293, A200332, W180100, W180099, W180091, W180103, S180104, A200181, A200168, A200190, W141451, E141314, W141436, D160246, E141315, E160329, W141434, E151091, W141456, W141457, E151088, E141313, D160278, W141449, W141458, W141432, E141327, and W141433. (**c**) The temperature zones distributed in China: A, tropical zone; B, subtropical zone; C, warm-temperate zone; D, mid-temperate zone; E, frigid zone; F, high plateau zone. The three populations of *C. militaris* strains collected for this research were mainly located in middle temperate (D), warm temperate (C), and subtropical (B) zones. (**d**) Cross validation error (CV-error) when K varies from 1 to 19 across the three populations. (**e**) Population structure as populations were set at 2 for all strains, in which brown and blue represent the two ancestries.

## 2. Materials and Methods

### 2.1. Collection of C. militaris Strains

In total, 43 strains of *C. militaris* (Table 1, Figure 1b) were collected over the vast geographic distribution of China, including middle temperate, warm temperate, and subtropical temperate zones (Figure 1c), and then they were stored in our in-house database, the Scientific Database of Edible and Medicinal Fungi in Institute of Microbiology, Guangdong Academy of Sciences, Guangzhou, China. The strain numbers, collection locations, latitude, longitude, and collection dates are listed in Table 1. In addition, the morphologic features are depicted in Figure 1b.

**Table 1.** *Cordyceps militaris* strains collected and their locations.

| Group and Temperate Zone | Seq. No. | Strain No. | Collection Location | Latitude | Longitude | Collection Date |
|---|---|---|---|---|---|---|
| subtropical temperate zone | S1 | A160421 | Xingdou Moutain, Hubei | 30°01′29″ | 109°06′10″ | 23 September 2016 |
| | S10 | A150549 | Huangsang, Hunan | 26°28′21″ | 110°06′23″ | 13 May 2015 |
| | S17 | M150611 | Tiantangzhai, Anhui | 31°07′19″ | 115°54′58″ | 12 August 2015 |
| | S29 | W141904 | Hailuogou, Sichuan | 29°34′27″ | 101°59′57″ | 17 September 2014 |
| | S32 | W180129 | Qingliang Peak, Zhejiang | 30°06′39″ | 118°53′28″ | 07 November 2018 |
| | S58 | A200293 | Tianma, Anhui | 31°09′32″ | 115°45′57″ | 14 September 2020 |
| | S59 | A200332 | Tianma, Anhui | 31°17′32″ | 115°41′07″ | 16 September 2020 |
| warm temperate zone | S6 | W180100 | Yaoxiang, Shandong | 36°19′43″ | 117°07′15″ | 15 August 2018 |
| | S15 | W180099 | Yaoxiang, Shandong | 36°19′43″ | 117°07′15″ | 15 August 2018 |
| | S24 | W180091 | Yaoxiang, Shandong | 36°19′43″ | 117°07′15″ | 15 August 2018 |
| | S27 | W180103 | Yaoxiang, Shandong | 36°19′43″ | 117°07′15″ | 15 August 2018 |
| | S28 | S180104 | Yaoxiang, Shandong | 36°19′35″ | 117°06′56″ | 15 August 2018 |
| | S48 | A200181 | Yaoxiang, Shandong | 36°19′38″ | 117°07′22″ | 18 August 2020 |
| | S50 | A200168 | Yaoxiang, Shandong | 36°19′45″ | 117°07′16″ | 18 August 2020 |
| | S54 | A200190 | Yaoxiang, Shandong | 36°19′49″ | 117°06′58″ | 18 August 2020 |
| middle temperate zone | S3 | W141451 | Zuojia, Jilin | 44°04′56″ | 126°04′17″ | 12 August 2014 |
| | S4 | E141314 | Zuojia, Jilin | 44°04′43″ | 126°04′12″ | 12 August 2014 |
| | S5 | W141436 | Zuojia, Jilin | 44°04′44″ | 126°04′13″ | 12 August 2014 |
| | S8 | D160246 | Zuojia, Jilin | 44°04′45″ | 126°04′14″ | 09 August 2016 |
| | S11 | E141315 | Zuojia, Jilin | 44°04′43″ | 126°04′12″ | 12 August 2014 |
| | S12 | E160329 | Jingyue Lake, Jilin | 43°46′32″ | 125°27′50″ | 10 August 2016 |
| | S14 | W141434 | Zuojia, Jilin | 44°04′44″ | 126°04′13″ | 12 August 2014 |
| | S16 | E151091 | Zuojia, Jilin | 44°04′46″ | 126°04′30″ | 29 August 2015 |
| | S18 | W141456 | Zuojia, Jilin | 44°05′04″ | 126°04′14″ | 12 August 2014 |
| | S20 | W141457 | Zuojia, Jilin | 44°05′04″ | 126°04′14″ | 12 August 2014 |
| | S22 | E151088 | Zuojia, Jilin | 44°04′52″ | 126°04′35″ | 29 August 2015 |
| | S23 | E141313 | Zuojia, Jilin | 44°04′43″ | 126°04′12″ | 12 August 2014 |
| | S25 | D160278 | Jingyue Lake, Jilin | 43°46′41″ | 125°28′02″ | 10 August 2016 |
| | S26 | W141449 | Zuojia, Jilin | 44°04′53″ | 126°04′15″ | 12 August 2014 |
| | S30 | W141458 | Zuojia, Jilin | 44°05′04″ | 126°04′14″ | 12 August 2014 |
| | S33 | W141432 | Zuojia, Jilin | 44°04′44″ | 126°04′13″ | 12 August 2014 |
| | S40 | E141327 | Zuojia, Jilin | 44°04′56″ | 126°04′18″ | 12 August 2014 |
| | S43 | W141433 | Zuojia, Jilin | 44°04′44″ | 126°04′13″ | 12 August 2014 |

### 2.2. Trans-Inoculation of Strains and Their DNA Extraction

Strains were trans-inoculated on a PDA medium (20% potato, 2% glucose, 2% agar, 0.3% $KH_2PO_4$, 0.15% $MgSO_4$, trace vitamin B1) and then grown at 25 °C for 7 days in a dark environment [30]. Then, mycelium was isolated and collected from the culture and ground into powder under liquid nitrogen to extract DNA.

Briefly, DNA extraction was performed with a QIAGEN® DNA extraction kit (Cat#13323, QIAGEN, Dusseldorf, Germany) according to the manufacturer's instructions. The purity of the extracted DNA was detected using a NanoDrop™One UV-Vis Detector (Thermo Fisher Scientific, Waltham, MA, USA) with OD260/280 in the range of 1.8-2.0 and OD260/230 in 2.0-2.2.

### 2.3. Construction of a DNA Library

To obtain proper DNA [36], ultrasonic waves were utilized to slice DNA sequences into randomized fragments. Then, fragments were amended by adding an A at the 3′-end. After the ligation of the sequencing adapter, magnetic beads were used to enrich the DNA sequences, and sequences of about 400 bp were absorbed, then amplified with PCR to construct a library. The constructed library was inspected, and then the qualified library was sequenced using an Illumina HiSeqTM platform with the method of Illumina PE150 and reads of 300 bp (paired-end) to target coverage of 30× [36]. The detailed flow is shown in Figure 1a.

### 2.4. Whole-Genome Re-Sequencing and SNP Calling

Raw data were downloaded from the Illumina HiseqTM platform and qualified [36], after which low-quality data were removed, and finally clean data were obtained. Therein, short reads were cleaned using Trimmomatic v.0.38 with a strict filtering process and then aligned to the *C. militaris* reference genome (CmilitarisCM01_v01) [4] using BWA v.0.7.15, and the obtained belongs of locations were BAM files. BAM files were corrected following the Best Practice of GATK, and SNPs and genotypes were called using HaplotypeCaller implemented in GATK v.4.1 [43]. SnpEff_v4_5 [44], Annovar 0.8 version [45], and genes of the reference genome were used to annotate the functions of SNPs. The obtained SNP markers were used for investigating the genetic diversity and evolutionary structure.

### 2.5. Population Structure Analyses

In order to elucidate the population structure, an admixture analysis was conducted with ADMIXTURE v.1.3.0 [46], and a principal component analysis (PCA) was performed with PCANGSD v.0.98 [47]. Also, the maximum likelihood algorithm was established for evolutionary trees with MEGAX. ADMIXTURE was run with predefined numbers of clusters (K) ranging from 1 to 19, each repeated 20 times. The K-value with the lowest cross-validation error was selected as the most likely number of putative genetic groups. PCANGSD was accomplished based on genotype likelihoods accounting for sequencing errors and uncertainty in genotype calls. Population structure analyses were carried out using common SNPs with missing rate < 0.1, minor allele frequency (MAF) > 5%, and correlation coefficient < 0.2 with any other SNPs in sliding windows of 50 SNPs.

### 2.6. Demographic History

The multiple sequentially Markovian coalescent (MSMC) was performed with MSMC v.2.00 to assess the cross-coalescence rate and to track variation in effective population size (Ne) over time using parameters reported [48]. The 2D joint-unfolded site frequency spectrum (SFS) was calculated with ANGSD v.0.935 based on intergenic sites, which are affected by selection least. Furthermore, the times of splitting events of the two were estimated according to MSMC2.

### 2.7. Population Genetic Statistics

ANGSD [49] was used to compute summary statistics including Pi, Tajima's D, and $F_{st}$ grounded on the folded SFS. All summary statistics were carried out in 10 kb nonoverlapping sliding windows. $F_{st}$ was estimated by averaging all pair-wise $F_{st}$ values between populations. Gene ontology (GO) was performed using GOWINDA [50] and Annovar [45]. LD between SNPs with a sliding window of 10 kb was computed using PLINK v.1.07 and then averaged over all pair-wise sites c. Only SNPs with MAF > 5% were included, and only windows with $\geq$ 10 SNPs were reserved for PLINK.

### 2.8. Protein Clustering and Analysis of Cytosine Deaminase of *C. militaris*

According to the annotated results, the SNPs of cytosine deaminase of *C. militaris* were selected and then extracted with VCFtools [51]. The variant nucleotide sequences were obtained by converting the SNPs by BCFtools with the genome region of the cytosine deaminase domain as reference, which was annotated above. Then, the protein sequences were obtained by translating the variant nucleotide sequences. After that, high confidence and accurate protein structure of each protein sequence were predicted by the AI algorithm RoseTTAFold [52] and filtered with average per-residue confidence metric predicted local-distance difference test (PLDDT) > 70. The paired structure alignment was performed based on the TM-score method with Foldseek [53] (van Kempen et al., 2023) to afford the overall structural matrix, which was then clustered, and the representatives of each cluster was presented by PyMol 2.1.0.

## 3. Results

### 3.1. Population Structure and Demographic History of the Three Populations of *C. militaris*

Forty-three strains (Figure 1b and Table 1) of *C. militaris* were collected from locations throughout China in different temperate zones (Figure 1c). Each sample was resequenced on an Illumina HiseqTM platform (paired-end 300 bp) to a target coverage 30×. Genetic structure analysis for a population may provide its origin clues and components. High-quality SNPs was analyzed and clustered with ADMIXTURE software to obtain the genetic structure. It was found that cross-validation error was the lowest when K was 2 during the scanning of K between the range of 1 to 19 (Figure 1d), suggesting that all the strains collected may originate from two common ancestors. Eight, seven, and eighteen individuals were assigned to the warm temperate, subtropical temperate, and middle temperate zone populations of *C. militaris* when K was set at 2, respectively. One individual showed a high admixture, which was excluded from downstream analyses (Figure 1e). Others were discarded and not considered since they failed to grow fruiting bodies.

Generally, PCA analysis was conducted on the basis of pure mathematics, in which several variables were transformed linearly from large variables. According to PCA analysis, all wild *C. militaris* strains were clustered into three populations, consisting of middle, warm temperate, and subtropical zone populations, except for two from the subtropical temperate zone (Figure 2a). Phylogenic trees are frequently used to elucidate and interpret the evolutionary distance and relationships between strains; strains are placed on the branches of a tree scheme, demonstrating evolutionary history and genetic relationships. The maximum likelihood algorithm was used to describe and establish the phylogenic tree of the collected populations (Figure 2b). All samples of the subtropical temperate zone population were separated from the warm temperate zone population except for S27, which was collected from Yaoxiang, Shandong Province. In terms of population structure, the PCA and maximum likelihood analyses further confirmed the separation patterns and clustering of the genetic differentiations supplied by the ADMIXTURE algorithm.

Overall, the three populations were all collected from locations in relatively low temperate climates, underlining the shade-loving feature of *C. militaris* and implying that the spread of *C. militaris* may be from north to south in China. On the other hand, the three populations presented different population sizes, with a pattern of subtropical zone > middle temperate zone > warm temperate zone (Figure S1a). In particular, the subtropical and middle temperate zone populations showed much larger population sizes than the warm temperate zone population, further implying the shade-loving feature and north-to-south migration.

Furthermore, all three populations experienced several bottlenecks and booms of different magnitudes and durations (Figure 2c). The middle temperate zone population boomed about 0.021 Myr between 0.025 Myr and 0.004 Myr ago, and then it bottlenecked about 0.001 Myr between 0.003 Myr and 0.0005 Myr. Then it was reduced to nearly zero, which may have been caused by over-exploitation due to its high value as medicine. The warm temperate zone population presented a much smaller population size in comparison to the other two populations. It boomed at 0.1 Myr ago and bottlenecked at 0.01 Myr ago, lasting about 0.09 Myr. It then boomed immediately at 0.01 Myr ago, and this ended at

0.001 Myr, lasting for 0.009 Myr. Then, it boomed and then declined to nearly zero, which may be attributed to over-exploitation. In detail, the subtropical temperate zone population experienced a short bottleneck between 0.05 Myr and 0.04 Myr years ago, lasting about 0.01 Myr, and then it boomed for about 0.04 Myr between 0.04 Myr and 0.0002 Myr ago. Interestingly, it declined evidently within the recent 0.0002 Myr, which may be have been caused by its over-exploitation. It was evident that the population sizes for the three populations in recent years were declining quickly, indicating the resources for this species suffered a bottleneck. These were consistent with modern breeding, which was fostered by the wild resource over-exploitation and the high value of this fungus as a traditional medicine in various Asia countries.

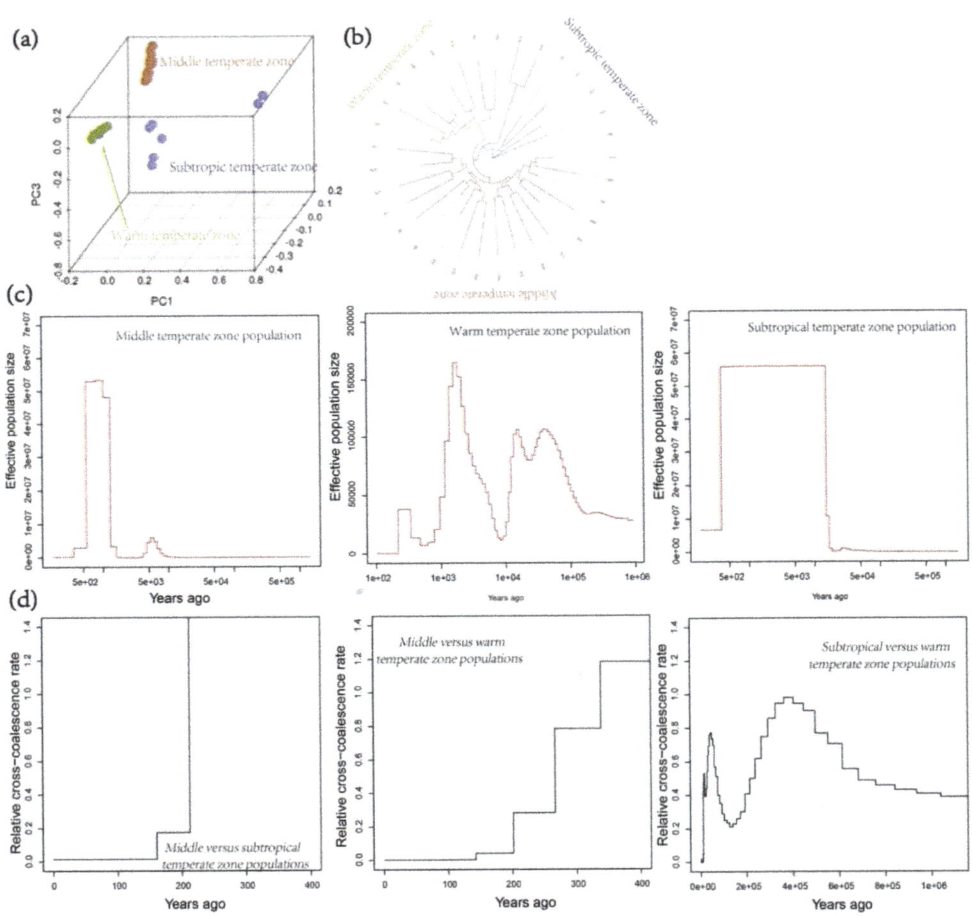


**Figure 2.** Population structure and demographic history of the *C. militaris* strains. (**a**) Principal component analysis (PCA) based on genome-wide single nucleotide polymorphisms (SNPs) is shown with the first three components (PCs). (**b**) Maximum likelihood tree based on SNP data. (**c**) Changes in the historic effective population size (Ne) of the *C. militaris* strains from the three temperate zones, estimated by multiple sequentially Markovian coalescent (MSMC2). (**d**) Divergence processes for pairs of *C. militaris* populations from the three temperate zones, inferred by MSMC2.

Demographic analyses using MSMC were conducted to reveal the divergence between the three populations of *C. militaris* (Figure 2d). Locally, the split in recent years between the middle and subtropical temperate zone populations occurred about 0.0002 Myr ago, and this occurred for the middle and warm temperate zone populations 0.00035 Myr ago. These events happened simultaneously with the declines in the sizes of the three populations. Specifically, the separating event occurred at 0.4 Myr ago for the subtropical and warm temperate zone populations. Over the whole relative cross—coalescence rate analysis, the warm population was first to start separating from the middle temperate population at 2.5 Myr and from the subtropical temperate population before 20 Myr, and it hybridized with the two in recent years and at about 7.5 Mry, experiencing several splits and hybridizations (Figure S1b). In the process, the population expansions between 0.1 Myr and 0.01 Myr ago were consistent with its relatively low hybridization with the subtropical temperate zone population, as reflected by the relative cross-coalescence rate analysis. Also, the recent splits between them were accompanied by steep population declines at almost the same time scales. Interestingly, it was found that the migration pattern of *C. militaris* may be from the north to south, provided by the earlier splitting times of middle *versus* warm temperate zone populations and the later splitting times of middle *versus* subtropical temperate zone populations. Possibly, the divergence between the subtropical temperate population and middle temperate population occurred earlier than that between the warm temperate population and subtropical temperate population, demonstrating further that the spread of *C. militaris* may be from the low temperate zone to the warm temperate zone. It gained its tolerance to high temperatures gradually.

### 3.2. Linkage Disequilibrium Analysis

For a population, the frequency of simultaneous inheritance of two genes at different loci being higher than random frequency is called linkage disequilibrium, providing the negative genetic element of a species. Generally, $r^2$ is obtained as linkage disequilibrium, through analyzing SNP combinations at the same chromosome and the linkages of SNPs in all samples. The linkage is high when it approaches 1. By fitting SNP distance and $r^2$, it would be discovered that the closer the SNP distance, the higher the $r^2$. A slow rate of $r^2$ decay indicates a high probability of SNP linkage. Generally, a species with a quick decay of $r^2$ may be original. In this research, $r^2$ values were calculated and are shown in Figure 3a. It was observed that strong LD ($r^2 > 0.3$) occurred for the three populations at short distances. The LD decays of the three populations were $r^2$ at 0.27, 026, and 0.23 for the middle, subtropical, and warm temperate zone populations, corresponding to distances of about 35 kb, 35 kb, and 10 kb. LD decay analysis suggested that the LD reaches very low values for distances greater than 150 kb ($r^2 < 0.10$). Linkage disequilibrium analysis revealed that *C. militaris* strains of middle temperate zone resemble the common ancestors, supporting that *C. militaris* in China may have originated from the north and then spread to the south.

### 3.3. Population Diversity of *C. militaris*

SNP distributions at each chromosome were presented (Figure 3b). Testing for deviations from Hardy–Weinberg equilibrium (HWE) offers fundamental information about genetic variation and evolutionary processes in natural populations. Overall, 176,310 sites were analyzed to obtain Hardy–Weinberg equilibrium for their alleles. Of those, 169,775 deviated from Hardy–Weinberg equilibrium (Figure 3c), illustrating that a large proportion of sites had undergone natural selection and also that natural selection has played an important role in the evolution and species shaping of *C. militaris*.

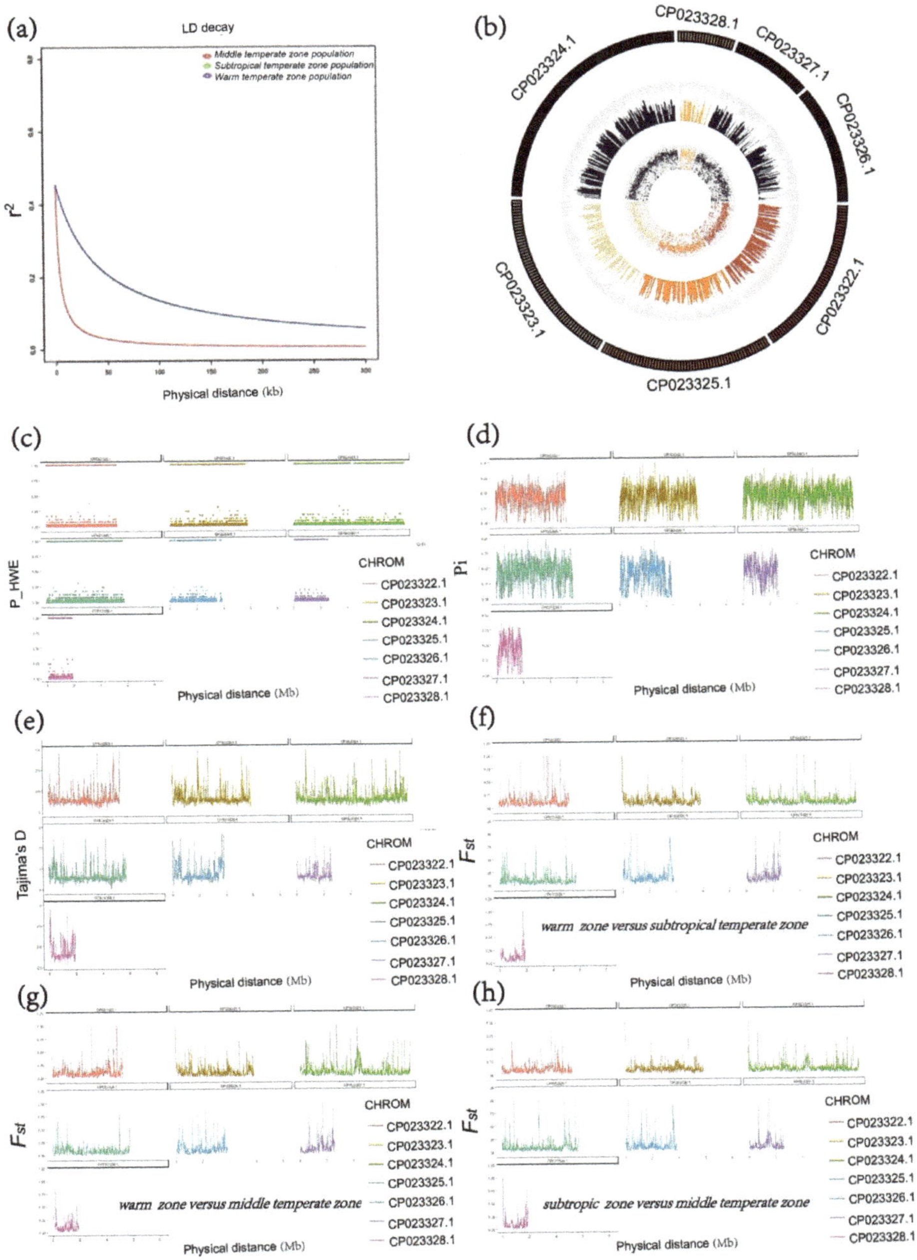


**Figure 3.** LD decay, SNP distributions and genome-wide differentiations occurred for or among the three populations of *C. militaris*. (**a**) LD decay plot for each population. SNP distance is on the X axis and $r^2$ on the Y axis. LD decay was at the SNP distance when $r^2$ decayed by half. (**b**) SNP distributions at each chromosome. (**c**) Hardy–Weinberg. (**d**) Pi. (**e**) Tajima's D. (**f**) Population differentiation ($F_{st}$) for warm *versus* subtropical temperate zone in 10 kb nonoverlapping windows. (**g**) Population differentiation ($F_{st}$) for warm *versus* middle temperate zone. (**h**) Population differentiation ($F_{st}$) for subtropical *versus* middle temperate zone. Different colors represent alternative chromosomes, and $F_{st}$ > 0.8 values were ascribed as $F_{st}$ islands.

Overall, the observed homozygosities of the three populations were at 0.932, 0.959, and 0.896 for the middle, subtropical, and warm temperate zone populations (Table 2). On the other hand, the observed heterozygosities of the three populations were 0.068, 0.041, and 0.104. Also, the average Pi within each population was at 0.261, 0.275, and 0.229. The genetic variation was heterogeneous across the genome for the three populations. The overall level of genome-wide polymorphisms of the three populations was estimated with nucleotide diversity Pi, which was in the range of 0–0.041, with an average of 0.016 and a max of 0.041 (Figure 3d). Altogether, the genome of *C. militaris* presented relatively stable and conserved features. Interestingly, there were 60 regions with Pi above 0.39, aggregating at chromosomes CP023325.1 (51) and CP023323.1 (9). The regions of relatively high nucleotide diversity at the two chromosomes may have been subjected to distinct selection mechanisms and/or demographic histories and then played a key role in adaption to the climate alterations in immigration or in the history of climate change. Tajima's D was calculated for the whole genome of the strains collected for the three populations (Figure 3e). The average Tajima's D score for them was found to be −0.943 with a standard deviation of 0.809, revealing that most elements of their genomes did not deviate significantly from neutrality. However, 127 sites or genes were flagged as not evolving neutrally, including 88 with a Tajima's D equal to or above 2 and 39 equal to or below −2.

**Table 2.** Parameters of population genetics of the collected *C. militaris* strains.

| Pop ID [a] | Private [b] | Num InDv [c] | Obs Het [d] | Obs Hom [e] | Exp Het [f] | Exp Hom [g] | Pi [h] |
|---|---|---|---|---|---|---|---|
| Middle | 21,786 | 18 | 0.068 | 0.932 | 0.253 | 0.747 | 0.261 |
| Subtropical | 12,924 | 8 | 0.041 | 0.959 | 0.258 | 0.742 | 0.275 |
| Warm | 4433 | 7 | 0.104 | 0.896 | 0.213 | 0.787 | 0.229 |

[a] Pop ID, population marker. [b] Private, featured SNP number of specific population. [c] Num Indv, average individual number of each locus included in population. [d] Obs Het, observed heterozygosity. [e] Obs Hom, observed homozygosity. [f] Exp Het, expected heterozygosity. [g] Exp Hom, expected homozygosity. [h] Pi, nucleotide polymorphisms.

Genetic diversity is the main component of biological diversity, for which species present their unique gene library, genetic organization, and then phenotype diversity. The genetic diversity of species and populations is formed during their long-term evolution and then serves in return as the prerequisite for their survival and evolutionary adaptation to climate alterations. To determine the pattern of sequence diversity between the three populations, we estimated relative sequence divergence, $F_{st}$, between the three populations by sequencing the individuals sampled from locations across China (Table 3). $F_{st}$ is approaching 1 when the speciation of two populations is high but approaching 0 when speciation is low. Consistent with the crystal population separation and then the structure, the average $F_{st}$ was at 0.109 in the analyses of the warm *versus* the subtropical temperate zone populations over 3,448,082 sites of the genome, 0.092 for the warm *versus* the middle temperate zone populations, and 0.083 for the subtropical *versus* the middle temperate zone populations (Table 3), given that they were within the species *C. militaris*. At a glance, a relatively high $F_{st}$ was observed for almost the whole genome except for a few regions with elevated $F_{st}$ on seven chromosomes (Figure 3f–h). At a finer scale, three sharp peaks ($F_{st}$ > 0.70) were found on chromosome CP023322.1, two on CP023323.1, four on CP023324.1, three on CP023325.1, four on CP023326.1, three on CP023327.1, and three on CP023328.1 for the comparison of the warm *versus* the subtropical temperate zone populations (Figure 3f). Overall, 22 $F_{st}$ peaks were observed between the warm and the subtropical temperate zone populations. Correspondingly, 21 $F_{st}$ peaks were obtained in the comparison of the warm versus the middle temperate zone populations, including two peaks on CP023322.1, three on CP023323.1, three on CP023324.1, three on CP023325.1, four on CP023326.1, three on

CP023327.1, and three on CP023328.1 (Figure 3g). On the other hand, 22 $F_{st}$ peaks were extracted through the comparison of the subtropical versus the middle temperate zone populations, consisting of three peaks on CP023322.1, two on CP023323.1, three on CP023324.1, five on CP023325.1, three on CP023326.1, three on CP023327.1, and three on CP023328.1 (Figure 3h). Thus, the $F_{st}$ peaks in the seven chromosomes represent genomic modules of divergences and differentiations between the three populations of *C. militaris*. Selective sweep was induced by the decline and sweep of the difference of adjacent nucleotides at selective loci under the stress of some strong positive natural selection. As a mutant is generated and then leads to an elevation of adaption to the environment, selective sweep may be produced. Natural selection benefits the survival of individuals with strong adaption, and as time goes on, the frequency of novel mutants for alleles increases gradually, and then linkages of neutral mutants and novel mutants may increase. Then, the domain of selective sweep in a genome forms a positive haplotype gradually, leading to a decrease. The $F_{st}$ and Pi selected thresholds were at 0.95 and 0.05, Tajima's D at 0.05 and 0.95, and these were correlated with each other to extract candidate regions and mutant loci in the regions. Selective sweep loci were identified (Figure 4a,b), and cytosine deaminase was selected as a candidate.

**Table 3.** Genetic differentiation between populations collected.

| $F_{st}$ [a] | **Warm Temperate Zone** | **Middle Temperate Zone** |
|---|---|---|
| Subtropical temperate zone | 0.109 | 0.083 |
| Warm temperate zone | | 0.092 |

[a] Genetic differentiation between populations is represented by $F_{st}$.

### *3.4. Representative Predicted Structures for Eight Deaminase Clades*

It was considered that structure determines function, which enabled the comparison and clustering of known or predicted protein structures, potentially classifying cytosine deaminases of *C. militaris* into functional clades and providing novel functions. Thus, an AI-assisted protein structure protein prediction, structure alignment, and clustering were combined to generate protein classification relationships among cytosine deaminases. From the strains of *C. militaris*, several non-synonymous cytosine deaminases were provided by calling SNPs and reconstructing sequences with the cytosine deaminase domain in the genome of *C. militaris* as reference. All the structures of the obtained protein sequences were predicted accurately with the AI algorithm RoseTTAFold. Also, multiple structural alignments (MSTAs) of the sequences were conducted, and then structural similarity matrices between these proteins were generated (Table S1), reflecting their overall structural correlations. According to the structure similarity matrices, clustering was carried out to classify them into eight unique structural clades (Figure 4c and Table S2), and the cytosine deaminases within each clade had distinct conserved protein structural domains, reflecting their diversity and then evolution at the structural and functional levels.

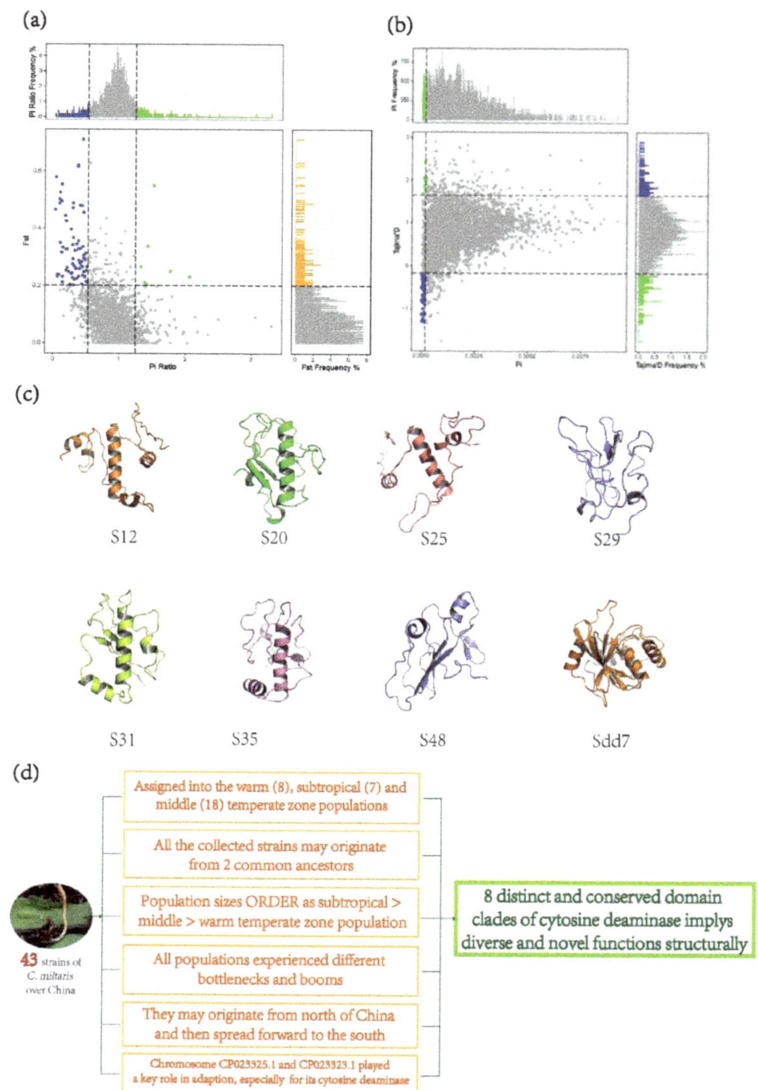


**Figure 4.** Correlations of diversity parameters and AI-assisted prediction and clustering of cytosine deaminases of collected *C. militaris* strains. (**a**) $F_{st}$ and Pi correlation. The abscissa coordinate of the middle scatter plot represents Pi ratio distribution and vertical coordinate $F_{st}$, in which blue scatters represent the Pi over 0.95 or below 0.05, and blue area was the selected; the right plot represents the distribution of $F_{st}$, wherein orange area was over 0.95; the above plot indicates the Pi ratio distribution, in which the blue area was below 0.05. (**b**) Pi and Tajima's D correlation. The abscissa coordinate of the middle scatter plot represents Pi distribution, and vertical coordinate Tajima's D, in which blue scatters represent the Tajima's D and Pi below 0.05, green scatters Tajima's D over 0.95 and Pi below 0.05, and blue or green areas were the selected; the right plot represents Tajima's D distribution, wherein the green area was below 0.05 and blue over 0.95; the above plot represents Pi distribution, in which the green area was below 0.05. (**c**) Representative predicted structures for eight deaminase clades. (**d**) Re-sequencing and AI-assisted structural predictions and alignments establish a new protein classification and functional mining method, providing a suite of cytosine deaminases of various activities and single- and double-stranded functions, potentially showing great probabilities as customed base editors, as therapeutics, or as tools for breeding novel species.

## 4. Discussion

In this paper, we reported the collection and re-sequencing, population structure, demographic history, diversity, and a cytosine deaminase of 43 wild strains of *C. militaris* collected from locations across China, offering a potential suite of single- and double-stranded deaminases of great promise as tunable base editors for therapeutic or agricultural breeding applications (Figure 4d). Genomic sequencing of fungi is rare since they present numerous genera and species and bisporous features in general. Therefore, the dynamics of their population sizes were seldom reported. The population size history of *C. militaris* is still poorly known, although it is essential for understanding the complex interactions and convolutions between it and its hosts and revealing the spread of their populations. It is likely to be difficult or impossible to obtain ancient DNA from *C. militaris* samples that would aid in dating its emergence. However, it would be particularly notable if present-day *C. militaris* genomic sequences could be used to robustly infer both the recent and ancient population size histories of this species. Population size changes that occurred hundreds of thousands of years ago affected the rates of coalescence and thus have left their signatures in the site frequency spectrum (SFS) of genomic sequences. In this research, the SFS is the distribution of allele frequencies in the sequences, randomly collected from the present-day population of *C. militaris*. Each SFS category contains a certain number of mutations of the same size. All three populations experienced several bottlenecks with different magnitudes and durations, especially in recent years. In order to conserve this species and its genetic diversity, recommendations for conservation parks should be established for this and other valuable species.

LD is an important tool in association studies as well as in studies aiming to evaluate genetic diversity. Thus, LD has been used in several studies to determine the diversity and history, signatures of selection, recombination rates, effective population size and other population events. Moreover, for high-resolution association mapping, it is also necessary to identify haplotype-block structures and a minimal set of polymorphisms. In this research, the LD pattern of the genome of *C. militaris* was evaluated. Furthermore, LD revealed that the middle temperate zone population resembled its common ancestors, implying that *C. militaris* in China may have migrated from the north to the south.

Darwin's evolutionary theory, presented in the book The Origin of Species, serves as a foundation of biology. However, the preservation and evolution of natural variation and selection within populations was still a puzzle at that time. About fifty years later, G.H. Hardy and W. Weinberg presented the proof of variation in a population mathematically, depicting that random mating results in stationary allele and genotype frequencies over generations. The genotypic frequencies were simply considered allele frequencies and called Hardy–Weinberg proportions (HWPs). In population genetics and evolutionary genetics, the proof of Hardy–Weinberg equilibrium [54] is regarded as the Law of Inertia of biology and is thus an important landmark. Hence, testing HWE is a routine and important procedure to infer the genetic basis of population evolution and to identify evidence for genetic associations. For the collected *C. militaris* strains, a large proportion of sites have undergone natural selection, playing an important role in this species' evolution and its species shaping. Over decades, considerable interest has been focused on detecting natural selection through sequencing. The neutrality test for allele frequency is popular, especially Tajima's D testing [55]. Numerous genes were determined to have undergone natural selection, such as lactase, trpv6, and the HLA immune complex. In this work, Tajima's D was estimated for the whole genome of the collected strains, revealing that most elements of their genomes have not deviated significantly from neutrality. However, several sites or genes were flagged as not evolving neutrally.

$F_{st}$ is defined as $(\pi_b - \pi_w)/(\pi_b + \pi_w)$, where $\pi_b$ (also known as $d_{xy}$) and $\pi_w$ are the absolute pairwise divergence between and within populations [56], respectively. An increase in $F_{st}$ can therefore be raised from an increase in $\pi_b$, a decline in $\pi_w$, or a synergetic behavior of these two. Genome scans of closely related populations or species have revealed "genome modules" as peaks of high relative sequence divergence ($F_{st}$) that stand out against a lower "sea" of divergence. Their causes remain unclear, but they have been suggested to contain key loci involved in local adaption and/or reproductive isolation. However, their significance for speciation or differentiation with or without gene flow between populations is a matter of debate. One hypothesis argued that gene flow is unimpeded across most of the genome, reducing inter-population diversity, except for loci under divergent selection and loci in close physical linkage to selected loci. Another hypothesis is that genomic modules reflect selective sweeps, where specific alleles are driven to high frequency, thus reducing within-population diversity. These two hypotheses are typically presented as alternatives, although they are not mutually exclusive—both barriers and selective sweeps of gene flows may play a role. Here, we determined how these processes were involved in controlling flora differences and divergences between subset populations of *C. militaris*. Consistent with the crystal population separation and then the structure, genetic differentiation was relatively high among the three populations, given that they were within a species *C. militaris*. The $F_{st}$ peaks in the seven chromosomes represent genomic modules [57] of divergences and differentiations between the three populations of *C. militaris*. This system has the advantage of being genetically tractable and having a hybrid zone that allows selection and gene flow to be analyzed in nature. However, these may be limited.

Elucidating the functions and mechanisms of genes and their translated proteins underlies modern biotechnology, which has been exploited functionally, such as through base editing [58], prime editing, epigenome editing, gene editing, and PROTACs [59], and has propelled the life sciences forward greatly, even defining an age of the life sciences. On the other hand, protein mining has been focused extensively in bio-enzyme refining and antibody and vaccine design, but it relies on sequences instead of structures from the past. However, structures define docking and interactions with others, as well as functions. In this work, the sequences of cytosine deaminase, which is an important nuclease functioning as a base editor for next-generation biotechnology, were provided from the collected *C. militaris* strains since they demonstrated importance in adaption. Specifically, cytosine deaminase is a significant component that may exert enzyme activity as a nuclease modifying bases on DNA and serve as an important tool in the next-generation biotechnology boom. Their structures were predicted accurately utilizing the AI algorithm RoseTTAFold [52,60], and then clustered subsequently based on predicted structure similarities [61]. In the future, cytidine deaminases will be engineered and may be efficient cytosine base editors packaged into a single adeno-associated virus. Also, some of them will be profiled and confirmed by editing plants, animals, and human cells. These discovered deaminases, based on AI-assisted structural predictions, may greatly expand the utility of base editors for therapeutic and agricultural applications. Accurate protein structure prediction and clustering may be generated on the basis of protein structural alignments, even without the use of contextual information such as conserved gene neighborhoods and domain architectures. When using structure-based clustering, different clades reflect unique structures, implying distinct catalytic functions and properties. Generally, structure-based clustering is much more robust and effective at sorting functional similarities than traditional 1D amino acid sequence-based clustering. AI-assisted 3D protein structures provide reliable clustering results and only require an amino acid sequence, making them a convenient and effective strategy for generating protein relationships and the discovery of novel functions.

## 5. Conclusions

In conclusion, we reported the collection and re-sequencing of 43 wild strains of *C. militaris* from sites across China, for which the population structure, demographic history, diversity, and a cytosine deaminase of *C. militaris* in China were examined. It was found that all the collected strains may originate from two common ancestors. Also, the individuals could be assigned to the warm, subtropical, and middle temperate zone populations, confirmed by PCA analysis and a phylogenic tree with a maximum likelihood algorithm. The three populations presented a population size pattern of subtropical zone > middle temperate zone > warm temperate zone, implying a north-to-south migration pattern. Furthermore, all three populations experienced bottlenecks and booms, especially in recent years. Strong LD occurred for the three populations at short distances. *C. militaris* strains of the middle temperate zone resemble the common ancestors, implying that *C. militaris* in China may have originated from northern China and then spread forward to the south, as supported by population size analysis. A large proportion of sites have undergone natural selection, and also, natural selection has played an important role in the evolution of *C. militaris* and its species shaping. The nucleotide diversity presented a relatively stable and conserved feature. Interestingly, the regions of high Pi were aggregated at the chromosomes CP023325.1 and CP023323.1, playing a key role in adaption to environmental alterations, especially the regions at cytosine deaminase. Genetic differentiation was relatively high among the three populations, given that they were within the species *C. militaris*. Relatively high $F_{st}$ was observed for almost the whole genome except for a few regions with elevated $F_{st}$ on the seven chromosomes. Thus, the $F_{st}$ peaks in the seven chromosomes represent genomic modules of divergences and differentiations between the three populations of *C. militaris*. According to the structure similarity matrices, clustering was carried out to classify cytosine deaminases into eight unique structural clades. The cytosine deaminases within each clade have distinct conserved protein structural domains, reflecting their diversity and evolution at the structural and functional levels. This work demonstrates how genomic variation leads to functional diversity.

**Supplementary Materials:** The following supporting information can be downloaded at: https://www.mdpi.com/article/10.3390/agriculture15070686/s1, Figure S1: (a) Overlap of the historic effective population size (Ne) for the *C. militaris* of the three temperate zones estimated by multiple sequentially Markovian coalescent (MSMC2). Ne was described as e.g., 5e+03 equals to $5 \times 10^3$. (b) Divergence processes for each pair of the *C. militaris* of the three temperate zones inferred by MSMC2 over a long time period; Table S1: Query and target identifier, TMscore, translation (3) and rotation vector = $(3 \times 3)$; Table S2: The protein structure clustering.

**Author Contributions:** Conceptualization, T.Y.; methodology, T.Y., Y.L. and M.C.; validation, H.H.; formal analysis, T.Y., Y.L. and M.C.; investigation, T.Y., Y.L., M.C., L.Z., X.W., H.G., H.H., Y.G., S.C., Y.X. and W.Z.; resources, Y.L.; writing—original draft preparation, T.Y.; writing—review and editing, T.Y.; visualization, T.Y.; supervision, H.H.; project administration, T.Y. and H.H.; funding acquisition, T.Y. and H.H. All authors have read and agreed to the published version of the manuscript.

**Funding:** This research was funded by National Key Research and Development Program for Food Nutrition and Safety (No. 2023YFF1104100, China), the Guangdong Provincial Science and Technology Fund Special Project (No. 210909154531306, China), the Key Research and Development Program of Guangdong Province (No. 2022B1111040002, China), the Science and Technology Program of Linzhi (No. LZZX-05, China), the Strategic Special Project for Rural Vitalization of Guangdong Province (No. 2022-WJS-00-001, China), the Natural Science Foundation of Guangdong (No. 2022A1515011066 and 2021A1515010960, China) and the Guangdong Province Sail Plan (No. 2017YT05S115, China).

**Institutional Review Board Statement:** Not applicable.

**Data Availability Statement:** The data that support the findings of this study are available from the corresponding author upon reasonable request.

**Conflicts of Interest:** The authors declare that there are no conflicts of interest.

## References

1. Tran, M.H.; Nguyen, T.M.; Huynh, V.B.; Sridhar, K.; Deshmukh, S.K.; Fung, S.-Y.; Mahadevakumar, S. Diversity evaluation of *Cordyceps* spp. in Bidoup Nui Ba, Lam Dong province, Vietnam. In *IOP Conference Series: Earth and Environmental Science, Proceedings of the 4th International Conference on Sustainable Agriculture and Environment, Ho Chi Minh City, Vietnam, 17–19 November 2022*; IOP Publishing Ltd.: Ho Chi Minh City, Vietnam, 2023; Volume 1155, p. 012003.
2. Tiwari, M.; Saraf, A.; Khelkar, T. Exploring the world of Cordyceps: Ecology, cultivation, biotechnology, and future horizons. In *Futuristic Trends in Biotechnology*; Iterative International Publishers (IIP), Selfypage Developers Pvt Ltd.: Novi, MI, USA, 2024; Volume 3, pp. 112–126.
3. Chen, W.; Han, Y. Taxonomy, phylogeny, and genetics of *Cordyceps*. In *Advances in Cordyceps Research*, 1st ed.; Sridhar, K., Deshmukh, S.K., Fung, S.-Y., Mahadevakumar, S., Eds.; CRC Press: Boca Raton, FL, USA, 2024; pp. 1–24.
4. Zheng, P.; Xia, Y.; Xiao, G.; Xiong, C.; Hu, X.; Zhang, S.; Zheng, H.; Huang, Y.; Zhou, Y.; Wang, S.; et al. Genome sequence of the insect pathogenic fungus *Cordyceps militaris*, a valued traditional Chinese medicine. *Genome Biol.* **2011**, *12*, R116. [CrossRef] [PubMed]
5. Mahadevakumar, S.; Sridhar, K.R. An overview of the phylogeny of *Cordyceps*. In *Advances in Cordyceps Research*, 1st ed.; Sridhar, K., Deshmukh, S.K., Fung, S.-Y., Mahadevakumar, S., Eds.; CRC Press: Boca Raton, FL, USA, 2024; pp. 1–21.
6. Roberson, R.W. Subcellular structure and behaviour in fungal hyphae. *J. Microsc.* **2020**, *280*, 75–85. [PubMed]
7. Woolley, V.C.; Teakle, G.R.; Prince, G.; de Moor, C.H.; Chandler, D. Cordycepin, a metabolite of *Cordyceps militaris*, reduces immune-related gene expression in insects. *J. Invertebr. Pathol.* **2020**, *177*, 107480. [PubMed]
8. Wang, Y.; Dong, Q.-Y.; Luo, R.; Fan, Q.; Duan, D.-E.; Dao, V.-M.; Wang, Y.-B.; Yu, H. Molecular phylogeny and morphology reveal cryptic species in the *Cordyceps militaris* complex from Vietnam. *J. Fungi* **2023**, *9*, 676. [CrossRef]
9. Wang, J.B.; Leger, R.S.; Wang, C. Advances in genomics of entomopathogenic fungi. *Adv. Genet.* **2016**, *94*, 67–105.
10. Lei, Y.; Hussain, A.; Guan, Z.; Wang, D.; Jaleel, W.; Lyu, L.; He, Y. Unraveling the mode of action of *Cordyceps fumosorosea*: Potential biocontrol agent against *Plutella xylostella* (*Lepidoptera*: *Plutellidae*). *Insects* **2021**, *12*, 179. [CrossRef]
11. Zhang, J.; Wen, C.; Duan, Y.; Zhang, H.; Ma, H. Advance in *Cordyceps militaris* (Linn) Link polysaccharides: Isolation, structure, and bioactivities: A review. *Int. J. Biol. Macromol.* **2019**, *132*, 906–914.
12. Avery, P.B.; Kumar, V.; Francis, A.; McKenzie, C.L.; Osborne, L.S. Compatibility of the predatory beetle, delphastus catalinae, with an entomopathogenic fungus, *Cordyceps fumosorosea*, for biocontrol of invasive pepper whitefly, *Aleurothrixus trachoides*, in Florida. *Insects* **2020**, *11*, 590. [CrossRef]
13. Kato, T.; Nishimura, K.; Misu, S.; Ikeo, K.; Park, E.Y. Changes of the gene expression in silkworm larvae and *Cordyceps militaris* at late stages of the pathogenesis. *Arch. Insect Biochem. Physiol.* **2022**, *111*, e21968. [CrossRef]
14. Zhang, Z.J.; Yin, Y.Y.; Cui, Y.; Zhang, Y.X.; Liu, B.Y.; Ma, Y.C.; Liu, Y.N.; Liu, G.Q. Chitinase is involved in the fruiting body development of medicinal fungus *Cordyceps militaris*. *Life* **2023**, *13*, 764. [CrossRef]
15. Lee, J.; Lee, H.; Lee, J.; Chang, P.S. Heterologous expression, purification, and characterization of a recombinant *Cordyceps militaris* lipase from Candida rugosa-like family in Pichia pastoris. *Enzym. Microb. Technol.* **2023**, *168*, 110254. [CrossRef] [PubMed]
16. Paterson, R.R.M. *Cordyceps*—A traditional Chinese medicine and another fungal therapeutic biofactory? *Phytochemistry* **2008**, *69*, 1469–1495. [CrossRef]
17. Yue, K.; Ye, M.; Zhou, Z.; Sun, W.; Lin, X. The genus *Cordyceps*: A chemical and pharmacological review. *J. Pharm. Pharmacol.* **2012**, *65*, 474–493. [CrossRef] [PubMed]
18. Cunningham, K.G.; Manson, W.; Spring, F.S.; Hutchinson, S.A. Cordycepin, a metabolic product isolated from cultures of *Cordyceps militaris* (Linn.) Link. *Nature* **1950**, *166*, 949. [CrossRef] [PubMed]
19. Xia, Y.; Luo, F.; Shang, Y.; Chen, P.; Lu, Y.; Wang, C. Fungal cordycepin biosynthesis is coupled with the production of the safeguard molecule pentostatin. *Cell Chem. Biol.* **2017**, *24*, 1479–1489.e1474. [CrossRef]
20. He, J.; Liu, S.; Tan, Q.; Liu, Z.; Fu, J.; Li, T.; Wei, C.; Liu, X.; Mei, Z.; Cheng, J.; et al. Antiviral potential of small molecules cordycepin, thymoquinone, and n6, n6-dimethyladenosine targeting SARS-CoV-2 entry protein ADAM17. *Molecules* **2022**, *27*, 9044. [CrossRef]
21. Liao, Y.; Ling, J.; Zhang, G.; Liu, F.; Tao, S.; Han, Z.; Chen, S.; Chen, Z.; Le, H. Cordycepin induces cell cycle arrest and apoptosis by inducing DNA damage and up-regulation of p53 in Leukemia cells. *Cell Cycle* **2015**, *14*, 761–771. [CrossRef]
22. Lan, T.; Yu, Y.; Zhang, J.; Li, H.; Weng, Q.; Jiang, S.; Tian, S.; Xu, T.; Hu, S.; Yang, G.; et al. Cordycepin ameliorates nonalcoholic steatohepatitis by activation of the amp-activated protein kinase signaling pathway. *Hepatology* **2021**, *74*, 686–703. [CrossRef]

23. Sirithep, K.; Xiao, F.; Raethong, N.; Zhang, Y.; Laoteng, K.; Hu, G.; Vongsangnak, W. Probing carbon utilization of *Cordyceps militaris* by sugar transportome and protein structural analysis. *Cells* **2020**, *9*, 401. [CrossRef]
24. Saito, M.; Xu, P.; Faure, G.; Maguire, S.; Kannan, S.; Altae-Tran, H.; Vo, S.; Desimone, A.; Macrae, R.K.; Zhang, F. Fanzor is a eukaryotic programmable RNA-guided endonuclease. *Nature* **2023**, *620*, 660–668. [CrossRef]
25. Balakireva, A.V.; Kuznetsova, N.V.; Petushkova, A.I.; Savvateeva, L.V.; Zamyatnin, A.A., Jr. Trends and prospects of plant proteases in therapeutics. *Curr. Med. Chem.* **2019**, *26*, 465–486. [PubMed]
26. Cui, Z.; Zeng, C.; Huang, F.; Yuan, F.; Yan, J.; Zhao, Y.; Zhou, Y.; Hankey, W.; Jin, V.X.; Huang, J.; et al. Cas13d knockdown of lung protease Ctsl prevents and treats SARS-CoV-2 infection. *Nat. Chem. Biol.* **2022**, *18*, 1056–1064. [CrossRef]
27. You, C.; Chen, H.; Myung, S.; Sathitsuksanoh, N.; Ma, H.; Zhang, X.-Z.; Li, J.; Zhang, Y.H.P. Enzymatic transformation of nonfood biomass to starch. *Proc. Natl. Acad. Sci. USA* **2013**, *110*, 7182–7187. [CrossRef] [PubMed]
28. Xu, X.; Zhang, W.; You, C.; Fan, C.; Ji, W.; Park, J.-T.; Kwak, J.; Chen, H.; Zhang, Y.-H.P.J.; Ma, Y. Biosynthesis of artificial starch and microbial protein from agricultural residue. *Sci. Bull.* **2023**, *68*, 214–223.
29. Huang, J.; Lin, Q.; Fei, H.; He, Z.; Xu, H.; Li, Y.; Qu, K.; Han, P.; Gao, Q.; Li, B.; et al. Discovery of deaminase functions by structure-based protein clustering. *Cell* **2023**, *186*, 3182–3195.e3114. [PubMed]
30. Feng, D.; Hu, H.; Yong, T.; Liu, Y.; Xiao, C.; Huang, L.; Xie, Y.; Wu, Q. Induction of sexual fruiting-body formation by pairing the opposite mating-type isolates of *Cordyceps militaris*. *Mycosystema* **2023**, *42*, 344–352.
31. Zu, Z.; Wang, S.; Zhao, Y.; Fan, W.; Li, T. Integrated enzymes activity and transcriptome reveal the effect of exogenous melatonin on the strain degeneration of *Cordyceps militaris*. *Front. Microbiol.* **2023**, *14*, 1112035. [CrossRef]
32. Wang, X.; Li, X.E.; Qiu, W.; Sa, F.; Feng, Y.; Ge, Y.; Yang, S.; Liu, Y.; Xie, J.; Zhang, W.; et al. Effects of mating-type ratio imbalance on the degeneration of *Cordyceps militaris* subculture and preventative measures. *PeerJ* **2024**, *12*, e17648. [CrossRef]
33. Lou, H.; Lin, J.; Guo, L.; Wang, X.; Tian, S.; Liu, C.; Zhao, Y.; Zhao, R. Advances in research on *Cordyceps militaris* degeneration. *Appl. Microbiol. Biotechnol.* **2019**, *103*, 7835–7841.
34. Yong, T.; Zhang, M.; Chen, D.; Shuai, O.; Chen, S.; Su, J.; Jiao, C.; Feng, D.; Xie, Y. Actions of water extract from *Cordyceps militaris* in hyperuricemic mice induced by potassium oxonate combined with hypoxanthine. *J. Ethnopharmacol.* **2016**, *194*, 403–411. [CrossRef]
35. Yong, T.; Chen, S.; Xie, Y.; Chen, D.; Su, J.; Shuai, O.; Jiao, C.; Zuo, D. Cordycepin, a characteristic bioactive constituent in *Cordyceps militaris*, ameliorates hyperuricemia through URAT1 in hyperuricemic mice. *Front. Microbiol.* **2018**, *9*, 58. [CrossRef] [PubMed]
36. Chai, L.; Li, J.; Guo, L.; Zhang, S.; Chen, F.; Zhu, W.; Li, Y. Genomic and transcriptome analysis reveals the biosynthesis network of cordycepin in *Cordyceps militaris*. *Genes* **2024**, *15*, 626. [CrossRef] [PubMed]
37. Dalbeth, N.; Gosling, A.L.; Gaffo, A.; Abhishek, A. Gout. *Lancet* **2021**, *397*, 1843–1855. [CrossRef]
38. Zhao, Z.; Luo, J.; Liao, H.; Zheng, F.; Chen, X.; Luo, J.; Chen, Y.; Zhao, K.; Zhang, S.; Tian, J.; et al. Pharmacological evaluation of a novel skeleton compound isobavachin (4′,7-dihydroxy-8-prenylflavanone) as a hypouricemic agent: Dual actions of URAT1/GLUT9 and xanthine oxidase inhibitory activity. *Bioorg. Chem.* **2023**, *133*, 106405. [CrossRef]
39. Shi, Y.; Wei, F.; Wang, G.-L.; Ma, S.-C.; Lin, R.-C. Identification of geographical origins of *Cordyceps* based on data of amino acids with self-organizing map neural network. *Zhongguo Zhong Yao Za Zhi* **2021**, *46*, 4765–4773.
40. Zeb, U.; Aziz, T.; Azizullah, A.; Zan, X.Y.; Khan, A.A.; Bacha, S.A.S.; Cui, F.J. Complete mitochondrial genomes of edible mushrooms: Features, evolution, and phylogeny. *Physiol. Plant* **2024**, *176*, e14363. [CrossRef]
41. Chiu, C.-P.; Hwang, T.-L.; Chan, Y.; El-Shazly, M.; Wu, T.-Y.; Lo, I.-W.; Hsu, Y.-M.; Lai, K.-H.; Hou, M.-F.; Yuan, S.-S.; et al. Research and development of *Cordyceps* in Taiwan. *Food Sci. Hum. Wellness* **2016**, *5*, 177–185.
42. Fu, X.; Wong, K.K.; Tseng, Y. Editorial: A new frontier for traditional medicine research-multi-omics approaches. *Front. Pharmacol.* **2023**, *14*, 1203097. [CrossRef]
43. Van der Auwera, G.A.; Carneiro, M.; Hartl, C.; Poplin, R.; del Angel, G.; Levy-Moonshine, A.; Jordan, T.; Shakir, K.; Roazen, D.; Thibault, J.; et al. From fastq data to high-confidence variant calls: The genome analysis toolkit best practices pipeline. *Curr. Protoc. Bioinform.* **2013**, *43*, 11.10.1–11.10.33. [CrossRef]
44. Cingolani, P.; Platts, A.; Wang, L.L.; Coon, M.; Nguyen, T.; Wang, L.; Land, S.J.; Lu, X.; Ruden, D.M. A program for annotating and predicting the effects of single nucleotide polymorphisms, SnpEff: SNPs in the genome of Drosophila melanogaster strain w1118; iso-2; iso-3. *Fly* **2012**, *6*, 80–92. [CrossRef]
45. Wang, K.; Li, M.; Hakonarson, H. ANNOVAR: Functional annotation of genetic variants from next-generation sequencing data. *Nucleic Acids Res.* **2010**, *38*, e164. [CrossRef] [PubMed]
46. Alexander, D.H.; Novembre, J.; Lange, K. Fast model-based estimation of ancestry in unrelated individuals. *Genome Res.* **2009**, *19*, 1655–1664. [PubMed]
47. Meisner, J.; Albrechtsen, A. Inferring population structure and admixture proportions in low-depth NGS data. *Genetics* **2018**, *210*, 719–731. [CrossRef]
48. Wang, Y.; Obbard, D.J. Experimental estimates of germline mutation rate in eukaryotes: A phylogenetic meta-analysis. *Evol. Lett.* **2023**, *7*, 216–226.

49. Korneliussen, T.S.; Albrechtsen, A.; Nielsen, R. ANGSD: Analysis of next generation sequencing data. *BMC Bioinform.* **2014**, *15*, 356.
50. Kofler, R.; Schlötterer, C. Gowinda: Unbiased analysis of gene set enrichment for genome-wide association studies. *Bioinformatics* **2012**, *28*, 2084–2085.
51. Danecek, P.; Auton, A.; Abecasis, G.; Albers, C.A.; Banks, E.; DePristo, M.A.; Handsaker, R.E.; Lunter, G.; Marth, G.T.; Sherry, S.T.; et al. The variant call format and VCFtools. *Bioinformatics* **2011**, *27*, 2156–2158. [CrossRef]
52. Baek, M.; DiMaio, F.; Anishchenko, I.; Dauparas, J.; Ovchinnikov, S.; Lee, G.R.; Wang, J.; Cong, Q.; Kinch, L.N.; Schaeffer, R.D.; et al. Accurate prediction of protein structures and interactions using a three-track neural network. *Science* **2021**, *373*, 871–876. [CrossRef]
53. Van Kempen, M.; Kim, S.S.; Tumescheit, C.; Mirdita, M.; Lee, J.; Gilchrist, C.L.M.; Söding, J.; Steinegger, M. Fast and accurate protein structure search with Foldseek. *Nat. Biotechnol.* **2024**, *42*, 243–246.
54. Wang, J.; Feng, L.; Mu, S.; Dong, A.; Gan, J.; Wen, Z.; Meng, J.; Li, M.; Wu, R.; Sun, L. Asymptotic tests for Hardy–Weinberg equilibrium in hexaploids. *Hort. Res.* **2022**, *9*, uhac104. [CrossRef]
55. Korneliussen, T.S.; Moltke, I.; Albrechtsen, A.; Nielsen, R. CalcuFlation of Tajima's D and other neutrality test statistics from low depth next-generation sequencing data. *BMC Bioinform.* **2013**, *14*, 289.
56. Tavares, H.; Whibley, A.; Field, D.L.; Bradley, D.; Couchman, M.; Copsey, L.; Elleouet, J.; Burrus, M.; Andalo, C.; Li, M.; et al. Selection and gene flow shape genomic islands that control floral guides. *Proc. Natl. Acad. Sci. USA* **2018**, *115*, 11006–11011. [CrossRef] [PubMed]
57. Malinsky, M.; Challis, R.J.; Tyers, A.M.; Schiffels, S.; Terai, Y.; Ngatunga, B.P.; Miska, E.A.; Durbin, R.; Genner, M.J.; Turner, G.F. Genomic islands of speciation separate cichlid ecomorphs in an East African crater lake. *Science* **2015**, *350*, 1493–1498.
58. Hu, J.; Sun, Y.; Li, B.; Liu, Z.; Wang, Z.; Gao, Q.; Guo, M.; Liu, G.; Zhao, K.T.; Gao, C. Strand-preferred base editing of organellar and nuclear genomes using CyDENT. *Nat. Biotechnol.* **2024**, *42*, 936–945.
59. Marei, H.; Tsai, W.-T.K.; Kee, Y.-S.; Ruiz, K.; He, J.; Cox, C.; Sun, F.T.; Penikalapati, S.; Dwivedi, P.; Choi, M.; et al. Antibody targeting of E3 ubiquitin ligases for receptor degradation. *Nature* **2022**, *610*, 182–189.
60. Wang, J.; Lisanza, S.; Juergens, D.; Tischer, D.; Watson, J.L.; Castro, K.M.; Ragotte, R.; Saragovi, A.; Milles, L.F.; Baek, M.; et al. Scaffolding protein functional sites using deep learning. *Science* **2022**, *377*, 387–394.
61. Mosalaganti, S.; Obarska-Kosinska, A.; Siggel, M.; Taniguchi, R.; Turoňová, B.; Zimmerli, C.E.; Buczak, K.; Schmidt, F.H.; Margiotta, E.; Mackmull, M.-T.; et al. AI-based structure prediction empowers integrative structural analysis of human nuclear pores. *Science* **2022**, *376*, eabm9506.

agriculture

MDPI

Article

# The Molecular Mechanism of Mycelial Incubation Time Effects on Primordium Formation of *Pleurotus tuoliensis* Through Transcriptome and Lipidomic Analyses

**Qi He [1,2,3], Chenyang Huang [1,2], Lijiao Zhang [1,2], Wei Gao [1,2] and Mengran Zhao [1,2,*]**

1 State Key Laboratory of Efficient Utilization of Arid and Semi-Arid Arable Land in Northern China, Institute of Agricultural Resources and Regional Planning, Chinese Academy of Agricultural Sciences, Beijing 100081, China; heqi_jlau@126.com (Q.H.); huangchenyang@caas.cn (C.H.); zhanglijiao@caas.cn (L.Z.); gaowei01@caas.cn (W.G.)

2 Key Laboratory of Microbial Resources, Ministry of Agriculture and Rural Affairs, Beijing 100081, China

3 Engineering Research Centre of Chinese Ministry of Education for Edible and Medicinal Fungi, Jilin Agricultural University, Changchun 130118, China

* Correspondence: zhaomengran@caas.cn (M.Z.)

**Abstract:** *Pleurotus tuoliensis* is a precious edible mushroom with a long cultivation cycle. Despite being cultivated in China for nearly 30 years, research on the molecular mechanisms underlying its primordium formation remains limited. In this study, the molecular mechanisms by which incubation time affects the primordium formation of *P. tuoliensis* were investigated using RNA-seq technology and lipid content detection. Our research revealed that the transcription of genes involved in lipid metabolism and lipid levels changed significantly during different incubation periods. Distinct differences were observed in gene transcription associated with signaling pathways, sphingolipid metabolism, fatty acid metabolism, and steroid biosynthesis in mycelia cultured for varying days and then stimulated by low temperature and light. These findings indicate that lipid accumulation and alterations in mycelial cell membrane components during incubation may affect the mycelial response to environmental signals, subsequently regulating primordium formation. This study revealed the crucial role of lipid metabolism during incubation in the primordium formation of *P. tuoliensis*, providing a novel perspective for investigating the molecular mechanism underlying fruiting body development.

**Keywords:** *Pleurotus tuoliensis*; incubation time; primordia; transcriptome; lipidomics

## 1. Introduction

*Pleurotus tuoliensis*, which used to be identified as *Pleurotus nebrodensis*, is a species of precious edible mushroom with a long cultivation cycle. The fruiting body is rich in lysine, arginine, vitamin D, mineral elements, and trace elements, which have high nutritional value [1]. The fruiting bodies contain numerous polysaccharides with health benefits, playing roles in protecting cardiovascular health, regulating immune activity, regulating blood lipids, and protecting the liver [2–5]. In view of their high economic value, many scholars have studied the breeding, cultivation technology, physiology, postharvest storage, and deep processing of *P. tuoliensis* [6–9].

*Pleurotus tuoliensis* is a medium-to-low-temperature fructification edible mushroom, and its optimum temperature for mycelial growth is generally 25 °C. Light and low-temperature are necessary to form primordia, and many studies have been conducted on the influence of environmental factors on primordium formation, such as blue light being favorable for the induction of primordium formation [10]. After 10 days of induction at 14 °C and 300–1000 lx with blue light, mycelia kink and finally form primordia [11]. Blue light affects the growth of primordia and fruiting bodies through glycolysis and pentose phosphate pathways [12]. In addition to environmental factors, the physiological state of mycelia can also affect the yield and quality of *P. tuoliensis*. The reproductive growth of edible mushrooms requires the consumption of many nutrients [13]. After spawn running is completed, mycelia can fully absorb and store nutrients from the substrate. Under appropriate environmental stimulation, mycelia undergo kinking and eventually form primordia. After 37 days of spawning in the dark, *P. tuoliensis* mycelia form primordia after 7 days of stimulation with low temperature at 17 °C and light [14].

In recent years, transcriptomics has been widely used to research the analysis of key genes and pathways involved in growth and development stages, pigment formation [15], and autolysis [16]. Transcriptome analysis of the mycelia and primordia of *Pleurotus ostreatus* revealed that 69 genes, which are involved mainly in energy metabolism, signal transduction, membrane proteins, and other pathways, were upregulated in the primordium stage [17]. Members of the $Zn_2Cys_6$-type zinc finger transcription factor family in *P. ostreatus* were identified via transcriptome analysis, and 13 candidate genes that play a role in fruiting body formation and the heat stress response were identified [18]. Transcriptome analysis searched for *Lentinula edodes* fruiting body-specific genes, such as aspartic protease, gamma-glutamyl transpeptidase, and cyclohexanone monooxygenase, which are involved in fruiting body maturation and the isolation of functional substances [19]. Transcriptome analyses revealed that blue light promoted the growth of *L. edodes* primordia and fruiting bodies, and DDR-48 heat shock protein genes involved in primordia morphogenesis were screened [20]. The maturity and development of transcriptomic technology further promoted the basic research of edible mushrooms.

Lipidomics has been applied in the research on edible mushrooms recently and can reveal complete molecular information about lipids in biological samples [21]. A lipidomics analysis of eight wild edible mushrooms identified lysophosphatidylethanolamine (16:1) and ceramide non-hydroxy fatty acid-dihydrosphingosine (d23:0–10:0) as potential biomarkers to distinguish different mushroom species [22]. Untargeted lipidomics analysis combined with chemometrics was applied to distinguish the edible mushrooms from poisonous mushrooms [23]. Lipidomic profiles and lipid dynamic changes during the growth of *Morchella sextelata* were analyzed, which revealed that glycerophospholipid metabolism was the major pathway involved in the growth of *M. sextelata* [24].

In this study, the molecular mechanisms underlying primordium formation in *P. tuoliensis* were investigated by assessing the effects of various mycelial incubation periods, sequencing the transcriptomes at different developmental stages, screening key genes and validating via RT-qPCR, and detecting the presence of related substances by lipidomics analysis. The present study further contributes to understanding the mechanism of primordium formation in *P. tuoliensis*.

## 2. Materials and Methods

### 2.1. Sample Collection

The *P. tuoliensis* CCMSSC 02607 strain was obtained from the Center for Mushroom Spawn Standards and Control of China (CCMSSC). The CCMSSC02607 strain was punched with a hole punch (φ7 mm), inoculated in PDA medium (15.6 g of potato dextrose agar powder dissolved in 400 mL of distilled water, sterilized at 121 °C for 20 min), and cultured at 25 °C in the dark for 4 days (T1) and 6 days (T3), respectively. Then, two different treatment groups were cultured at 15 °C with alternate light (12 h light, 12 h dark) for 13 days (T2 and T4) (Table 1). Mycelia and primordia were collected and stored at −80 °C after being snap-frozen in liquid nitrogen for subsequent experiments.

**Table 1.** Samples and treatment conditions.

| Sample | Culture Condition |
|---|---|
| T1 | Culture at 25 °C for 4 days in the dark |
| T2 | Cultured at 25 °C for 4 days in the dark, then cultured at 15 °C with alternate light for 13 days |
| T3 | Culture at 25 °C for 6 days in the dark |
| T4 | Cultured at 25 °C for 6 days in the dark, then cultured at 15 °C with alternate light for 13 days |

*2.2. Library Preparation and RNA-Seq*

Four treatments were set up with three replicates in each treatment, with a total of 12 samples in transcriptome sequencing. Transcriptome sequencing was carried out at Beijing Biomarker Technologies Co., Ltd. Total RNA was extracted from the mycelia and primordia of *P. tuoliensis*. Purity, concentration, and integrity of the RNA samples were tested to guarantee the use of qualified samples for transcriptome sequencing. mRNAs were enriched with magnetic beads with oligo (dT), which were randomly disrupted with fragmentation buffer. The first cDNA strand was then synthesized with six-base random hexamers, and the second cDNA strand was synthesized by adding buffer, dNTPs, RNase H, and DNA polymerase I. The resulting cDNA fragments were purified using AMPure XP beads and then subjected to end repair through the addition of a single "A" base and the ligation of Illumina multiplex barcode adapters, followed by the separation of the fragments via gel purification using AMPure XP beads. Finally, the cDNA libraries were constructed via PCR to create a total of 12 libraries.

After library construction, the effective library concentration (library effective concentration > 2 nmol/L) was accurately quantified via q-PCR to ensure library quality. After library inspection, the different libraries were pooled according to the target downstream data volume and sequenced via the Illumina platform.

Before performing data analysis, we ensured that reads were of high enough quality to ensure the accuracy of subsequent analyses. Reads containing joints and filtered low-quality reads were removed to obtain high-quality clean data. HISAT2 (v 2.0.4) software was used to obtain the RNA sequencing experimental reads for an efficient comparison system. The reads on the pair were assembled via StringTie to construct more complete transcripts and better assess expression.

*2.3. Identification of Differentially Expressed Genes (DEGs)*

To identify DEGs, the expression levels of transcripts were calculated via the fragments per kilobase of transcript per million fragments mapped (FPKM) method [25]. The fold change indicates the ratio of expression between two samples (groups). The false discovery rate (FDR) was obtained by correcting for the differential significance of the *p*-value. The differential expression gene was selected according to DESeq2_edgeR, and fold change ≥ 2.0 and an FDR < 0.01 were used as screening criteria.

The functional annotation of the DEGs was performed via the non-redundant protein sequence database (NR) (https://www.ncbi.nlm.nih.gov/ (accessed on 25 December 2020)), gene ontology (GO) (https://www.geneontology.org/ (accessed on 25 December 2020)), cluster of orthologous groups (COG) (https://www.ncbi.nlm.nih.gov/research/cog-project/ (accessed on 25 December 2020)), eukaryotic orthologous groups (KOG) (https://ftp.ncbi.nlm.nih.gov/pub/COG/KOG/ (accessed on 25 December 2020)), the protein family database (Pfam) (http://pfam.xfam.org/ (accessed on 25 December 2020)), Kyoto encyclopedia of genes and genomes (KEGG) (https://www.kegg.jp/kegg/kegg1.html (accessed on 25 December 2020)), Swiss-Prot (https://www.expasy.org/resources/uniprotkb-swiss-prot (accessed on 25 December 2020)), and other databases to obtain annotation information for the genes.

*2.4. Reverse Transcription-Quantitative PCR (RT-qPCR) Analysis*

Twenty-five DEGs identified by RNA-seq were selected for RT-qPCR validation. RNA extraction was performed using the E.Z.N.A® Plant RNA Kit from Omega Bio-Tek, Inc., Norcross, GA, USA. The RNA concentrations were measured via an Agilent 2100 Bioanalyzer. The RNA was used as a template to obtain cDNA using the HiScript® II 1st Strand cDNA Synthesis Kit from Nanjing Vazyme Biotech Co., Ltd., Nanjing, China. The qPCR primers for the target genes were designed via the online website IDT (https://sg.idtdna.com/Scitools/Applications/RealTimePCR/Default.aspx (accessed on 13 April 2024)), as shown in Table S1. The expression of the target gene was examined via RT-qPCR. RT-qPCR amplification was performed as follows: 95 °C for 30 s, 40 cycles of 95 °C for 10 s, and 60 °C for 30 s, followed by a 72 °C extension for 30 s. The relative gene expression was calculated via the $2^{-\Delta\Delta CT}$ method, with *β-actin* used as an internal reference gene [26].

*2.5. Lipid Metabolomic Analysis*

To analyze lipid differences after different mycelial incubation periods, mycelia were collected after culture in the dark from 4 d to 6 d. For lipid analysis, the lipid extracts from three biological replicates were separated via an Orbitrap Exploris 120 mass spectrometer (Thermo Fisher Scientific, Waltham, MA, USA) at Biomarker Technologies Co., Ltd., Beijing, China. Finally, based on the mass spectrometry analysis results, a comparison of the mycelium lipid composition and quantification was conducted using ultra-high-performance liquid chromatography quadruple time-of-flight mass spectrometry (UHPLC-QTOF/MS) [27].

*2.6. Statistical Analysis*

In this study, each experiment was performed three times. Statistical analyses were performed using SPSS Statistics 17.0 (SPSS Inc., Chicago, IL, USA) with the data generated above. Multiple comparisons were conducted via one-way analysis of variance (ANOVA). When the $p$-value was $< 0.05$, results were considered statistically significant. Figures were plotted with GraphPad Prism version 8 (GraphPad Software Inc., San Diego, CA, USA).

**3. Results**

*3.1. Incubation Time Affects Primordium Formation*

After 4 days of mycelial incubation at 25 °C in the dark (T1) and then stimulation by light and cold subsequently for 13 days (T2), the mycelia kinked but failed to form primordia. However, when the mycelia were incubated for a duration of 6 days (T3) and subjected to the same stimulation for 13 days (T4), they were able to form primordia after hyphal knot formation. The results indicated that incubation time is one of the important factors affecting the formation of primordia in *P. tuoliensis* (Figure 1).

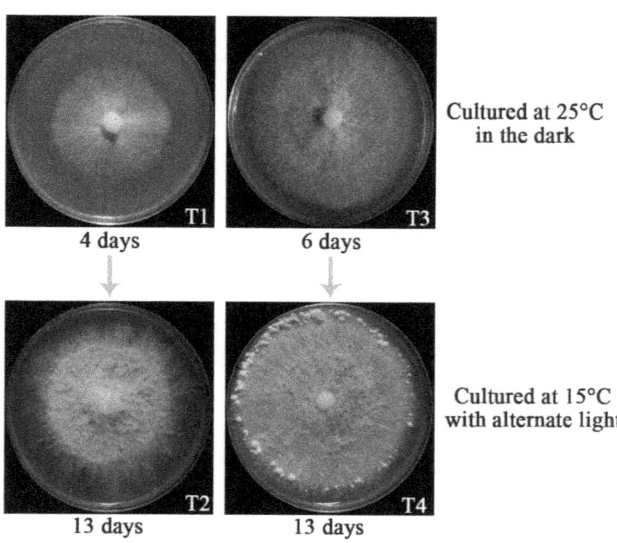


**Figure 1.** Effects of incubation time on primordium formation. T1: Cultured at 25 °C for 4 days in the dark; T2: cultured at 25 °C for 4 days in the dark and then stimulated at 15 °C with alternate light for 13 days; T3: cultured at 25 °C for 6 days in the dark; T4: cultured at 25 °C for 6 days in the dark and then stimulated at 15 °C with alternate light for 13 days.

### 3.2. *Transcription of Genes Involved in Lipid Metabolism Significantly Changed During the Incubation Period*

To investigate the mechanism by which primordium formation is affected by various incubation periods (T1 vs. T3), genes whose expressions changed during incubation were screened (Table S2). There were 242 DEGs affected by incubation time, of which 188 were upregulated, and 54 were downregulated (Figure S1). KEGG analyses showed that the upregulated DEGs were significantly enriched in glycosphingolipid biosynthesis-globo series, galactose metabolism, sphingolipid metabolism, glycerolipid metabolism, pentose and glucuronate interconversion, and starch and sucrose metabolism (Figure 2a), whereas the downregulated DEGs were significantly enriched in protein processing in the endoplasmic reticulum, fatty acid biosynthesis, and fatty acid metabolism (Figure 2b). These results suggest that the expression patterns of genes involved in lipid metabolism significantly changed during the incubation period when the mycelia grew at 25 °C in the dark.

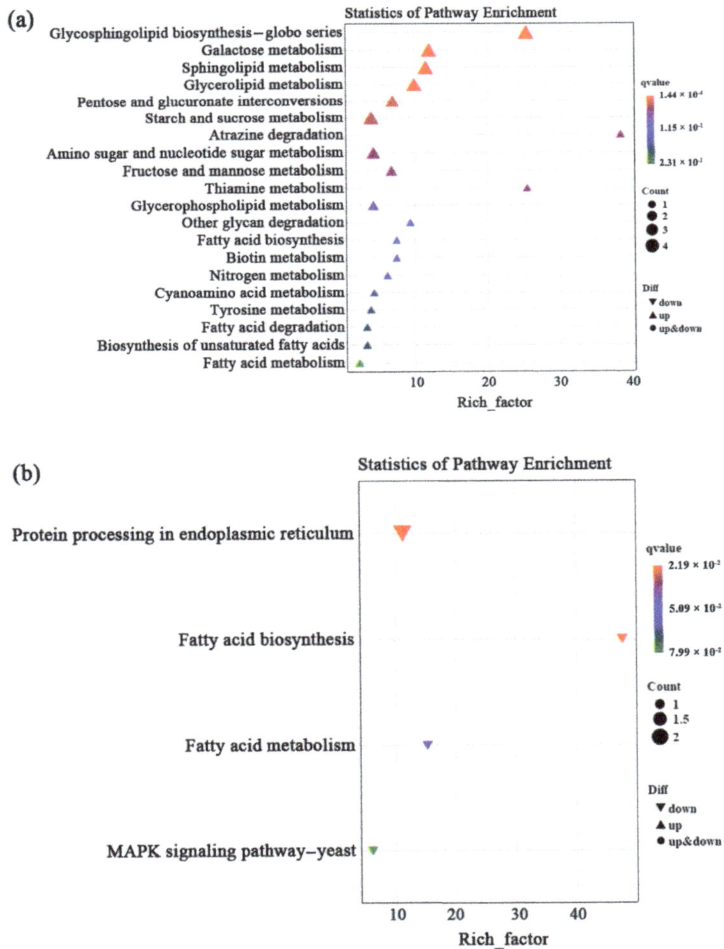


**Figure 2.** KEGG functional enrichment analysis of DEGs affected by incubation time (T1 vs. T3). (**a**) KEGG functional enrichment analysis of upregulated DEGs affected by incubation time. (**b**) KEGG functional enrichment analysis of downregulated DEGs affected by incubation time.

GO analyses showed that the upregulated DEGs were significantly enriched in the biological processes of carbohydrate metabolism, phospholipid biosynthesis, proteolysis, and thiazole biosynthesis and the molecular functions of raffinose alpha-galactosidase activity and hydrolase activity (Figure 3a), whereas downregulated DEGs were related mainly to the biological processes of oxidation-reduction, response to stress, oxalate metabolism, cellular metabolism, organic substance metabolism, and primary metabolism; the molecular functions of flavin adenine dinucleotide binding, oxidoreductase, D-arabinono-1,4-lactone oxidase, enoyl-[acyl-carrier-protein] reductase, L-gulonolactone oxidase, aspartic-type endopeptidase, prenyltransferase, nutrient reservoir, holo-[acyl-carrier-protein] synthase, oxidoreductase, and protein domain-specific binding; and the cellular components of the fatty acid synthase complex and membrane (Figure 3b). The results were similar to those from the KEGG analyses.

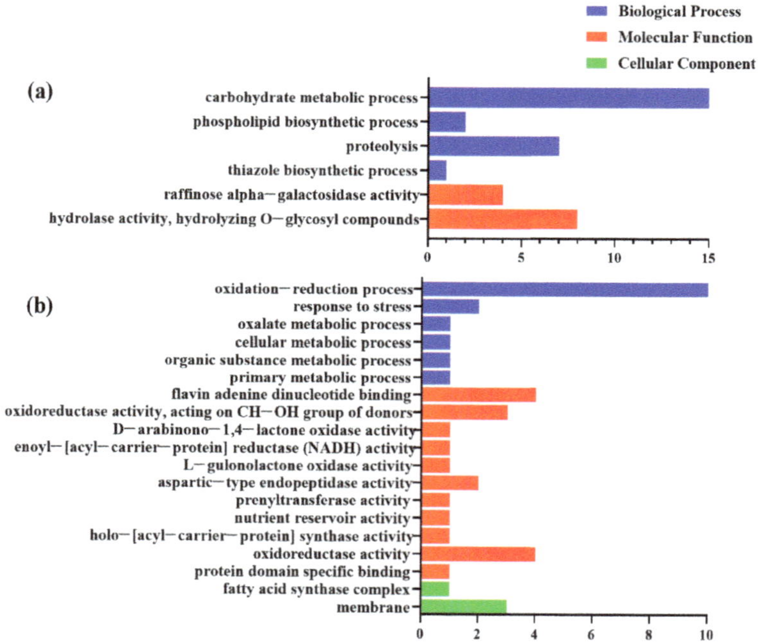


**Figure 3.** GO functional enrichment analysis of DEGs affected by incubation time (T1 vs. T3). (**a**) GO functional enrichment analysis of upregulated DEGs affected by incubation time in the biological process and molecular functions categories. (**b**) GO functional enrichment analysis of downregulated DEGs affected by incubation time in the biological process molecular function and cellular component categories. Only the significant GO terms ($q$-value < 0.05) are shown.

Heatmap analyses demonstrated that among these 13 DEGs enriched significantly in KEGG terms, only one gene coding for fatty acid synthase was significantly downregulated, whereas the other 12 DEGs were significantly upregulated (Figure 4a). A total of 58 DEGs were enriched significantly in GO terms, of which the significantly downregulated genes encoded fatty acid synthase, NADP-binding protein, and NADPH2 dehydrogenase, whereas 20 significantly upregulated genes encoded galactan endo-β-1,3-galactanase, mannan endo-1,4-β-mannosidase, UPF0271 protein, alcohol dehydrogenase, propanol-preferring, thiamine thiazole synthase, D-xylose reductase, L-glyceraldehyde reductase, bacterial leucyl aminopeptidase, carboxypeptidase D, 3-oxoacyl-[acyl-carrier protein] reductase, β-mannosidase, β-D-xylosidase 4, D-xylulose reductase, Gly-Xaa carboxypeptidase, phosphatidylserine decarboxylase, and α-galactosidase (Figure 4b). Some of the key DEGs selected via KEGG analysis were the same as those selected via GO analysis.

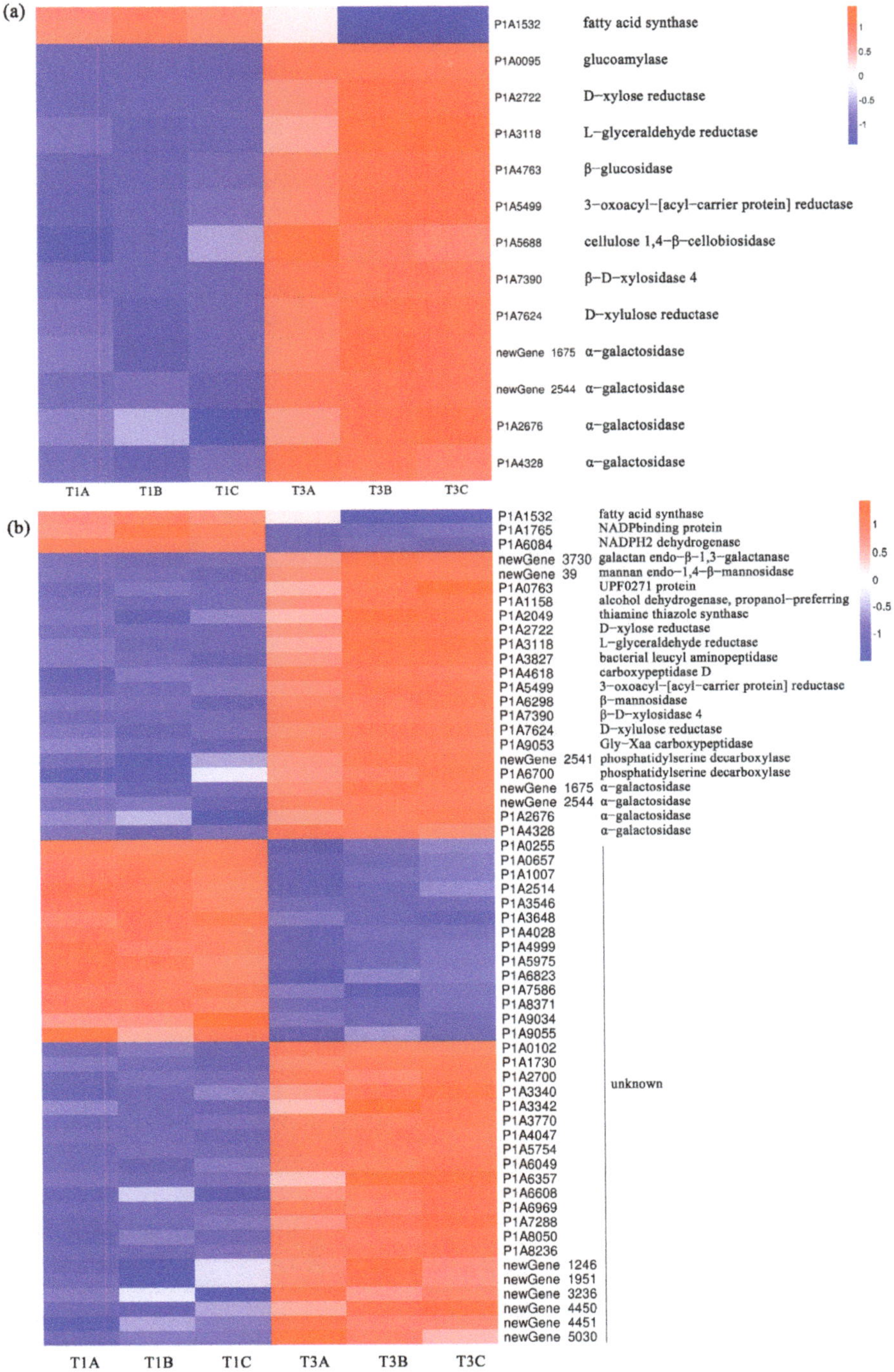


**Figure 4.** Heatmap analyses of DEGs affected by incubation time. (**a**) Heatmap analyses of the 13 DEGs enriched significantly in KEGG terms. (**b**) Heatmap analyses of the 58 DEGs enriched significantly in GO terms. A, B, and C represent 3 repetitions.

### 3.3. *The Lipid Components Changed Significantly During the Incubation Period*

Because transcription analysis revealed that lipid metabolism was most active during the mycelial culture period, the changes in the relevant lipid components during the incubation period were further analyzed. The results showed that the triglycerides (Figure 5a), unsaturated free fatty acids (Figure 5b), sphingosine (Figure 5i), and some types of phospholipids, including phosphatidic acid (Figure 5d), phosphatidylglycerol (Figure 5e), phosphatidylinositol (Figure 5f), and phosphatidylethanolamine (Figure 5h), significantly decreased in T3, as compared to T1. However, the saturated fatty acids and phosphatidylcholine (Figure 5g) significantly increased during the incubation period. The proportion of unsaturated free fatty acids was much greater than that of saturated fatty acids (Figure 5c). Phosphatidylglycerol and phosphatidylcholine were the main components of the phospholipids involved in mycelial growth in the dark.

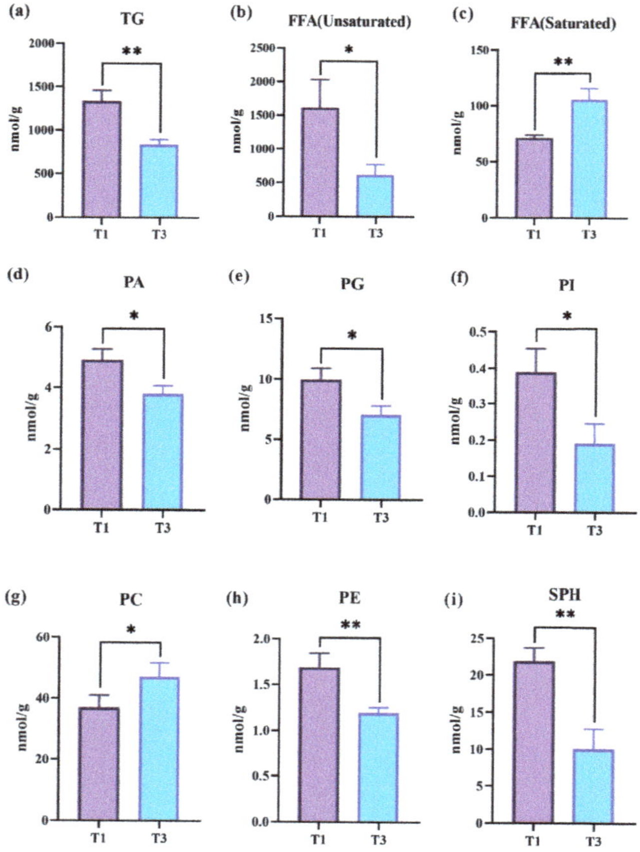


**Figure 5.** Changes in the levels of lipid components during mycelia culture in the dark. (**a**) Triglyceride (TG); (**b**) unsaturated free fatty acids (FFAs); (**c**) saturated free fatty acids (FFAs); (**d**) phosphatidic acid (PA); (**e**) phosphatidylglycerol (PG); (**f**) phosphatidylinositol (PI); (**g**) phosphatidylcholine (PC); (**h**) phosphatidylethanolamine (PE); (**i**) sphingosine (SPH); (* indicates $p < 0.05$, and ** indicates $p < 0.01$, according to Student's $t$ test).

### 3.4. *The Effect of Environmental Stimulation on Metabolic Pathways in Mycelia of Various Incubation days*

Free fatty acids and sphingosine are precursors to the synthesis of sphingolipids, which are essential structural substances of cell membranes and signaling molecules for cell fate determination. Transcription data showed that the process of the response to stress is significantly downregulated during mycelial incubation in the dark. Therefore, we compared the metabolic pathway differences after mycelia cultured for various incubation days were stimulated by light and low temperature. A total of 2660 DEGs changed in the mycelia cultured for 4 days in the dark followed by stimulation (T1 vs. T2); however, in mycelia cultured for 6 days followed by environmental stimulation, the transcriptions of 2143 genes were changed (T3 vs. T4) (Figure S2). GO classification showed that the mycelia cultured for different incubation periods presented marked differences in the transcript levels of genes involved in location, cellular component organization, membranes, transporter activity, and nucleic acid binding transcription factor activity (Figure 6a,d). KEGG enrichment analyses indicated that mycelial culture for 4 d followed by environmental stimulation induced the upregulation of 1264 DEGs that were significantly enriched in the nitrogen, starch, sucrose, and glycerolipid metabolic pathways and the monobactam-, steroid-, sesquiterpenoid-, triterpenoid-, and longevity-regulating pathways (Figure 6b), whereas the 1396 DEGs with decreased expressions were related mainly to ribosomes; aminoacyl-tRNA biosynthesis; the degradation of valine, leucine, and isoleucine; tryptophan metabolism; etc. (Figure 6c). A total of 870 DEGs, whose expressions were increased by environmental stimulation in mycelia cultured for 6 days in the dark, were significantly enriched in tyrosine metabolism, amino sugar and nucleotide sugar metabolism, glycerophospholipid metabolism, and phenylalanine and tryptophan metabolism (Figure 6e), whereas a total of 1273 DEGs whose expressions were downregulated were involved mainly in the metabolism of glycine, serine, threonine, cysteine, methionine, and histidine; the biosynthesis of valine, leucine, and isoleucine; and the metabolism of fructose and mannose (Figure 6f). These results demonstrated that changes in the number and transcriptional level of DEGs during incubation affected the mycelial response to environmental stimuli.

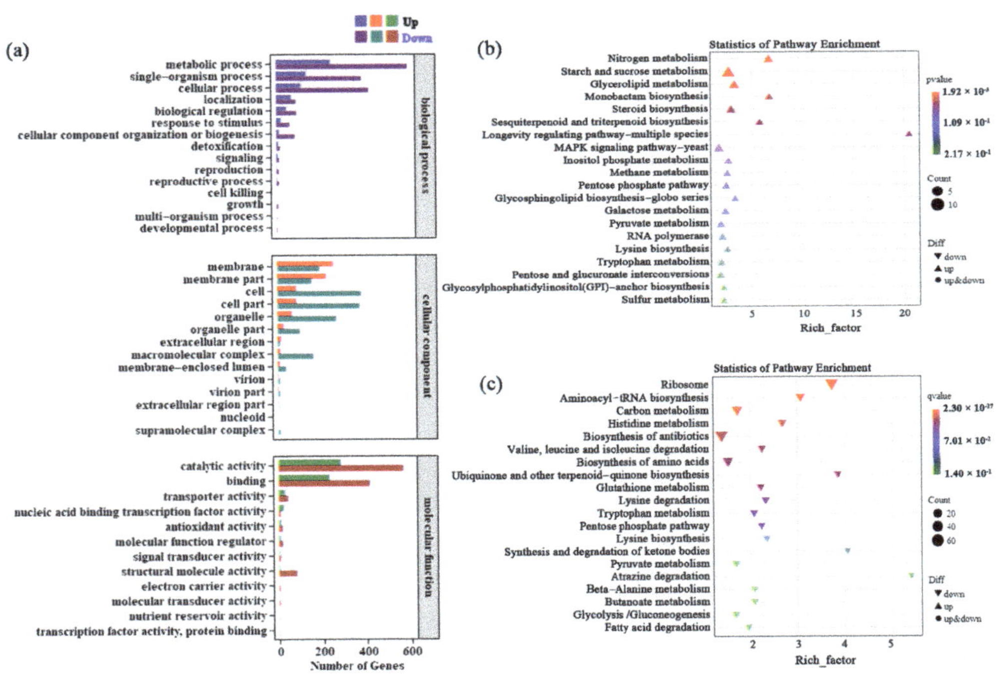


**Figure 6.** *Cont.*

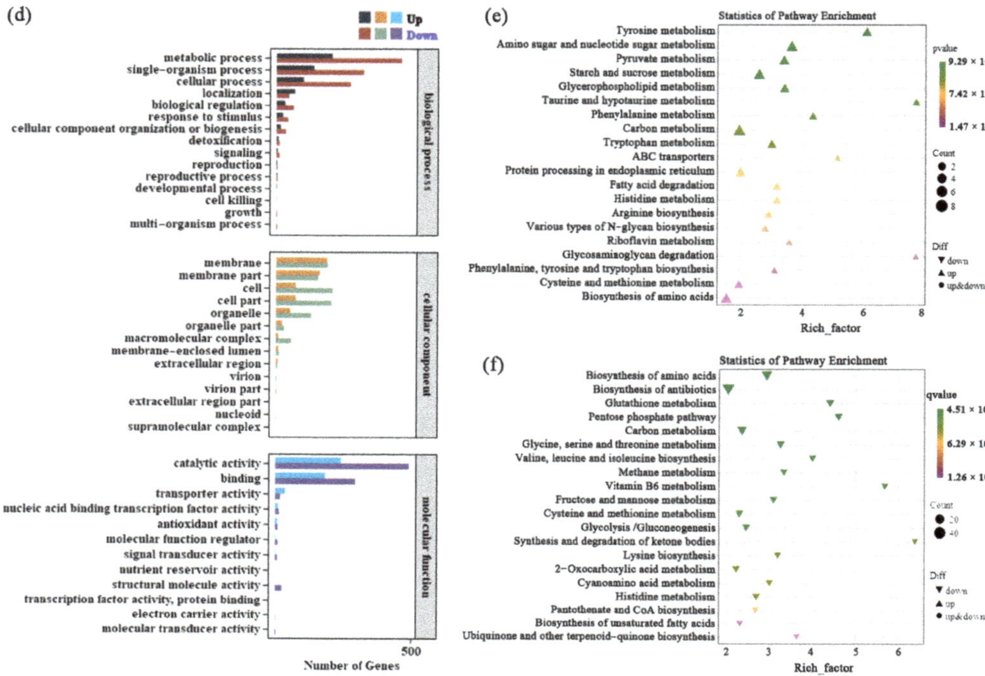

**Figure 6.** Metabolic pathways of mycelia that went through various incubation periods and then stimulation. (**a**) GO functional enrichment analysis of DEGs identified between T1 and T2. (**b**) KEGG functional enrichment analysis of upregulated DEGs identified between T1 and T2. (**c**) KEGG functional enrichment analysis of downregulated DEGs identified between T1 and T2. (**d**) GO functional enrichment analysis of DEGs identified between T3 and T4. (**e**) KEGG functional enrichment analysis of upregulated DEGs identified between T3 and T4. (**f**) KEGG functional enrichment analysis of downregulated DEGs identified between T3 and T4.

Heatmap analyses revealed that the DEGs in the longevity-regulating pathway and MAPK signaling pathway were significantly upregulated in the mycelia that were cultured for 4 days and then stimulated (T1 vs. T2). An opposite change in gene transcription of the transcriptional enhancer factor Tec1 was observed in mycelia cultured for 6 days (T3 vs. T4); however, other genes encoding the transcription factor RLM1, Sho1 osmosensor, protein kinase A, and another transcriptional enhancer factor Tec1 presented no obvious transcriptional changes (Figure 7a). The DEGs enriched in lipid metabolism according to the KEGG database were selected for analysis. The genes encoding delta-9 desaturase, delta-12 desaturase, and 3-oxoacyl-[acyl-carrier protein] reductase, which are related to the biosynthesis of unsaturated fatty acids in mycelia cultured for 6 days and then stimulated (T3 vs. T4), were significantly downregulated; however, the genes encoding phosphatidylserine decarboxylase, phosphatidylserine synthase 2, phospholipase D, and lysophospholipase I, which are enriched in glycerophospholipid metabolism, were significantly upregulated. Most genes encoding alcohol dehydrogenase, aldehyde dehydrogenase ($NAD^+$), glutaryl-CoA dehydrogenase, acetyl-CoA acetyltransferase, and enoyl-CoA hydratase, which are involved in fatty acid degradation, were significantly downregulated in the mycelia cultured for 4 days and then stimulated (T1 vs. T2), whereas the genes involved in steroid biosynthesis, including C-4 methyl sterol oxidase and squalene monooxygenase, were significantly upregulated (Figure 7b).

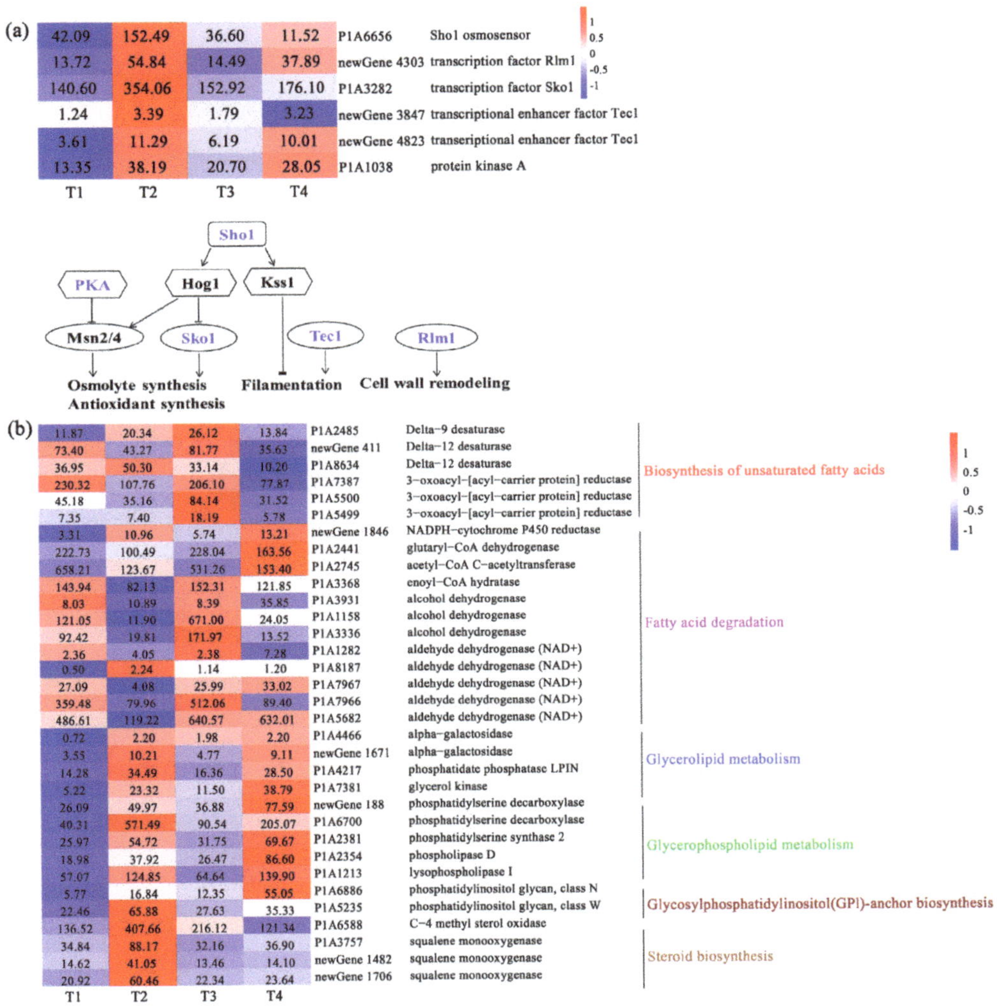

**Figure 7.** Heatmap analyses for the DEGs identified in mycelia that went through various incubation days and then stimulation. (**a**) Analysis of the DEGs in the longevity-regulating pathway and the MAPK signaling pathway. (**b**) DEGs enriched in lipid metabolism were selected on the basis of the KEGG database. The number indicates the FPKM value.

*3.5. Validation of Key DEGs*

To confirm the quality of RNA-seq data and the expression patterns of DEGs in mycelia that were cultured for different incubation periods followed by light and low-temperature stimulation, six DEGs involved in signaling pathways and 14 DEGs involved in lipid metabolism were selected to validate their expression trends via RT-qPCR (Figure 8). The results from RT-qPCR validation showed that the expression trends of most selected genes were consistent with those detected by transcriptome sequencing, except for four genes encoding squalene monooxygenase (newGene 1706), transcriptional enhancer factor Tec1 (newGene 4823), SHO1 osmosensor (P1A1038), and protein kinase A (newGene 3847), whose expression trends in the mycelia cultured for 6 days and then stimulated were opposite to those demonstrated by transcriptome analyses. These results suggest that our RNA-seq data are reliable.

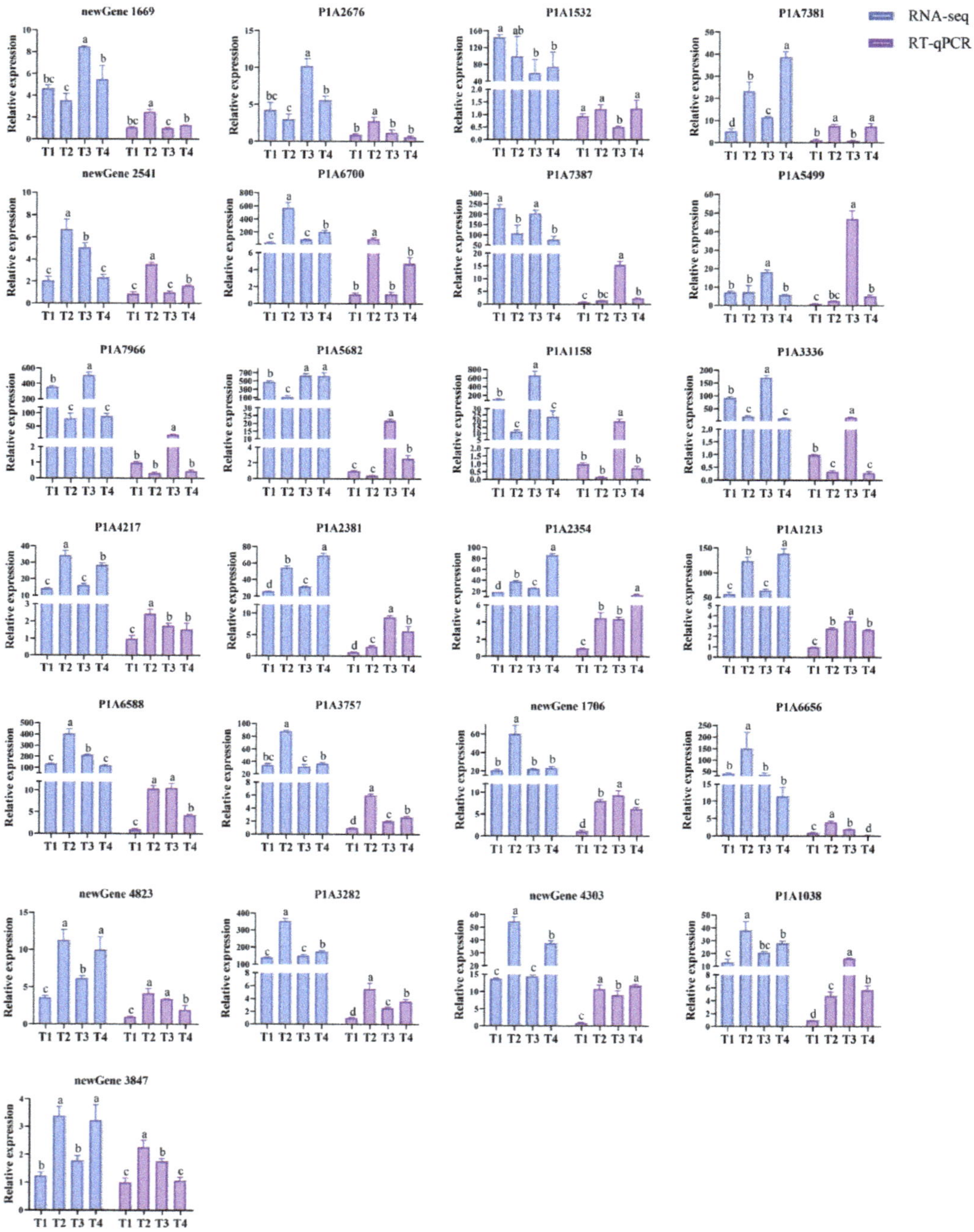


**Figure 8.** Gene expression validation of 25 DEGs by RT-qPCR. Different letters indicate significant differences between the groups ($p < 0.05$, according to Duncan's test).

## 4. Discussion

*Pleurotus tuoliensis* is a species of low-temperature fructifying edible mushroom. Mycelia can only form primordia after accumulating sufficient nutrients during the growth phase and then being subjected to low temperature and light stimulation [14]. In this study, we investigated the correlation between the duration of mycelial incubation at 25 °C in the dark and the primordium formation of *P. tuoliensis* on the basis of analyses of the transcriptome and targeted metabolome.

### 4.1. Lipid Metabolism Was Most Active During the Incubation Period

Our studies suggested that lipid metabolism, including glycosphingolipid biosynthesis-globo series, glycerolipid, and sphingolipid metabolism, were active during the incubation period. Studies on *Hypsizygus marmoreus* revealed that the glycerolipid metabolic pathway was upregulated during the hyphal knot phase [28], which is consistent with the results of the present study. The decreased expression of fatty acid synthase during mycelial incubation in the dark from 4 days to 6 days and decreased levels of triacylglycerol and free fatty acid demonstrated the slowdown of fatty acid and triacylglycerol biosynthesis. Previous studies have suggested that triglyceride and phospholipid levels are significantly reduced after the silencing of *fas*1, which can affect growth and development [29]. Phosphatidylserine decarboxylase plays important roles in phospholipid homeostasis, filamentous fungal growth, and morphogenesis [30]. Our study revealed significantly increased transcriptional levels of phosphatidylserine decarboxylase genes during mycelial growth in the dark and in mycelial response to environmental stimulation. The deletion of *FgPsd*2 resulted in significant reductions in the mycelial growth rate and conidial number of *Fusarium graminearum*. *FgPsd2* mutants grew better when they were supplied with exogenous ethanolamine, which demonstrated that the PE content produced by *FgPsd2* was essential for mycelial growth [31].

Targeted lipidomic analysis revealed that the lipid composition, which included triglycerides, free fatty acids, phosphatidylcholine, phosphatidylglycerol, and sphingosine, significantly changed during the incubation period. Recent studies suggest that the TAG stores in lipid droplets release FAs that act as signals to influence the transcriptional control of gene expression [32]. The process of synthesizing or degrading TAG produces lipid intermediates, such as phosphatidic acid and diacylglycerol, which may serve as activators or inhibitors of signaling pathways [33]. The present study revealed that phosphatidylglycerol and phosphatidylcholine were the main phospholipids involved in mycelial growth in the dark. The level of phosphatidylcholine gradually increased during the stage of mycelial growth and was the most abundant phospholipid. Previous studies have shown that phosphatidylcholine is usually the most abundant phospholipid in eukaryotic cells, comprising about 40–50% of the total cellular phospholipids [34]. Phosphatidylcholine is predicted to participate in organellar and cellular biogenesis, as well as the formation of vesicles for the transport of proteins and lipids within cells [35]. GO enrichment analyses also revealed that DEGs whose expressions changed during incubation were enriched mainly in integral components of the membrane, extracellular regions, and membrane parts in the cellular component category. Sphingosines belong to a structurally diverse group of lipids, which are abundant in membranes. In our study, a significantly greater level of sphingosine was detected in mycelia cultured for 4 days than in that cultured for 6 days. Numerous studies have shown that sphingosine can regulate the phosphorylation of many proteins in the MAPK pathway, in addition to its inhibitory effects on protein kinase C [36].

### 4.2. Lipids Are Components of the Eukaryotic Cellular Membrane and Influence the Mycelial Response to Environmental Stimuli

Lipids not only maintain the homeostasis of the intracellular environment and participate in the formation of cell membranes but also act as signaling molecules in the eukaryotic response to physiological signals and multiple types of stress [37,38]; thus, we hypothesized that the change in lipid metabolism during the incubation period would cause differences in the mycelial response to environmental stimulation, thus affecting primordium formation. The accumulation of TAG can activate the stress pathways of the transcription factors Msn2/4 and superoxide dismutase and significantly prolong lifespan [39]. In the present research, the expression of protein kinase A, which is involved in the longevity-regulating pathway, was significantly increased after the mycelia were cultured for 4 days and then stimulated with light and low temperature. RNA-seq data showed that the transcription levels of three kinds of transcription factors and the osmosensor SHO1, which is involved in the MAPK signaling pathway, significantly changed during the response to stimulation of mycelia of various incubation periods. MAPK activation results in the modification of transcription factors, allowing cellular responses to adapt to external stimuli. Sho1 is an osmosensor in the HKR1 subbranch of the HOG pathway. High external osmolarity induces structural changes in Sho1, which lead to Hog1 activation. Numerous non-osmotic stressors are also known to activate the HOG pathway, including cold stress, the inhibition of glycosylphosphatidylinositol anchor synthesis, and the inhibition of sphingolipid synthesis [40]. *Sho1* mutants are defective in the filamentous growth Kss1 MAPK signaling pathway, which might indicate that Sho1 serves a scaffold function in the filamentous growth pathway [41]. Sko1, a transcriptional repressor of the ATF/CREB family, is doubly regulated by Hog1 and PKA to regulate the transcription of a subset of stress-responsive genes upon osmotic stress. Compared with wild-type *Candida albicans*, the *sko1* mutant of *C. albicans* has an increased ability to filament [42]. The transcription factor Tec1 is an activator of G1 cyclin and vegetative adhesin genes. In *Saccharomyces cerevisiae*, Tec1, which is involved in the Kss1 MAPK signaling pathway, is necessary for morphological changes leading to pseudohyphal growth [43]. In *Ustilago maydis*, Tec1 was found to be a crucial factor for normal mating and basidiocarp development [44]. In our study, Tec1, Sho1, and PKA presented opposite transcriptional patterns in mycelia incubated for different periods followed by environmental stimulation, suggesting that they have different influences on osmolyte synthesis, antioxidant synthesis, and filamentation. Rlm1 is a downstream target of the MAPK signaling pathway. The absence of RLM1 alters the cell wall content, specifically the chitin and mannan layers [45]. The present study demonstrated that the upregulated DEGs in mycelia that were cultured for 6 days in the dark and then stimulated were significantly enriched in amino sugar and nucleotide sugar metabolism, and the downregulated DEGs were significantly enriched in fructose and mannose metabolism, which were different from the responses of the mycelia cultured for 4 days to environmental stimulation.

Our results showed that if mycelial incubation time in the dark is insufficient, glycerolipid metabolism and steroid biosynthesis would be strengthened after the mycelia are subjected to environmental stimulation, which would have a passive influence on primordium formation. Glycerolipid metabolism is critical for the homeostasis of cellular lipid stores and membranes. Glycerol kinase is a key enzyme in the glycerol metabolic pathway. Mogly1 is associated with carbon source utilization and glycerol catabolism in *Magnaporthe oryzae*, and the single knockout mutant *Mogly1* reduces aerial mycelium [46]. Steroids play indispensable roles in the process of eukaryotic endocytosis by regulating membrane fluidity and permeability. Deletion of the gene encoding squalene monooxygenase results in an ergosterol decrease and inhibits growth in *Sporisorium scitamineum* [47]. The ectopic overexpression of *AaSMO1* causes an increase in total sterol content and improves germination and growth in subsequent generations [48]. Steroid biosynthesis is an oxygen-dependent process, and half of the enzymatic reactions require molecular oxygen [49]. Increased steroid biosynthesis increases oxygen consumption.

## 5. Conclusions

Our research indicated that various incubation times resulted in changes in the levels of lipid components and the expression of genes involved in lipid metabolism, which have an influence on the primordium formation of *P. tuoliensis* (Figure 9). The present study contributes to a better understanding of the molecular mechanisms underlying the primordium formation of *P. tuoliensis*. The functional validation of key genes was further performed to elucidate the regulatory pathways, and low-temperature treatment increased the sphingolipid content involved in the growth and development of *P. tuoliensis*.

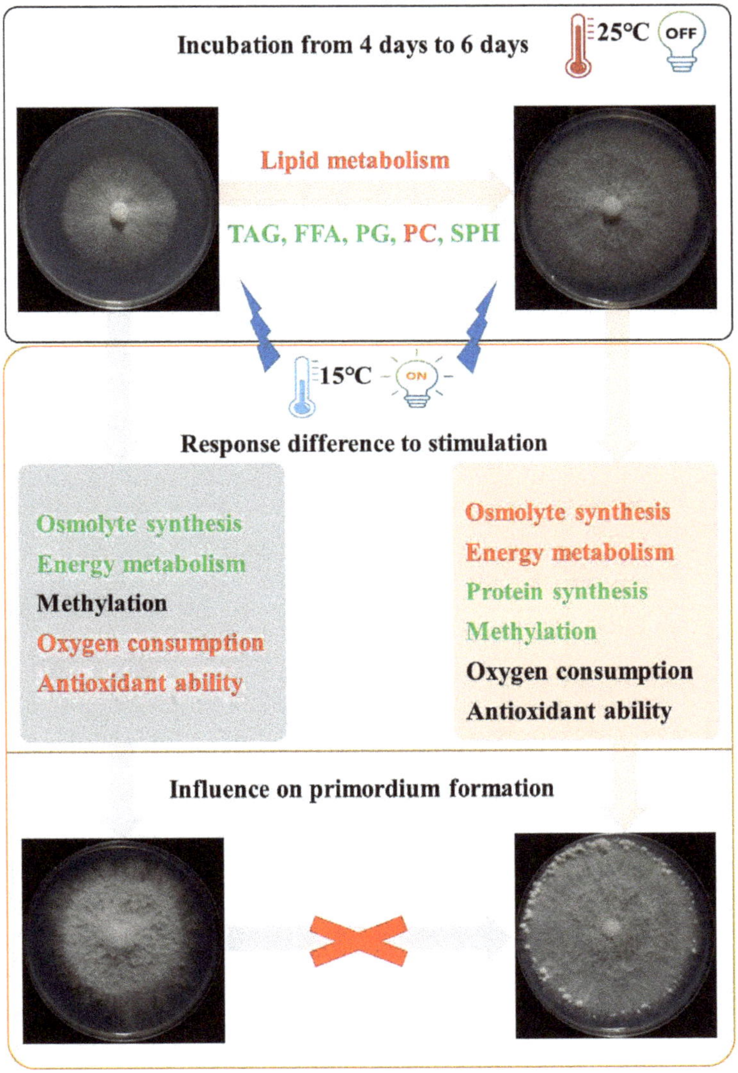


**Figure 9.** Schematic representation of the molecular mechanisms by which incubation time affects the primordium formation of *P. tuoliensis*. Triacylglycerol (TAG), free fatty acids (FFA), phosphatidylglycerol (PG), phosphatidylcholine (PC), and sphingosine (SPH). The red words in the box indicate the upregulated pathways. Green words indicate the downregulated pathways.

**Supplementary Materials:** The following supporting information can be downloaded at: https://www.mdpi.com/article/10.3390/agriculture14122277/s1, Table S1. Primers for qRT-PCR of 25 DEGs; Table S2. Statistical analysis of the number of sequencing reads; Figure S1. Volcano plot analysis of DEGs with different incubation times; Figure S2. The DEGs were changed in the mycelia cultured for 4 days in the dark and then stimulated (T1 vs. T2). The DEGs were changed in the mycelia cultured for 6 days in the dark and then stimulated (T3 vs. T4).

**Author Contributions:** Data curation, Q.H.; funding acquisition, C.H. and M.Z.; resources, W.G.; supervision, L.Z.; writing—original draft, Q.H.; writing—review and editing, C.H. and M.Z. All authors have read and agreed to the published version of the manuscript.

**Funding:** This research was funded by the National Key R&D Program of China (2022YFD1200600) and the earmarked fund for the China Agriculture Research System (CARS-20), and it was also supported by the National Natural Science Foundation of China (32002110).

**Institutional Review Board Statement:** Not applicable.

**Data Availability Statement:** The original contributions presented in the study are included in the article/Supplementary Materials. Further inquiries can be directed to the corresponding author.

**Conflicts of Interest:** The authors declare no conflicts of interest.

## References


1. Usman, M.; Murtaza, G.; Ditta, A. Nutritional, medicinal, and cosmetic value of bioactive compounds in button mushroom (*Agaricus bisporus*): A review. *Appl. Sci.* **2021**, *11*, 5943. [CrossRef]
2. Wang, C.L.; Cui, H.Y.; Wang, Y.R.; Wang, Z.F.; Li, Z.J.; Chen, M.H.; Li, F.J. Bidirectional immunomodulatory activities of polysaccharides purified from *Pleurotus nebrodensis*. *Inflammation* **2014**, *37*, 83–93. [CrossRef] [PubMed]
3. Yan, B.J.; Jing, L.Y.; Wang, J. A polysaccharide (PNPA) from *Pleurotus nebrodensis* offers cardiac protection against ischemia-reperfusion injury in rats. *Carbohydr. Polym.* **2015**, *133*, 1–7. [CrossRef]
4. Xu, N.; Gao, Z.; Zhang, J.J.; Jing, H.J.; Li, S.S.; Ren, Z.Z.; Wang, S.X.; Jia, L. Hepatoprotection of enzymatic-extractable mycelia zinc polysaccharides by *Pleurotus eryngii* var. *tuoliensis*. *Carbohydr. Polym.* **2017**, *157*, 196–206. [CrossRef] [PubMed]
5. Ren, Z.Z.; Li, J.; Xu, N.; Zhang, J.J.; Song, X.L.; Wang, X.X.; Gao, Z.; Jing, H.J.; Li, S.S.; Zhang, C.; et al. Anti-hyperlipidemic and antioxidant effects of alkali-extractable mycelia polysaccharides by *Pleurotus eryngii* var. *tuolensis*. *Carbohydr. Polym.* **2017**, *175*, 282–292. [CrossRef] [PubMed]
6. Kong, W.W.; Huang, C.Y.; Chen, Q.; Zou, Y.J.; Zhang, J.X. Nitric oxide alleviates heat stress-induced oxidative damage in *Pleurotus eryngii* var. *tuoliensis*. *Fungal Genet. Biol.* **2012**, *49*, 15–20. [CrossRef]
7. Dai, Y.T.; Su, W.Y.; Yang, C.T.; Song, B.; Li, Y.; Fu, Y.P. Development of novel polymorphic EST-SSR markers in Bailinggu (*Pleurotus tuoliensis*) for crossbreeding. *Genes* **2017**, *8*, 325. [CrossRef] [PubMed]
8. Li, R.R.; Zheng, Q.W.; Lu, J.L.; Zou, Y.; Lin, J.F.; Guo, L.Q.; Ye, S.Q.; Xing, Z.M. Chemical composition and deterioration mechanism of *Pleurotus tuoliensis* during postharvest storage. *Food Chem.* **2021**, *338*, 127731. [CrossRef]
9. Du, F.; Qu, J.B.; Hu, Q.X.; Yuan, X.F.; Yin, G.H.; Wang, L.; Zou, Y.J. Maximizing the value of Korshinsk peashrub branches by the integration of *Pleurotus tuoliensis* cultivation and anaerobic digestion of spent mushroom substrate. *Renew. Energy* **2021**, *179*, 679–686. [CrossRef]
10. Xie, C.; Gong, W.; Zhu, Z.; Yan, L.; Hu, Z.; Peng, Y. Comparative transcriptomics of *Pleurotus eryngii* reveals blue-light regulation of carbohydrate-active enzymes (CAZymes) expression at primordium differentiated into fruiting body stage. *Genomics* **2018**, *110*, 201–209. [CrossRef]
11. Liu, W.; Cai, Y.L.; He, P.X.; Chen, L.F.; Bian, Y.B. Comparative transcriptomics reveals potential genes involved in the vegetative growth of *Morchella importuna*. *3 Biotech* **2019**, *9*, 81. [CrossRef] [PubMed]
12. Wang, H.; Tong, X.D.; Tian, F.H.; Jia, C.W.; Li, C.T.; Li, Y. Transcriptomic profiling sheds light on the blue-light and red-light response of oyster mushroom (*Pleurotus ostreatus*). *AMB Express* **2020**, *10*, 10. [CrossRef] [PubMed]
13. Carrasco, J.; Zied, D.C.; Pardo, J.E.; Preston, G.M.; Pardo-Giménez, A. Supplementation in mushroom crops and its impact on yield and quality. *AMB Express* **2018**, *8*, 146. [CrossRef] [PubMed]
14. Du, F.; Ti, N.; Hu, Q.X.; Zou, Y.J.; Ye, D.; Zhang, H.J. A comparative transcriptome analysis reveals physiological maturation properties of mycelia in *Pleurotus tuoliensis*. *Genes* **2019**, *10*, 703. [CrossRef] [PubMed]
15. Im, J.H.; Park, C.H.; Shin, J.H.; Oh, Y.L.; Oh, M.; Paek, N.C.; Park, Y.J. Effects of light on the fruiting body color and differentially expressed genes in *Flammulina velutipes*. *J. Fungi* **2024**, *10*, 372. [CrossRef] [PubMed]
16. Guo, H.B.; Zhang, Z.F.; Wang, J.Q.; Wang, S.Y.; Yang, J.K.; Xing, X.Y.; Qi, X.J.; Yu, X.D. Transcriptome analysis of genes associated with autolysis of *Coprinus comatus*. *Sci. Rep.* **2022**, *12*, 2476. [CrossRef] [PubMed]
17. Qi, Y.C.; Sun, X.K.; Zhang, M.K.; Wen, Q.; Qiu, L.Y.; Shen, J.W. Identification of up-regulated transcripts during *Pleurotus ostreatus* primordium stage and characterization of *PoALDH1*. *J. Basic Microbiol.* **2018**, *58*, 1071–1082. [CrossRef]
18. Hou, Z.H.; Chen, Q.; Zhao, M.R.; Huang, C.Y.; Wu, X.L. Genome-wide characterization of the Zn(II)$_2$Cys$_6$ zinc cluster-encoding gene family in *Pleurotus ostreatus* and expression analyses of this family during developmental stages and under heat stress. *PeerJ* **2020**, *8*, e9336. [CrossRef]
19. Song, H.Y.; Kim, D.H.; Kim, J.M. Comparative transcriptome analysis of dikaryotic mycelia and mature fruiting bodies in the edible mushroom *Lentinula edodes*. *Sci. Rep.* **2018**, *8*, 8983. [CrossRef] [PubMed]
20. Kim, J.Y.; Kim, D.Y.; Park, Y.J.; Jang, M.J. Transcriptome analysis of the edible mushroom *Lentinula edodes* in response to blue light. *PLoS ONE* **2020**, *15*, e0230680. [CrossRef] [PubMed]
21. Han, X.; Gross, R.W. The foundations and development of lipidomics. *J. Lipid Res.* **2022**, *63*, 100164. [CrossRef]

22. Yang, F.; Zhao, M.; Zhou, L.; Zhang, M.; Liu, J.; Marchioni, E. Identification and differentiation of wide edible mushrooms based on lipidomics profiling combined with principal component analysis. *J. Agric. Food Chem.* **2021**, *69*, 9991–10001. [CrossRef]
23. Yao, J.; Zhou, L.; Hu, Y.; Zhao, M.; Ma, Y.; Liu, J.; Marchioni, E. Combining untargeted lipidomics analysis and chemometrics to identify the edible and poisonous mushrooms (*Pleurotus cornucopiae* vs. *Omphalotus japonicus*). *J. Agric. Food Chem.* **2023**, *71*, 8220–8229. [CrossRef]
24. Xie, G.; Chen, M.; Yang, Y.; Xie, Y.; Deng, K.; Xie, L. Comprehensive untargeted lipidomics study of black morel (*Morchella sextelata*) at different growth stages. *Food Chem.* **2024**, *451*, 139431. [CrossRef]
25. Trapnell, C.; Williams, B.A.; Pertea, G.; Mortazavi, A.; Kwan, G.; van Baren, M.J.; Salzberg, S.L.; Wold, B.J.; Pachter, L. Transcript assembly and quantification by RNA-Seq reveals unannotated transcripts and isoform switching during cell differentiation. *Nat. Biotechnol.* **2010**, *28*, 511–515. [CrossRef] [PubMed]
26. Wang, L.N.; Wu, X.L.; Gao, W.; Zhao, M.R.; Zhang, J.X.; Huang, C.Y. Differential expression patterns of *Pleurotus ostreatus* catalase genes during developmental stages and under heat stress. *Genes* **2017**, *8*, 335. [CrossRef] [PubMed]
27. Zhang, S.; Lu, X.; Hu, C.; Li, Y.; Yang, H.; Yan, H.; Fan, J.; Xu, G.; Abnet, C.C.; Qiao, Y. Serum metabolomics for biomarker screening of esophageal squamous cell carcinoma and esophageal squamous dysplasia using gas chromatography-mass spectrometry. *ACS Omega* **2020**, *5*, 26402–26412. [CrossRef] [PubMed]
28. Zhang, J.J.; Ren, A.; Chen, H.; Zhao, M.W.; Shi, L.; Chen, M.J.; Wang, H.; Feng, Z.Y. Transcriptome analysis and its application in identifying genes associated with fruiting body development in basidiomycete *Hypsizygus marmoreus*. *PLoS ONE* **2015**, *10*, e0123025. [CrossRef] [PubMed]
29. Kuhnlein, R.; Garrido, D.; Rubin, T.; Poidevin, M.; Maroni, B.; Le Rouzic, A.; Parvy, J.-P.; Montagne, J. Fatty acid synthase cooperates with glyoxalase 1 to protect against sugar toxicity. *PLoS Genet.* **2015**, *11*, e1004995.
30. Takagi, K.; Kikkawa, A.; Iwama, R.; Fukuda, R.; Horiuchi, H. Type II phosphatidylserine decarboxylase is crucial for the growth and morphogenesis of the filamentous fungus *Aspergillus nidulans*. *J. Biosci. Bioeng.* **2021**, *131*, 139–146. [CrossRef]
31. Tang, L.; Chi, H.W.; Li, W.D.; Zhang, L.; Zhang, L.Y.; Chen, L.; Zou, S.S.; Liu, H.X.; Liang, Y.C.; Yu, J.F.; et al. FgPsd2, a phosphatidylserine decarboxylase of *Fusarium graminearum*, regulates development and virulence. *Fungal Genet. Biol.* **2021**, *146*, 103483. [CrossRef]
32. Heier, C.; Kühnlein, R.P. Triacylglycerol metabolism in *Drosophila melanogaster*. *Genetics* **2018**, *210*, 1163–1184. [CrossRef] [PubMed]
33. Coleman, R.A.; Mashek, D.G. Mammalian triacylglycerol metabolism: Synthesis, lipolysis, and signaling. *Chem. Rev.* **2011**, *111*, 6359–6386. [CrossRef] [PubMed]
34. Howe, A.G.; McMaster, C.R. Regulation of phosphatidylcholine homeostasis by Sec14. *Can. J. Physiol. Pharmacol.* **2006**, *84*, 29–38. [CrossRef] [PubMed]
35. Deng, Y.; Zhu, H.; Wang, Y.; Dong, Y.; Du, J.; Yu, Q.; Li, M. The endoplasmic reticulum–plasma membrane tethering protein Ice2 controls lipid droplet size via the regulation of phosphatidylcholine in *Candida albicans*. *J. Fungi* **2024**, *10*, 87. [CrossRef] [PubMed]
36. Spiegel, S.; Milstien, S. Sphingosine-1-phosphate: An enigmatic signalling lipid. *Nat. Rev. Mol. Cell Biol.* **2003**, *4*, 397–407. [CrossRef]
37. Spiegel, S.; Cuvillier, O.; Edsall, L.; Kohama, T.; Menzeleev, R.; Olivera, A.; Thomas, D.; Tu, Z.; Van Brocklyn, J.; Wang, F. Roles of sphingosine-1-phosphate in cell growth, differentiation, and death. *Biochemistry* **1998**, *63*, 69–73.
38. Luberto, C.; Kraveka, J.M.; Hannun, Y.A. Ceramide regulation of apoptosis versus differentiation: A walk on a fine line. Lessons from neurobiology. *Neurochem. Res.* **2002**, *27*, 609–617. [CrossRef] [PubMed]
39. Handee, W.; Li, X.; Hall, K.W.; Deng, X.; Li, P.; Benning, C.; Williams, B.L.; Kuo, M.H. An energy-independent pro-longevity function of triacylglycerol in Yeast. *PLoS Genet.* **2016**, *12*, e1005878. [CrossRef]
40. Saito, H.; Posas, F. Response to hyperosmotic stress. *Genetics* **2012**, *192*, 289–318. [CrossRef]
41. Tatebayashi, K.; Yamamoto, K.; Nagoya, M.; Takayama, T.; Nishimura, A.; Sakurai, M.; Momma, T.; Saito, H. Osmosensing and scaffolding functions of the oligomeric four-transmembrane domain osmosensor Sho1. *Nat. Commun.* **2015**, *6*, 6975. [CrossRef]
42. Alonso-Monge, R.; Román, E.; Arana, D.M.; Prieto, D.; Urrialde, V.; Nombela, C.; Pla, J. The Sko1 protein represses the yeast-to-hypha transition and regulates the oxidative stress response in *Candida albicans*. *Fungal Genet. Biol.* **2010**, *47*, 587–601. [CrossRef]
43. Gavrias, V.; Andrianopoulos, A.; Gimeno, C.J.; Timberlake, W.E. Saccharomyces cerevisiae TEC1 is required for pseudohyphal growth. *Mol. Microbiol.* **1996**, *19*, 1255–1263. [CrossRef]
44. León-Ramírez, C.G.; Sánchez-Arreguin, J.A.; Cabrera-Ponce, J.L.; Martínez-Soto, D.; Ortiz-Castellanos, M.L.; Aréchiga-Carvajal, E.T.; Salazar-Chávez, M.F.; Sánchez-Segura, L.; Ruiz-Herrera, J. Tec1, a member of the TEA transcription factors family, is involved in virulence and basidiocarp development in *Ustilago maydis*. *Int. Microbiol.* **2021**, *25*, 17–26. [CrossRef] [PubMed]
45. Oliveira-Pacheco, J.; Alves, R.; Costa-Barbosa, A.; Cerqueira-Rodrigues, B.; Pereira-Silva, P.; Paiva, S.; Silva, S.; Henriques, M.; Pais, C.; Sampaio, P. The role of *Candida albicans* transcription factor *RLM1* in response to carbon adaptation. *Front. Microbiol.* **2018**, *9*, 1127. [CrossRef]
46. Fan, G.L.; Zhao, A.; Chen, L.B.; Yang, J.; Lu, G.D. The function analysis of glycerol kinases genes in the metabolism of carbon source in *Magnaporthe oryzae*. *Genom. Appl. Biol.* **2016**, *35*, 2716–2723.
47. Cai, Y.; Zhang, Y.; Bao, H.; Chen, J.; Chen, J.; Shen, W. Squalene monooxygenase gene *SsCI80130* regulates *Sporisorium scitamineum* mating/filamentation and pathogenicity. *J. Fungi* **2022**, *8*, 470. [CrossRef] [PubMed]

48. Singh, A.; Jindal, S.; Longchar, B.; Khan, F.; Gupta, V. Overexpression of *Artemisia annua* sterol C-4 methyl oxidase gene, *AaSMO1*, enhances total sterols and improves tolerance to dehydration stress in tobacco. *Plant Cell Tissue Organ Cult.* **2014**, *121*, 167–181. [CrossRef]
49. Hoshino, Y.; Gaucher, E.A. Evolution of bacterial steroid biosynthesis and its impact on eukaryogenesis. *Proc. Natl. Acad. Sci. USA* **2021**, *118*, e2101276118. [CrossRef] [PubMed]

*agriculture*

*Article*

# Study of the Degradation and Utilization of Cellulose from *Auricularia heimuer* and the Gene Expression Level of Its Decomposition Enzyme

**Xianqi Shan [1], Fangjie Yao [1,2,*], Lixin Lu [2], Ming Fang [2], Jia Lu [3] and Xu Sun [1]**

1 Engineering Research Centre of Chinese Ministry of Education for Edible and Medicinal Fungi, Jilin Agricultural University, Changchun 130118, China; shanxianqi2022@163.com (X.S.); sunxu0512@163.com (X.S.)
2 Laboratory of the Genetic Breeding of Edible Mushroom, College of Horticulture, Jilin Agricultural University, Changchun 130118, China; lixinl@jlau.edu.cn (L.L.); fangming@jlau.edu.cn (M.F.)
3 Hangzhou Academy of Agricultural Sciences, Hangzhou 310024, China; lujia920316@163.com
* Correspondence: yaofj@jlau.edu.cn

**Abstract:** *Auricularia heimuer* is a wood-rotting edible mushroom, and with the continuous development of the industry, the research on its grass-rotting cultivation is becoming more and more important. In this study, *A. heimuer* was cultivated using herbaceous substrate (reed) completely replacing the traditional woody substrate (oak), and the correlation between the relative expression of cellulase gene, cellulase activity, cellulose degradation and yield of different strains of *A. heimuer* were studied by combining qRT-PCR technology at different growth stages. The results showed that the cellulose degradation were positively correlated with the yield of reed and sawdust substrate at two growth stages, and were positively correlated with three cellulase activities. The relative expression of four cellulase genes were positively correlated with enzyme activity. There were inter-strain differences in the expression of the enzyme genes, which were basically consistent with the trend of the enzyme activity of the strains; *g5372* and *g7270* were more actively expressed in the mycelium period, while *g9664* and *g10234* were more actively expressed in the fruiting period. The results of SEM showed that the mycelium of A15 and A125 were different in their ability to degrade and utilize lignocellulose in reed substrate. The parental hybridization test further verified that qRT-PCR could be used as a rapid method to evaluate the cellulose degradation ability of *A. heimuer* strains. Seven strains (A12, A15, A184, A224, Z6, Z12, and Z18) with high cellulose degradation ability were screened. This study provides a reference for further understanding the role of *A. heimuer* cellulase genes in the degradation and metabolism of cellulose and for breeding new varieties more suitable for herbaceous substrate cultivation.

**Keywords:** *Auricularia heimuer*; real-time quantitative PCR; cellulase gene; cellulase activity; hybrid breeding; molecular-assisted breeding

## 1. Introduction

*Auricularia heimuer* [1] is an edible and medicinal fungus [2]. It is widely distributed naturally in China and has a long history of cultivation. It has now become the second largest edible mushroom species in China [3]. *A. heimuer* is a typical wood-rotting edible mushroom, which mainly uses wood chips of broad-leaved trees as the substrate for cultivation and domestication, and obtains nutrients and energy by decomgenerating lignin, cellulose and hemicellulose in wood residues in the form of secreting extracellular enzymes. However, with the gradual halt of natural forests, sawdust resources are increasingly tight, and prices are rising, which has become a major factor limiting the development of the *heimuer* industry [4]. At the same time, wetland lakes have a large amount of reed straw (herbaceous substrate) that is not well utilized and discarded, and it has a high cellulose content and a low lignin content, which makes it a good substrate for wood-rotting edible-mushroom grass-rotting cultivation research [5,6].

Cellulase is one of the many extracellular enzymes secreted by edible mushrooms. According to their catalytic methods, which can be divided into Endo-β-1, 4-glucanase (EC, 3.2.1.4), Exo-β-1, 4-glucanase (EC, 3.2.1.91) and β-1, 4-glucosidase (EC, 3.2.1.21) [7,8]. They are mixtures of glycoside hydrolases (GHs) that completely degrade cellulose into glucose [9]. Specifically, there are three main types of enzymes, carboxymethyl cellulase (CMCase), filter-paper cellulase (FPase) and β-glucosidase (β-Gase), and the three enzymes work together to complete the decomposition of cellulose components in the substrate [10,11]. Therefore, the level of enzyme activity will directly affect the growth of mycelia and the formation and development of fruiting bodies, and will also have a huge impact on yield [12]. Cellulase mainly degrades cellulose in a cultivated substrate, and its activity is closely related to the growth and development of edible mushrooms [13]. Fan et al. [14] found that different carbon and nitrogen sources would lead to changes in extracellular enzyme activity of *Hericium erinaceus*, and there was a positive correlation between extracellular enzyme activity and mycelial dry weight. In general, the higher the activity of extracellular enzymes in mycelia, the stronger the ability of substrate degradation [15], the faster the growth rate of edible mushroom mycelium [16], and the higher its yield [17]. The extracellular enzyme activity secreted by the mycelia of edible mushrooms is closely related to the regulation of enzyme genes [18,19], and there is also a strong internal correlation between the expression level of enzyme genes, enzyme activity and substrate degradation ability [20]. Therefore, it is necessary to study the relative expression of enzyme genes by means of real-time quantitative PCR. [18,21,22]. Real-time quantitative PCR can theoretically be used to evaluate the relative expression of the cellulase gene of the *A. heimuer* strain growing on reed substrate, but no relevant studies have been reported. However, Bai et al. [23] studied the relationship between *Morchella esculenta* antioxidant enzymes and their gene expression at different temperatures by using real-time quantitative PCR technology. Mao [24] studied the expression levels of 11 laccase-coding genes (*lac1-11*) in *Volvariella Volvacea* at different developmental stages of fruiting bodies by using real-time quantitative PCR. Using qRT-PCR, Zhang [25] showed that 18S rRNA, β-TUB, EF1-a and 28S rRNA could be used as internal reference genes of different strains of *A. heimuer*. Xiang et al. [26] found in the screening and analysis of *Lentinula edodes* genome data that RpI4 was the most stable internal reference gene in all experimental conditions. These studies provided a theoretical basis for real-time quantitative PCR to evaluate the relative expression of the cellulase gene of the *A. heimuer* strain growing in reed substrate. Therefore, it is possible to study the expression level of the cellulase gene and the relationship between the cellulase activity of *A. heimuer* grown on different substrates, by qRT-PCR [18,27,28].

Based on the scientific research concept of "Research on the cultivation of wood-rotting edible-mushroom grass-rotting" proposed by the team [4,5], the laboratory has used corn cob and reed to partially or completely replace wood chips in the cultivation experiment of *A. heimuer* in the early stage, and studied the influence of different proportions of cultivation substrate on the degradation rate of extracellular enzymes and lignocellulose of the *A. heimuer* strain [29,30]. However, there were few studies on the relative expression of the cellulase gene in the *A. heimuer* strain and its correlation with cellulase activity and cellulose degradation in different cultivation substrates. Through previous laboratory studies on the *A. heimuer* genome [31] and transcriptome [32], key genes controlling cellulose degradation have been identified, and the ability of strains to degrade cellulose in the substrate could be evaluated by gene expression; it could lay a foundation for the rapid auxiliary breeding of the *A. heimuer* strain more suitable for herbaceous substrate (agricultural waste such as reed straw and corn cob) cultivation in the future.

## 2. Materials and Methods

### 2.1. Materials

#### 2.1.1. Test Strain

The 28 *A. heimuer* strains used in this experiment were provided by the *A. heimuer* Variety Improvement Post of the National Modern Edible Mushroom Industry technical system. It is now preserved in the College of Horticulture, Jilin Agricultural University, including cultivated strains and wild strains. Specific information is shown in Table 1.

**Table 1.** Strain names and origins of germplasm of the preserved fungi *A. heimuer.*

| Serial Number | Storage ID | Type | Origin of Resource |
|---|---|---|---|
| 1 | JAUAH002 | Cultivated | Hubei Province |
| 2 | JAUAH004 | Cultivated | Beijing |
| 3 | JAUAH012 | Cultivated | Hubei Province |
| 4 | JAUAH013 | Cultivated | Heilongjiang Province |
| 5 | JAUAH014 | Cultivated | Heilongjiang Province |
| 6 | JAUAH015 | Cultivated | Heilongjiang Province |
| 7 | JAUAH016 | Cultivated | Heilongjiang Province |
| 8 | AUAH017 | Cultivated | Heilongjiang Province |
| 9 | AUAH019 | Cultivated | Jilin Province |
| 10 | AUAH020 | Cultivated | Shanghai |
| 11 | AUAH124 | Wild | Heilongjiang Province |
| 12 | JAUAH125 | Wild | Heilongjiang Province |
| 13 | JAUAH127 | Wild | Jilin Province |
| 14 | JAUAH132 | Wild | Heilongjiang Province |
| 15 | JAUAH134 | Wild | Heilongjiang Province |
| 16 | JAUAH139 | Wild | Heilongjiang Province |
| 17 | JAUAH184 | Wild | Jilin Province |
| 18 | JAUAH224 | Wild | Heilongjiang Province |
| 19 | JAUAH282 | Wild | Jilin Province |
| 20 | JAUAH308 | Wild | Yunnan Province |
| 21 | JAUAH314 | Wild | Yunnan Province |
| 22 | JAUAH336 | Wild | Yunnan Province |
| 23 | JAUAH345 | Wild | Yunnan Province |
| 24 | JAUAH356 | Wild | Yunnan Province |
| 25 | JAUAH496 | Wild | Jilin Province |
| 26 | JAUAH593 | Cultivated | Jilin Province |
| 27 | JAUAH596 | Cultivated | Jilin Province |
| 28 | JAUAH599 | Cultivated | Jilin Province |

#### 2.1.2. Sample Preparation and Preservation

The 28 strains of *A. heimuer* were activated with PDA plates, then transferred to a new PDA plate culture. The tested strains were inoculated on a glass plate containing reed and wood chips, incubated at 25 °C for 25 days, and the mycelium was collected and stored at −80 °C. At the same time, fruiting bodies were obtained by a fruiting test, stored at −80 °C, and used to study the relative expression of cellulase gene in different strains at different stages. During the fruiting experiment, the tested strains were stored at −80 °C at the full-bag stage of the mycelium and fruiting-body maturity stage for the determination of cellulose content in cultivation materials and the preparation of crude enzyme solution. The medium used in the experiment was the following: the formula of PDA solid medium was peeled potato 200 g (boiled filter juice), glucose 20 g, AGAR powder 10 g, and distilled water 1.0 L, pH natural. The formula of sodium carboxymethyl cellulose Congo red culture medium was KCl 0.5 g, $MgSO_4 \cdot 7H_2O$ 0.5 g, $KH_2PO_4$ 1.0 g, $(NH_4)_2SO_4$ 2.0 g, sodium carboxymethyl cellulose 10 g, AGAR 10 g, and distilled water 1.0 L, pH natural [33]. The formula of the reed culture medium was reed 88%, wheat bran 10%, lime 1%, gypsum 1%, and water content about 55%. The formula of the wood chips culture medium (CK) was reed 88%, wheat bran 10%, lime 1%, gypsum 1%, and water content about 55%.

### 2.2. Methods

#### 2.2.1. Preliminary Screening of Cellulase Production Capacity of *A. heimuer* Strain

The 5 mm diameter strains were cut with a hole punch at the edge of the activated plate colony and inoculated in the center of sodium carboxymethyl cellulose Congo red medium. After 7 days of constant temperature culture at 25 °C, the hydrolytic ring D and colony diameter d were measured by staining with 1 g/L Congo red for 30 min and decolorization with 1 mol/L NaCl for 30 min, and the D/d value was calculated. The ratio of the two was used as an index to measure the cellulose degradation ability of the strain [33,34]. Three repetitions were set for each process.

#### 2.2.2. *A. heimuer* Cellulase Gene Mining and Primer Design

According to the previous whole-genome sequencing analysis of *A. heimuer* and transcriptome sequencing analysis at different stages based on different substrate conditions [32], enrichment analysis and screening of differential genes involved in cellulose metabolic pathways were performed, and combined with annotation of the gene database of carbohydrate-active enzymes (CAZymes), and 5 key coding genes related to cellulose degradation were identified. They were *g5372*, *g6295*, *g7270*, *g9664* and *g10234* [35]. Combined with NCBI-BLAST-blastx gene information annotation, it was confirmed that *g5372* and *g9664* were related to encoding cellobiose hydrolase (exo-glucanase), *g7270* was related to encoding endo-glucanase, and *g10234* was related to encoding β-glucosidase [36]. In combination with the whole-genome annotation results of *A. heimuer* and the sequences of 5 cellulase genes, NCBI-BLAST-Primer was used for primer design, requiring the amplified product to span introns, be 80–300 bp long, and contain 45–55% GC % [37]. At the same time, *EF1-a* was selected as the internal reference gene according to previous laboratory results [25,26].

#### 2.2.3. Determination of Relative Expression of *A. heimuer* Cellulase Gene

##### Extraction of Total RNA and Synthesis of cDNA

The frozen sample was ground into a fine powder in liquid nitrogen, and 0.1000 g sample was extracted by the Trizol method [38]. RNA purity and concentration were measured using a nano-spectrophotometer, and RNA integrity was detected by agarose gel electrophoresis. Total RNA was reverse-transcribed into cDNA using the TRNA AT-341 reverse transcription kit (Transgen Biotech, Beijing, China).

##### Real-Time Quantitative PCR

Real-time amplification response detection in 96-well plates was performed using SYBR Green. The reaction system consisted of 1 μL cDNA template, 0.4 μL amplification primers, 0.4 μL passive Reference Dye (50×, 10 μL Top Green qPCR supermix and 7.8 μL RNase-free water. The final volume was 20 μL. Reaction conditions were predenaturation at 95 °C for 2 min, denaturation at 95 °C for 15 s, annealing at 60 °C for 20 s, and extension at 72 °C for 26 s, a total of 40 cycles. A melt curve analysis was generated by heating the amplicon from 60 °C to 95 °C to confirm that each reaction produced a single product. Thermal cycling was performed to collect fluorescence data.

##### Calculation of Gene Relative Expression

Ct values of cellulolytic enzyme genes were obtained by qRT-PCR, and the relative expression levels of genes on two substrates in the mycelium stage and fruiting-body mature stage were calculated by the formula $F = 2^{-\Delta\Delta Ct}$ by comparing Ct values. In formula $F = 2^{-\Delta\Delta Ct}$, ΔΔCt = (Ct value of the target gene in the test group − Ct value of the reference gene in the test group) − (Ct value of the target gene in the control group − Ct value of the reference gene in the control group). The control group in this formula is the following: the strain with the lowest relative expression of each enzyme gene calculated by using sawdust substrate as the control in the previous experiment was used as the control group to calculate the relative expression of each enzyme gene between different strains. The replication of the study was a technical replication.

#### 2.2.4. Cultivation Seed Preparation

This was carried out using a 17 × 33 cm polypropylene folding bag as an edible-mushroom bag, and each bag contained about 1 kg of wet material. The bags were sterilized at 121 °C for 4 h in an autoclave. After the edible-mushroom bags were sterilized out of the pot, they were transferred to the cooling room to cool down to room temperature, and the original species (secondary species) were used for inoculation. The two substrate formulations for each strain were ten bags of replicates.

#### 2.2.5. Fruiting Test

After the bag was full of mycelium, the *A. heimuer* mechanical piercing machine was used for piercing after 6–7 days of post-ripening, and the method of piercing the ear through small holes was adopted [39]. After the mycelia at the perforated area recovered and turned white 3–5 days later, the bags were transferred to the mushroom-extraction chamber for the ear-extraction test. Water spray 3–5 times a day, ventilation 2–3 times a day, and see dry and wet dry and wet alternately, kept the humidity of the mushroom greenhouse about 85–95%. When the ear piece was fully unfolded, the edge was slightly shrunken and rolled in, and the villi on the back of the ear piece were erect, which indicated that the fruiting body of *A. heimuer* was mature, and it could be harvested.

Yield determination: the actual yield of each edible-mushroom bag was converted to the mass of dry ear produced per 100 kg of dry material (kg).

#### 2.2.6. Hybrid Breeding

General Steps of Hybrid Breeding

Select suitable parents ➨ Monospore isolation ➨ Hybrid pairing ➨ Microscopic examination ➨ Tube propagation ➨ Primary screening ➨ Rescreening ➨ Fruiting test.

Monospore Isolation

Plate dilution separation method: the collected spores were picked up with a vaccination needle on a super-clean workbench and loaded into a pre-prepared plastic plate containing 5 mL sterile water, and thoroughly mixed to obtain the original spore suspension. A pipette gun to was used absorb 100 µL spore suspension and to transfer it into a small test tube filled with 900 µL sterile water; this was thoroughly mixed to obtain a spore suspension diluted $10^{-2}$ times, and dilution was continued until $10^{-5}$ times the spore suspension was obtained. A pipette gun was used to take about 100 µL of the spore diluent from $10^{-5}$, $10^{-4}$, $10^{-3}$, and $10^{-2}$ dilution times, these were transferred to the new PDA-plate-culture medium with a spore diluent coated evenly, and were then put in a constant temperature incubator at 25 °C. The spore germination on the plate was checked at regular intervals, and the small colonies of single spores were transferred to the new PDA-plate-culture medium for culture and to wait for hybridization.

Monospore Hybridization

One piece of the mononuclear mycelia from each of the two parental spores to be hybridized were picked up by inoculation needle and placed at both ends of fresh PDA solid medium, with a distance of about 1.5–2 cm between the two, and cultured in a constant-temperature incubator at 25 °C. When the mononuclear mycelia of the two parents came into contact, they were transferred to a new PDA medium with a cover slide inserted at an angle for further culture. After the mycelia climbed on the slide, the mycelia were stained with a drop of 1% Congo red solution and a drop of 10% KOH solution after mixing, and they were left for 30 min before the microscopic examination (Objective lens 40×, eyepiece 10× began. If the mycelium had a lock-like joint structure observed during microscopic examination, it was a successfully hybridized dikaryotic mycelium, which could be transferred to a new PDA plate medium for culture and use [40].

2.2.7. Determination of *A. heimuer* Cellulase Activity and Degradation Amount

Determination of Cellulase Activity

Preparation of crude enzyme solution: samples were taken at the full-bag period of mycelium and the first fruiting-body collection period, and 5 g fresh samples were placed in a small beaker containing 45 mL distilled water at a distance of 3–5 cm below the feed surface. The samples were extracted on a shaker at 15 °C for 4 h and centrifuged at 4000 r/min for 10 min. The supernatant, i.e., crude enzyme solution, was stored in a refrigerator at 4 °C for later use [29].

Carboxymethyl cellulase enzyme activity determination: 0.6 mL sodium 0.5% carboxymethyl cellulose solution (prepared with a PH 4.6, 0.1 mol /L acetate buffer) was added into a 10 mL tube, then 0.2 mL enzyme solution was added. This was mixed well, and held in a water bath at 50 °C for 30 min. A total of 0.6 mL DNS reagent was added immediately after removal, and was boiled for 5 min. This was removed and allowed to stand, with distilled water was fixed to 10 mL. The $OD_{520}$ value was measured by an enzymoleter, the inactivated enzyme solution was boiled as a blank control (only 1–2 times), the glucose standard curve was substituted for calculation, with 3 replicates for each group [29,41].

Filter-paper cellulase-enzyme activity determination: 0.25 mL of enzyme liquid and 0.75 mL of 0.05 mol /L of citric acid buffer at PH 4.6 were added to each of the four numbered test tubes, and they were mixed them well and preheated in a 50 °C water bath for 5–10 min. Then 25 mg of filter-paper strip was added to each tube, and they were held in a 50 °C water bath for 1 h. A total of 0.75 mL DNS reagent was immediately added to stop the enzyme reaction, and was shaken well, boiling water bath for 5 min, and then taken out and fixed to 10 mL. The $OD_{540}$ value was measured by enzymoleter, and the inactivated enzyme solution was boiled as a blank control (only 1–2 times); and the glucose standard curve was substituted for calculation, with 3 replicates for each group [42,43].

β-glucosidase enzyme activity determination: 0.75 mL of 0.5% salicylate citrate buffer was added to each of the four numbered test tubes, and 0.75 mL DNS reagent was added to test tube No. 1 to deactivate the enzyme activity as a blank control (only 1–2 times). Four test tubes were preheated in a 50 °C water bath for 5–10 min, then 0.25 mL of enzyme solution was added to each, and they were kept warm in a 50 °C water bath for 30 min. Immediately after removal, 0.75 mL of DNS reagent was added to each of the test tubes No. 2, No. 3 and No. 4, to terminate the enzyme reaction. This was mixed well, and the color was developed in a boiling water bath for 5 min, and then taken out and fixed to 10 mL. The $OD_{540}$ value was measured by an enzyme-labeled instrument and calculated by glucose standard curve. The amount of enzyme required to produce 1 umol glucose per hour from the substrate was defined as one unit of enzyme activity (U) [44,45].

Cellulose Content Determination

Samples of the tested strains were taken from the edible-mushroom bag at the mycelium full-bag period and the fruiting-body mature period, and the fresh cultivated materials were mixed and put into the oven to dry, and each of them was taken three times that is, three times of repetition, and the unused cultivated materials of reed and wood chips were used as the control group. The cellulose content of the cultivated material was determined by the cellulose (CLL) content-detection kit method of Beijing Box Sheng-gong Technology Co., Ltd., Beijing, China.

2.2.8. SEM Observation

Strains A15 and A125, which had the highest and lowest yield on the reed substrate, were selected as the experimental group, and the reed cultivation material at the maturation stage of its fruiting body was sampled and observed by scanning electron microscope, while the wet reed material with unused mycelium was selected as the control group. The samples to be tested were stored at 4 °C in the refrigerator. The scanning electron microscopy was completed by Wuhan Xavier Biotechnology Co., Ltd., Wuhan, China.

2.2.9. Statistical Analysis

The experimental results were expressed as mean ± standard deviation ($n \geq 3$). SPSS 26.0 and Origin-Pro 2022 software were used for statistical analysis and mapping. SPSS 26.0 was used to conduct one-way ANOVA and Duncan's multiple range test to determine the significant differences between samples, SPSS 26.0 and Origin-Pro 2022 were used to analyze the correlation between the data, and the correlation results were represented by Pearson correlation coefficient ($p < 0.05$, $p < 0.01$).

## 3. Results

### *3.1. Analysis Preliminary Screening Results of Cellulase Production Capacity of A. heimuer Strain*

According to the characteristics of cellulose use by extracellular enzymes secreted by mycelium, the strains were screened by the Congo red plate method of carboxymethyl cellulose sodium medium [34,46,47]. Under the same conditions, 28 *A. heimuer* test strains were cultured in plates, and the growth rate was measured by colony diameter(d). The larger the diameter, the faster the growth rate of the strain on the plates. Carboxymethyl cellulose was degraded in areas where cellulase was produced, so that it could not be stained by Congo red, that is, a transparent ring(D) was formed, and the width of the transparent ring indicated the enzyme-producing capacity of different strains. As shown in Table 2, the colony diameters of 28 strains ranged from 0.78–3.10 cm, and D/d ranged from 0.77–1.62, with significant differences among strains. Strains with a colony diameter greater than 1.90 cm and a ratio greater than 0.80 were screened for subsequent tests [34]. By combining the two sets of data, 13 strains were obtained, accounting for about 46% of the total tested strains.

**Table 2.** Screening results of cellulase production capacity of *A. heimuer* strain.

| Sample ID | Colony Diameter (cm) | D/d | Sample ID | Colony Diameter (cm) | D/d |
|---|---|---|---|---|---|
| A14 | $3.10 \pm 0.05\ ^{a}$ | $1.00 \pm 0.00\ ^{c–f}$ | A127 | $1.90 \pm 0.08\ ^{d–g}$ | $1.27 \pm 0.01\ ^{a–e}$ |
| A314 | $2.99 \pm 0.10\ ^{a,b}$ | $0.82 \pm 0.01\ ^{f}$ | A4 | $1.86 \pm 0.07\ ^{e–g}$ | $1.32 \pm 0.24\ ^{a–d}$ |
| A125 | $2.85 \pm 0.19\ ^{a–c}$ | $1.04 \pm 0.04\ ^{c–f}$ | A17 | $1.85 \pm 0.08\ ^{e–g}$ | $1.06 \pm 0.30\ ^{b–f}$ |
| A2 | $2.83 \pm 0.16\ ^{a–c}$ | $0.94 \pm 0.05\ ^{c–f}$ | A16 | $1.77 \pm 0.07\ ^{e–g}$ | $1.08 \pm 0.11\ ^{b–f}$ |
| A12 | $2.83 \pm 0.22\ ^{a–c}$ | $0.93 \pm 0.12\ ^{c–f}$ | A599 | $1.73 \pm 0.08\ ^{f,g}$ | $0.78 \pm 0.09\ ^{f}$ |
| A596 | $2.77 \pm 0.19\ ^{a–c}$ | $0.99 \pm 0.00\ ^{c–f}$ | A282 | $1.69 \pm 0.30\ ^{f–h}$ | $1.27 \pm 0.25\ ^{a–e}$ |
| A15 | $2.77 \pm 0.15\ ^{a–c}$ | $0.91 \pm 0.13\ ^{d–f}$ | A20 | $1.68 \pm 0.23\ ^{f–h}$ | $1.03 \pm 0.21\ ^{c–f}$ |
| A593 | $2.74 \pm 0.08\ ^{a–c}$ | $0.93 \pm 0.01\ ^{c–f}$ | A345 | $1.60 \pm 0.13\ ^{g–i}$ | $1.04 \pm 0.08\ ^{c–f}$ |
| A336 | $2.72 \pm 0.38\ ^{a–c}$ | $0.77 \pm 0.12\ ^{f}$ | A124 | $1.42 \pm 0.12\ ^{g–k}$ | $0.80 \pm 0.02\ ^{f}$ |
| A13 | $2.55 \pm 0.29\ ^{a–c}$ | $0.79 \pm 0.02\ ^{f}$ | A356 | $1.08 \pm 0.13\ ^{h–l}$ | $0.88 \pm 1.69\ ^{e,f}$ |
| A184 | $2.53 \pm 0.19\ ^{a–d}$ | $1.12 \pm 0.03\ ^{b–f}$ | A132 | $1.01 \pm 0.14\ ^{i–l}$ | $1.50 \pm 0.28\ ^{a,b}$ |
| A496 | $2.37 \pm 0.15\ ^{b–e}$ | $0.82 \pm 0.08\ ^{f}$ | A139 | $0.97 \pm 0.08\ ^{j–l}$ | $1.27 \pm 0.07\ ^{a–e}$ |
| A134 | $2.24 \pm 0.56\ ^{c–f}$ | $0.88 \pm 0.14\ ^{e,f}$ | A308 | $0.91 \pm 0.00\ ^{k,l}$ | $1.37 \pm 0.05\ ^{a–c}$ |
| A224 | $1.90 \pm 0.15\ ^{g–j}$ | $1.62 \pm 0.22\ ^{a}$ | A19 | $0.78 \pm 0.07\ ^{l}$ | $1.32 \pm 0.12\ ^{a–d}$ |

Note: The data in the table are mean ± standard deviation, and different lowercase letters in the same name column indicate significant differences between strains ($p < 0.05$). The type of average comparison test was Duncan's multiple range test.

### *3.2. Screening and Analysis of Cellulase Gene Primers*

First, the test strains were used for PCR amplification of the designed five cellulase gene primers, and the PCR products were analyzed by agarose gel electrophoresis. Finally, qualified primers of genes *g5372*, *g7270*, *g9664* and *g10234* were initially screened: the products were all single bands with non-specific amplification, indicating strong primer specificity (Figure 1A), and no qualified primer was screened for *g6295*. qRT-PCR was performed on the selected qualified primers using different strains and samples at different growth stages, and the melting curves obtained were all single melting peaks, which further verified the strong specificity of the primers (Figure 1B). The primer sequences of the four cellulase genes obtained by screening are shown in Table 3.

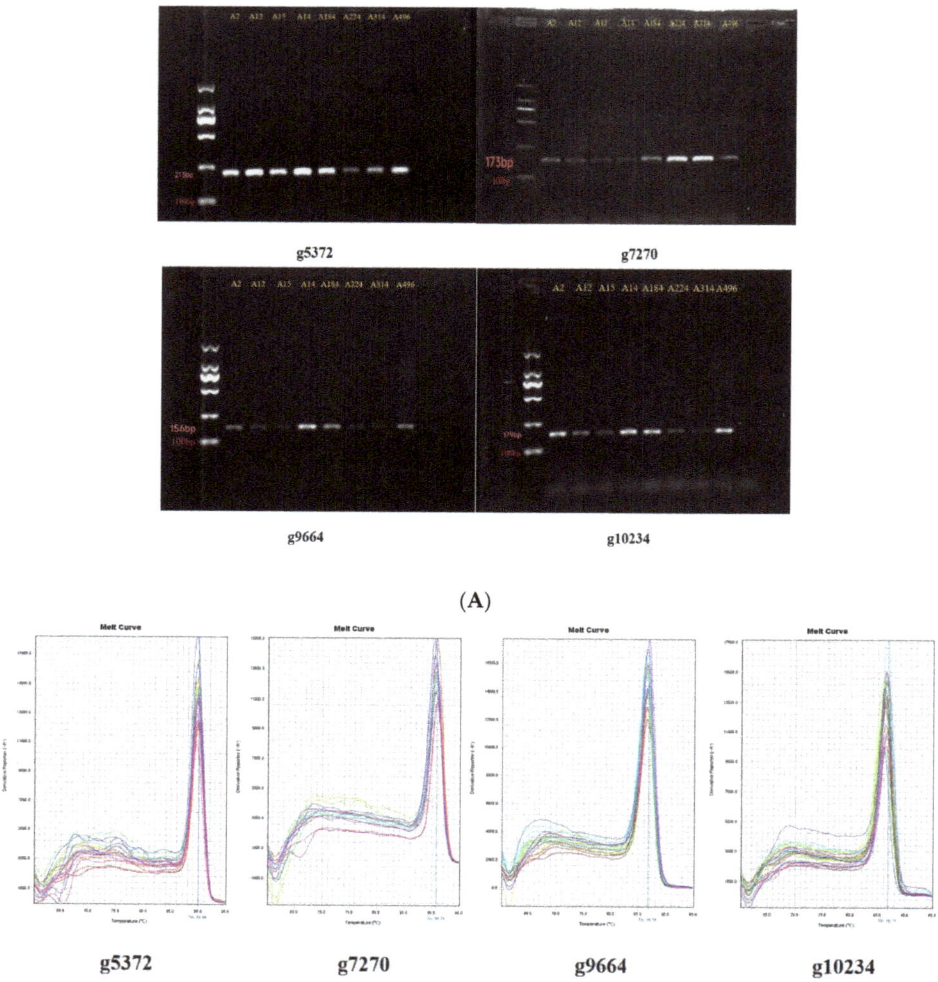


**Figure 1.** The amplicon length and specificity of candidate reference genes. (**A**): Amplified fragments of candidate reference genes shown by agarose gel electrophoresis with ethidium bromide staining; (**B**): melting curves generated by qRT-PCR.

**Table 3.** Sequence of cellulase gene primers.

| Gene ID | Primer Sequences (5′–3′) | Amplicon Size (bp) |
|---|---|---|
| *g5372* | CCTCGCCGTTAATGAACTTGATGTC<br>AGTACGAGATGTTCAACCTCCTGA | 213 bp |
| *g7270* | GAAAGTGCTGGGGTTGTTCTTG<br>ATACTAACTGTGTGACGGACAACG | 173 bp |
| *g9664* | CTCCTGGAGAGCGAATCAAAATACG<br>GTTGGTCGAGAACTTGGACATACC | 156 bp |
| *g10234* | GTCCTTGAATTGCTGCATGAGAAG<br>CACTACATCAACAACGAGCAAGAG | 179 bp |

## 3.3. Analysis of Cellulose Degradation and Yield in Different Strains

### 3.3.1. Determination and Correlation Analysis of Cellulose Degradation and Yield in Different Strains

The correlation between the yield of fruit bodies and cellulose degradation in two growth stages (mycelium full- bag period and fruiting period) of 13 selected strains on the reed substrate and the sawdust substrate was analyzed. Results are as shown in Table 4: the yield of the strain on the reed substrate was positively correlated with the amount of cellulose degradation in the two growth stages, and the correlation coefficients were 0.387 and 0.506. The yield on the sawdust substrate was positively correlated with the degradation amount of cellulose in the two growth stages, and the correlation coefficients were 0.310 and 0.444. The degradation of cellulose in the reed substrate and the sawdust substrate at the full-bag period was positively correlated with the degradation of cellulose at the fruiting period, and the correlation coefficients were 0.532 and 0.610 ($p < 0.05$), respectively. The yield of each strain could reflect the degradation amount of cellulose well, to some extent, and vice versa. In general, the strain with higher yield had higher cellulose degradation.

**Table 4.** Correlation analysis of degradation amount and yield in two growth stages of different strains.

| | RS-CDR1 | RS-CDR2 | RS-Yield | SS-CDR1 | SS-CDR2 | SS-Yield |
|---|---|---|---|---|---|---|
| RS-CDR1 | 1.000 | | | | | |
| RS-CDR2 | 0.532 | 1.000 | | | | |
| RS-Yield | 0.387 | 0.506 | 1.000 | | | |
| SS-CDR1 | 0.219 | 0.726 ** | 0.361 | 1.000 | | |
| SS-CDR2 | 0.285 | 0.534 | 0.645 * | 0.610 * | 1.000 | |
| SS-Yield | 0.588 * | 0.455 | 0.774 ** | 0.310 | 0.444 | 1.000 |

Note: RS: reed substrate, SS: sawdust substrate. RS-CDR is the degradation amount of cellulose in reed substrate, and SS-CDR is the degradation amount of cellulose in sawdust substrate. The number 1 represents the full-bag period of mycelium and 2 represents the fruiting period. * Correlation is significant at the 0.05 level (2-tailed). ** Correlation is significant at the 0.01 level (2-tailed).

The yields of the tested strains are shown in Table 5. There were significant differences in the yields of different strains on the reed substrate, and there were also significant differences between the reed substrate and the sawdust substrate. The reed-substrate yield was between 0.67 and 3.98 kg, and the sawdust-substrate yield was between 1.55 and 4.28 kg. There were no significant difference in the yield of strains A15, A134 and A184 in the two substrates. The yield of A15 in the reed substrate was the highest, and that of A224 in the sawdust substrate was the highest. The cellulose degradation capacity of different strains was further measured (Figure 2). There were differences in the ability of different strains to utilize the two substrates, and the cellulose degradation capacity of more than half of the strains in the reed substrate exceeded 40% at the fruiting-body maturity stage. The degradation of cellulose in A12 (58.35%), A15 (48.15%), and A496 (47.96%) was relatively high, and the degradation of cellulose in A12 and A496 was higher than that in the sawdust substrate. The adaptability of A12 and A496 to the reed substrate was relatively better, and the degradation of cellulose were basically consistent with the yield trend of each strain. The ability of different strains to degrade cellulose varied, but the overall trends were similar. The degradation of cellulose in the culture substrate gradually increased with the growth of mycelia, and reached the highest point after entering the reproductive stage, especially in the process of fruiting-body harvesting, which indicated that the degradation of cellulose provided abundant nutrients for the growth of *A. heimuer*.

**Table 5.** Yield characteristics of different strains.

| Sample ID | Yield (kg) | |
|---|---|---|
| | Reed Substrate | Sawdust Substrate |
| A2 | 2.10 ± 0.01 $^{k,l}$ | 2.55 ± 0.05 $^{g,h}$ |
| A12 | 2.83 ± 0.11 $^{e,f}$ | 3.02 ± 0.02 $^{d}$ |
| A14 | 0.75 ± 0.28 $^{r}$ | 2.27 ± 0.03 $^{i,j}$ |
| A15 | 3.98 ± 0.18 $^{b}$ | 3.90 ± 0.14 $^{b}$ |
| A125 | 0.73 ± 0.04 $^{r}$ | 1.55 ± 0.21 $^{p}$ |
| A127 | 1.15 ± 0.02 $^{q}$ | 2.11 ± 0.01 $^{k,l}$ |
| A134 | 1.84 ± 0.10 $^{n}$ | 1.90 ± 0.11 $^{m,n}$ |
| A184 | 2.18 ± 0.19 $^{j,k}$ | 2.25 ± 0.16 $^{i,j}$ |
| A224 | 2.30 ± 0.03 $^{i}$ | 4.28 ± 0.02 $^{a}$ |
| A314 | 1.12 ± 0.03 $^{q}$ | 1.68 ± 0.02 $^{o}$ |
| A496 | 2.01 ± 0.12 $^{l,m}$ | 2.75 ± 0.04 $^{f}$ |
| A593 | 2.61 ± 0.01 $^{g}$ | 3.33 ± 0.22 $^{c}$ |
| A596 | 2.48 ± 0.02 $^{h}$ | 2.90 ± 0.13 $^{e}$ |

Note: Different lowercase letters indicate the significant difference between different strains and different substrates ($p < 0.05$). The type of average comparison test was Duncan's multiple range test.

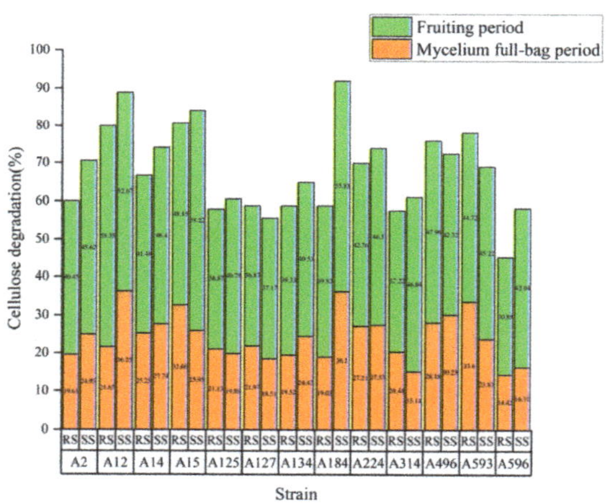


**Figure 2.** Degradation amount of cellulose in two growth stages of different strains. RS: Reed substrate. SS: Sawdust substrate.

### 3.3.2. Analysis of SEM Results

The reed-cultivation materials of strain A15 with the highest yield of reed substrate and strain A125 with the lowest yield were selected for the observation by scanning electron microscopy (SEM). The results showed, as shown in Figure 3, that the surface structure of the reed before mycelial degradation and utilization (Figure 3A1,A2) were compact and orderly, and the lignocellulose was intertwined to form a complete and dense structure. After degradation and utilization by extracellular enzymes secreted by mycelia, the surface structure of the reed became rough, the surface wax layer were destroyed, and numerous grooves have appeared; the wood fiber was cut, and the internal structure became loose and porous, destroying the original complex structure of the natural lignocellulose [33]. In addition, according to Figure 3B1,B2,C1,C2, compared with strain A15, which had the highest yield, although the reed substrate of A125 was also degraded and utilized by mycelium, it was more clearly seen that the surface of the reed substrate was not as thoroughly degraded as that of A15, and the reed was not fully degraded and utilized. It was proved that the two strains had different cellulose degradation ability, which could also explain the reason why the yield of A15 was significantly higher than that of A125.

**Figure 3.** Surface structure and morphology of reed used for *A. heimuer* mycelium degradation. (**A1**,**A2**) indicate the surface structural morphology of the reed substrate before mycelium utilization under 100 and 500 times microscope, respectively. (**B1**,**B2**) indicate the surface structural morphology of reed substrate during the fruiting period of strains A15 under 100 and 500 times microscope, respectively. (**C1**,**C2**) indicate the surface structural morphology of reed substrate during the fruiting period of strains A125 under 100 and 500 times microscope, respectively.

### *3.4. Study on Cellulase Activity of Different Strains*

#### 3.4.1. Correlation Analysis of Cellulase Activity and Cellulose Degradation in Different Strains

The biodegradation of lignocellulose in the culture substrate requires the synergistic action of various extracellular enzymes, in which mycelia secretes cellulase to degrade cellulose in the substrate. In general, the activity of cellulase in strains directly determines the utilization of cellulose degradation. Therefore, the activity of 3 cellulases in 13 strains were measured, and the correlation with the degradation of cellulose in the reed substrate were analyzed. As shown in Table 6, the activities of three cellulases in the two growth stages were positively correlated with the amount of cellulose degradation. The correlation coefficients between CMCase enzyme activity and the amount of degradation were all significant or highly significant positive correlations, with the highest correlation coefficient of 0.820 ($p < 0.01$). All positive correlations were found between β-Gase enzyme activity and the amount of degradation, with the highest correlation coefficient of 0.538. The correlation coefficients between the FPase enzyme activity and the amount of degradation were all positive correlations, with some of them significant positive correlations, with the highest correlation coefficient of 0.734 ($p < 0.01$). In addition, the positive correlation between the three cellulase activities also confirmed that cellulose degradation was the result of their cooperation. Therefore, the high activity of CMCase, FPase and β-Gase enzymes was conducive to the degradation of cellulose in the reed substrate.

**Table 6.** Correlation analysis between cellulose degradation and cellulase activity.

| | CMCase1 | CMCase2 | β-Gase1 | β-Gase2 | FPase1 | FPase2 | RS-CDR1 | RS-CDR2 |
|---|---|---|---|---|---|---|---|---|
| CMCase1 | 1.000 | | | | | | | |
| CMCase2 | 0.614 * | 1.000 | | | | | | |
| β-Gase1 | 0.506 | 0.751 ** | 1.000 | | | | | |
| β-Gase2 | 0.488 | 0.808 ** | 0.872 ** | 1.000 | | | | |
| FPase1 | 0.790 ** | 0.517 | 0.429 | 0.398 | 1.000 | | | |
| FPase2 | 0.435 | 0.663 * | 0.816 ** | 0.898 ** | 0.311 | 1.000 | | |
| RS-CDR1 | 0.820 ** | 0.592 * | 0.350 | 0.320 | 0.734 ** | 0.191 | 1.000 | |
| RS-CDR2 | 0.622 * | 0.694 ** | 0.350 | 0.538 | 0.355 | 0.470 | 0.532 | 1.000 |

Note: CMCase 1 indicates the enzyme activity of CMCase mycelium full-bag period, and CMCase 2 indicates the enzyme activity of the fruiting period. The same goes for β-Gase and FPase. * Correlation is significant at the 0.05 level (2-tailed). ** Correlation is significant at the 0.01 level (2-tailed).

### 3.4.2. Analysis of Enzyme Activity of Carboxymethyl Cellulase (CMCase) in Different Strains

As shown in Figure 4, the enzyme activity of carboxymethyl cellulase at different periods and on different substrates was different among the strains. The enzyme activity was low in the full-bag period of mycelium and high in the fruiting period, which was consistent with the trend of degradation amount. In the full-bag period of mycelium, three strains (A14, A496, A596) showed no significant difference from the sawdust substrate, and five strains (A15, A125, A134, A224, A593) were highly significant higher than the sawdust substrate. At the fruiting period, five strains (A2, A125, A127, A134, A314) showed no significant difference from the sawdust substrate. Four strains (A12, A15, A184, A593) had relatively high enzyme activity in the reed substrate during the fruiting period, all of which exceeded 100 U/g. The CMCase enzyme activity of strains A12, A15, A184, A224 and A593 on the reed substrate was relatively high, and the enzyme activity of strains A15, A224 and A593 was significantly higher than that of the sawdust substrate in the mycelium period, while there was no significant difference between strains A2 and the sawdust substrate in the fruiting period. The enzyme activity of the A125, A127 and A134 strains was relatively low, and their degradation capacity and yield were also average.

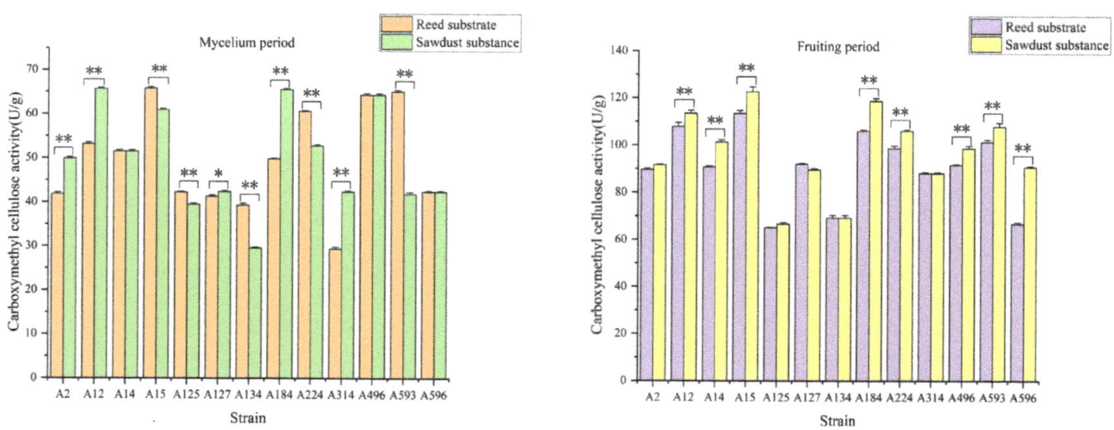


**Figure 4.** Activity of carboxymethyl cellulase at the period of full-bag of mycelium (**left**) and fruiting-body maturation (**right**). * indicates the significance of difference at $p < 0.05$ level, ** indicates the significance of difference at $p < 0.01$ level. The significant difference was expressed by Duncan's multiple range test.

### 3.4.3. Enzyme Activity Analysis of Filter-Paper Cellulase (FPase) from Different Strains

The filter-paper cellulase reflects the synergistic effect of exodextranase, endodextranase and β-glucosidase. As shown in Figure 5, at the full-bag period of mycelium, there were six strains (A12, A125, A127, A134, A224, and A593) that showed no significant difference from the sawdust substrate, one strain (A2) that was significantly higher than the sawdust substrate, and six strains (A14, A15, A184, A314, A496, and A596) that were highly significant higher than the sawdust substrate. At the fruiting period, there were five strains (A2, A134, A314, A593, and A596) with no significant difference from the sawdust substrate, and one strain (A12) with highly significant difference from the sawdust substrate. The activity of the reed substrate enzyme was relatively high in the five strains (A12, A15, A184, A224, and A593), all of which were over 150 U/g. Strains A12, A14, A15, A184, A224, and A593 had relatively high FPase enzyme activity in the two growth stages; from the perspective of FPase enzyme activity, they might be more suitable for growth in the reed substrate.

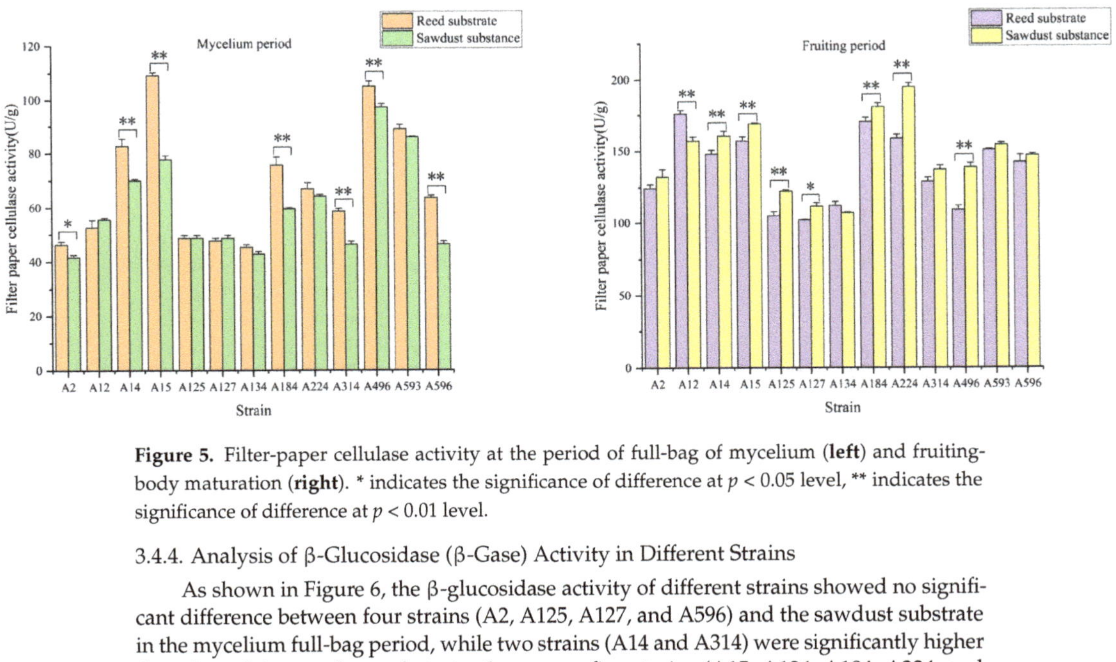


**Figure 5.** Filter-paper cellulase activity at the period of full-bag of mycelium (**left**) and fruiting-body maturation (**right**). * indicates the significance of difference at $p < 0.05$ level, ** indicates the significance of difference at $p < 0.01$ level.

### 3.4.4. Analysis of β-Glucosidase (β-Gase) Activity in Different Strains

As shown in Figure 6, the β-glucosidase activity of different strains showed no significant difference between four strains (A2, A125, A127, and A596) and the sawdust substrate in the mycelium full-bag period, while two strains (A14 and A314) were significantly higher than that of the sawdust substrate; there were five strains (A15, A134, A184, A224, and A593) that were highly significant higher than the sawdust substrate. In the fruiting period, there were two strains (A12 and A184) with no significant difference from the sawdust substrate, and one strain (A2) with significant difference from the sawdust substrate. The activity of the reed substrate enzyme was relatively high in four strains (A12, A15, A184, and A224), all of which were over 400 U/g. In the two growth stages, strains A12, A15, A184, A224 and A593 had relatively high enzyme activity, while strains A125, A127, A134, A496 and A596 had relatively low enzyme activity. The regularity of β-Gase enzyme activity in the reed substrate was similar to that of CMCase and FPase. The higher the enzyme activity, the stronger the mycelia could degrade and utilize the reed substrate. Through the determination of the activity of three cellulases, it could be seen from the perspective of enzyme activity that the strains A2, A12, A15, A184, A224, A593, etc., were more suitable for growth on the reed substrate, and they could be referred as the next breeding objects for grass-rotting.

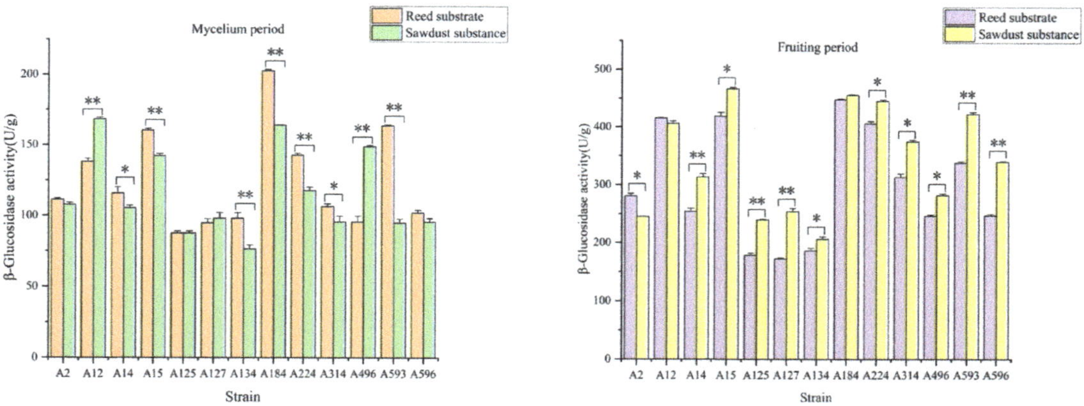


**Figure 6.** β-glucosidase activity at the period of full-bag of mycelium (**left**) and fruiting-body maturation (**right**). * indicates the significance of difference at $p < 0.05$ level, ** indicates the significance of difference at $p < 0.01$ level.

### *3.5. Study on the Expression Level of Cellulase Gene in Different Strains*

#### 3.5.1. Correlation Analysis of Cellulase Activity and Relative Expression of Enzyme Genes in Different Strains

*A. heimuer* mycelium secretes cellulase to degrade cellulose in the substrate, and the higher the enzyme activity, the higher the cellulose degradation amount. However, the activity of cellulase is usually regulated by cellulase genes, and is closely related to gene expression levels. Therefore, the relative expression levels of cellulase genes (*g5372*, *g7270*, *g9664* and *g10234*) of different strains in the reed substrate at different growth stages were measured by qRT-PCR, and correlation analyses were conducted between the three cellulase activities of the strains measured in the reed substrate. As shown in Figure 7, there was a positive correlation between enzyme gene expression and enzyme activity in the two growth stages on the whole, and some of them showed a significant positive correlation, with the highest correlation coefficient being 0.80 ($p < 0.01$). This indicated that the higher the transcriptional expression level of the cellulase gene of the strain, the higher its cellulase activity, and the stronger its ability to degrade and utilize cellulose in the substrate.

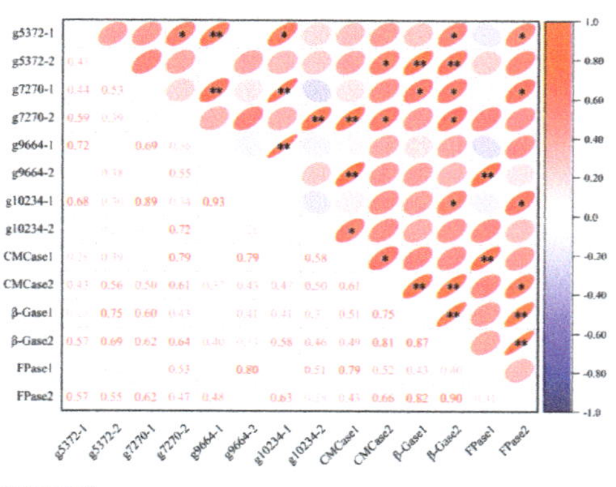


**Figure 7.** Correlation analysis between cellulase activity and relative expression of enzyme genes in different strains. *g5372*-1 indicates the relative gene expression level at the mycelium period of *g5372*, and *g5372*-2 indicates the relative gene expression level at the fruiting period of *g5372*. The same applies to *g7270*, *g9664*, and *g10234*.

#### 3.5.2. Analysis of Cellulase Gene Expression in Different Strains

The relative expression levels of four cellulase genes of thirteen *A. heimuer* strains on the reed substrate at two growth stages are shown in Figures 8 and 9, and the relative expression levels of four cellulase genes at two growth stages were different among strains. In the mycelium period (Figure 8), the relative expression of *g5372* was higher in strains A2, A224 and A12, which was significantly higher than for the other strains. The relative expression of *g7270* was higher in A12, A184, A224 and A496, and the lowest in A127, A314 and A596. The relative expression of *g9664* was the highest in A2 and A12, and medium in A15, A184, A224 and A496. The relative expression of *g10234* was higher in A12 and A184 and significantly higher than in the other strains. In the fruiting period (Figure 9), the relative expression of *g5372* was the highest in strains A184 and A224, and medium in strains A12, A15, A127, A496 and A593. The relative expression of *g7270* was significantly higher in A12, A15, A224, A496 and A593 than in the other strains. The relative expression of *g9664* was significantly higher in A15, A184, A224, A496 and A593 than in the other strains. The relative expression of *g10234* was significantly higher in A2, A15, A224 and A593 than that of the other strains. Overall, seven strains of *g5372* had higher relative gene expression in the mycelium period than in the fruiting period. The expression of 11 strains of *g7270* in the mycelium period was higher than that in the fruiting period. The expression levels of all 13 strains of *g9664* in the fruiting period were higher than those in the mycelium period. The expression of 11 strains of *g10234* in the fruiting period was higher than that in the mycelium period. *g5372* and *g7270* were more actively expressed in the mycelium period, while *g9664* and *g10234* were more actively expressed in the fruiting period.

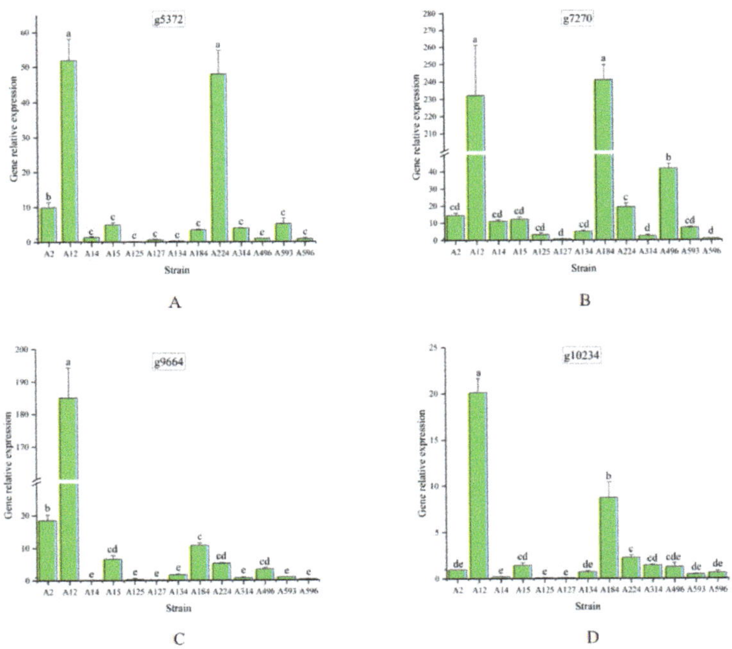


**Figure 8.** Expression of cellulase gene in mycelium period of different strains. The (**A–D**) in the graphs indicate the relative expression of genes *g5372*, *g7270*, *g9664*, and *g10234* in the test strains, in that order. Different lowercase letters in the figure indicate the significant difference between different strains ($p < 0.05$). The significant difference was expressed by Duncan's multiple range test.

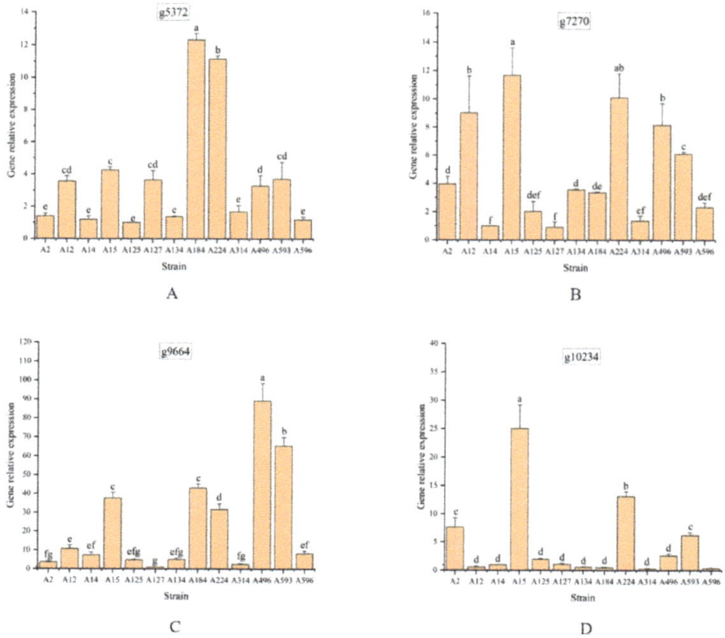


**Figure 9.** Expression of cellulase gene in fruiting period of different strains. The (**A–D**) in the graphs indicate the relative expression of genes *g5372*, *g7270*, *g9664*, and *g10234* in the test strains, in that order. Different lowercase letters in the figure indicate the significant difference between different strains ($p < 0.05$). The significant difference was expressed by Duncan's multiple range test.

### *3.6. Study on Hybrid Cellulose Degradation and Utilization Ability*

#### 3.6.1. Analysis of Relative Expression of Cellulase Gene, Cellulose Degradation and Yield of Hybrid

In order to further verify the correlation between the relative expression of enzyme genes, the degradation amount of cellulose and the yield, hybrid breeding experiments were carried out on *A. heimuer* strains A12, A15, A184 and A224, which had been screened in the first half of this paper and had relatively strong cellulose utilization ability. Three mononuclear strains (spores) were obtained from each of the four parental strains for round mating, and a total of 41 hybrids were obtained; then the pouch (300 g) fruiting test (five bags repeated for each strain) was carried out, and finally 26 hybrids with normal fruiting were obtained.

The expression of two enzyme genes, yield and cellulose degradation of the hybrid and parent in hmycelium period were measured (Table 7, Figure 10). The expression level of *g9664* in six hybrids was higher than in the four parents. In *g10234*, there were also six hybrids with higher expression levels than the four parents, and the hybrids with higher expression levels generally had higher yields in the reed substrate and the sawdust substrate. As can be seen from Table 7, the reed-substrate yield of most hybrids was lower than that of the sawdust substrate, and the cellulose degradation amount in the two growth stages was generally consistent with that shown in Figure 10. Specifically, the yield of eight hybrids in the reed substrate exceeded that of the parent A184, among which hybrid Z12 was the hybrid offspring of parent A184 and A15, and its yield was 1.78 kg, which was the same as that of the parent A15. The expression of two enzyme genes of Z12 was the highest among all the hybrids, while Z6 was the hybrid offspring of A184 and A224, and its yield was 2.21 kg, more than its yield on the sawdust substrate (1.55 kg) and exceeded the yields of the four parents, making it a superparental strain, and Z6 and Z12 were more adaptable to the reed substrate. The yield of 15 hybrids in the sawdust substrate exceeded that of the parent A184; Z18 was the hybrid offspring of parent A184 and parent

A15, with a yield of 3.04 kg, which exceeded the yields of the four parents, and was a superparental strain There was no significant difference between reed-substrate yield and sawdust-substrate yield in 11 of the 26 hybrid strains.

**Table 7.** Relative expression of cellulase gene and yield of hybrids.

| Hybrid Number | Gene Relative Expression | | Parental Hybrid Number | Yield (kg) | |
|---|---|---|---|---|---|
| | *g9664* | *g10234* | | Reed Substrate | Sawdust Substrate |
| Z1 | $0.86 \pm 0.15^{i,j}$ | $5.20 \pm 0.66^{f,g}$ | A15-1*A224-3 | $0.82 \pm 0.02^{k,l}$ | $1.39 \pm 0.06^{h–j}$ |
| Z2 | $0.83 \pm 0.23^{i,j}$ | $5.33 \pm 0.52^{f,g}$ | A15-1*A224-1 | $1.28 \pm 0.14^{h–j}$ | $1.17 \pm 0.17^{j–l}$ |
| Z3 | $0.79 \pm 0.04^{i,j}$ | $0.25 \pm 0.04^{g}$ | A224-3*A12-1 | $0.18 \pm 0.04^{p–r}$ | $0.15 \pm 0.05^{p}$ |
| Z4 | $0.32 \pm 0.04^{i,j}$ | $1.74 \pm 0.12^{g}$ | A224-1*A12-1 | $0.28 \pm 0.04^{o–r}$ | $0.64 \pm 0.28^{o}$ |
| Z5 | $1.38 \pm 0.24^{i,j}$ | $1.20 \pm 0.09^{g}$ | A184-1*A224-2 | $1.02 \pm 0.03^{j,k}$ | $1.18 \pm 0.04^{j–l}$ |
| Z6 | $38.20 \pm 3.69^{d}$ | $6.37 \pm 2.29^{f,g}$ | A184-2*A224-2 | $2.21 \pm 0.28^{b–d}$ | $1.55 \pm 0.05^{f,g}$ |
| Z7 | $1.58 \pm 0.10^{i,j}$ | $0.63 \pm 0.06^{g}$ | A184-3*A12-1 | $0.17 \pm 0.03^{p–r}$ | $1.07 \pm 0.08^{k–n}$ |
| Z8 | $4.78 \pm 1.11^{i,j}$ | $19.52 \pm 1.12^{d}$ | A184-2*A15-1 | $0.45 \pm 0.07^{m–o}$ | $2.11 \pm 0.10^{c,d}$ |
| Z9 | $2.00 \pm 0.18^{i,j}$ | $0.79 \pm 0.10^{g}$ | A184-2*A224-3 | $0.72 \pm 0.03^{l}$ | $1.14 \pm 0.14^{j–l}$ |
| Z10 | $1.15 \pm 0.10^{i,j}$ | $0.21 \pm 0.06^{g}$ | A184-2*A15-3 | $1.50 \pm 0.05^{e–h}$ | $0.87 \pm 0.14^{m–o}$ |
| Z11 | $0.91 \pm 0.14^{i,j}$ | $2.27 \pm 0.33^{g}$ | A184-3*A12-3 | $0.07 \pm 0.01^{r}$ | $0.11 \pm 0.00^{p}$ |
| Z12 | $272.35 \pm 14.73^{b}$ | $246.59 \pm 2.91^{a}$ | A184-2*A15-2 | $1.78 \pm 0.10^{d,e}$ | $2.37 \pm 0.01^{b}$ |
| Z13 | $1.00 \pm 0.00^{i,j}$ | $1.00 \pm 0.00^{g}$ | A224-3*A12-2 | $0.10 \pm 0.00^{q,r}$ | $0.87 \pm 0.28^{m–o}$ |
| Z14 | $0.24 \pm 0.05^{j}$ | $0.23 \pm 0.06^{g}$ | A15-2*A224-3 | $0.69 \pm 0.04^{l,m}$ | $0.93 \pm 0.04^{l–n}$ |
| Z15 | $3.17 \pm 0.65^{i,j}$ | $1.24 \pm 0.17^{g}$ | A15-2*A224-1 | $1.29 \pm 0.06^{h,i}$ | $0.81 \pm 0.10^{n,o}$ |
| Z16 | $5.95 \pm 0.74^{h–j}$ | $11.42 \pm 0.53^{e,f}$ | A15-2*A224-2 | $1.44 \pm 0.06^{e–i}$ | $1.79 \pm 0.01^{e}$ |
| Z17 | $13.31 \pm 1.12^{f,g}$ | $0.77 \pm 0.12^{g}$ | A184-2*A12-2 | $0.34 \pm 0.06^{o–q}$ | $1.47 \pm 0.04^{g,h}$ |
| Z18 | $7.79 \pm 2.18^{g–i}$ | $1.30 \pm 0.16^{g}$ | A184-1*A15-1 | $1.61 \pm 0.01^{e–g}$ | $3.04 \pm 0.01^{a}$ |
| Z19 | $16.70 \pm 1.45^{e,f}$ | $5.30 \pm 0.82^{f,g}$ | A184-3*A15-1 | $0.59 \pm 0.01^{l–n}$ | $2.33 \pm 0.02^{b,c}$ |
| Z20 | $0.99 \pm 0.19^{i,j}$ | $0.84 \pm 0.10^{g}$ | A184-1*A12-3 | $0.17 \pm 0.07^{p–r}$ | $1.59 \pm 0.06^{e,f}$ |
| Z21 | $4.17 \pm 0.25^{i,j}$ | $0.56 \pm 0.05^{g}$ | A184-3*A15-2 | $0.69 \pm 0.06^{l,m}$ | $1.13 \pm 0.02^{j–m}$ |
| Z22 | $1.33 \pm 0.158^{i,j}$ | $2.64 \pm 0.26^{g}$ | A184-1*A12-3 | $0.42 \pm 0.03^{n–p}$ | $0.98 \pm 0.11^{l–n}$ |
| Z23 | $21.74 \pm 0.93^{e}$ | $6.51 \pm 0.78^{f,g}$ | A184-3*A224-3 | $1.00 \pm 0.16^{j,k}$ | $1.84 \pm 0.01^{e}$ |
| Z24 | $0.22 \pm 0.08^{j}$ | $1.17 \pm 0.14^{g}$ | A184-1*A224-3 | $0.68 \pm 0.10^{l,m}$ | $1.63 \pm 0.03^{e,f}$ |
| Z25 | $2.96 \pm 0.32^{i,j}$ | $4.61 \pm 1.28^{f,g}$ | A184-2*A12-1 | $1.20 \pm 0.14^{i,j}$ | $1.28 \pm 0.03^{i–k}$ |
| Z26 | $2.05 \pm 0.31^{i,j}$ | $46.21 \pm 0.27^{c}$ | A184-2*A12-3 | $0.27 \pm 0.01^{o–r}$ | $0.17 \pm 0.01^{p}$ |
| A12 | $545.40 \pm 8.60^{a}$ | $95.20 \pm 4.22^{b}$ | A12 | $1.38 \pm 0.14^{f–i}$ | $1.63 \pm 0.30^{e,f}$ |
| A15 | $11.59 \pm 2.09^{f–h}$ | $6.29 \pm 0.63^{f,g}$ | A15 | $1.78 \pm 0.28^{d,e}$ | $2.07 \pm 0.10^{d}$ |
| A184 | $46.39 \pm 9.93^{c}$ | $249.91 \pm 23.61^{a}$ | A184 | $1.03 \pm 0.03^{j,k}$ | $1.13 \pm 0.03^{j–m}$ |
| A224 | $6.24 \pm 0.12^{h–j}$ | $16.31 \pm 2.03^{d,e}$ | A224 | $1.21 \pm 0.01^{i,j}$ | $2.88 \pm 0.20^{a}$ |

Note: All lowercase letters in the same column of data in the table show significant differences at the 0.05 level.
* Indicates hybridization of mononuclear mycelium from two parents. The type of average comparison test was Duncan's multiple range test.

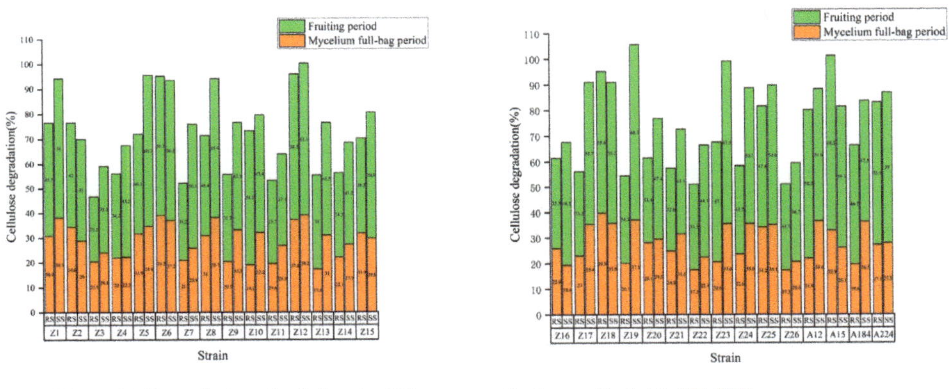


**Figure 10.** Degradation of cellulose in different growth stages of hybrids (Z1–Z26).

3.6.2. Correlation Analysis of Relative Expression of Cellulase Gene, Cellulose Degradation and Yield in Hybrid

Correlation analysis was conducted between the relative expression of enzyme genes in the mycelium period of the hybridand parentalreed substrates, and cellulose degradation and yield in the two growth stages of the reed and sawdust substrate (Table 8). It can be seen from the data in the table that the amount of cellulose degradation at the full-bag period and the amount of cellulose degradation at the fruiting period of the reed substrate were highly significantly positively correlated with the yield of the reed substrate, and the correlation coefficients were 0.573 and 0.623 ($p < 0.01$). The degradation amount of cellulose at the full-bag period and at the fruiting period of the sawdust substrate were highly significantly positively correlated with the yield of the sawdust substrate, and the correlation coefficients were 0.468 and 0.660 ($p < 0.01$). In addition, the relative expression levels of *g9664* and *g10234* were highly significantly positively correlated (r = 0.532, $p < 0.01$), and the expression levels of *g9664* and *g10234* were positively correlated with the cellulose degradation levels of the reed substrate at the full-bag period and the fruiting period. There was a significant positive correlation between the expression level of *g9664* and the degradation amount of cellulose at the fruiting period (r = 0.433, $p < 0.05$), while the correlation between the expression level of *g10234* and the degradation amount of cellulose at the fruiting period was small (r = 0.350). The expression levels of the two enzyme genes were positively correlated with the yield of the reed substrate (r = 0.317, r = 0.274).

**Table 8.** Correlation analysis of hybrid-cellulase gene relative expression, cellulose degradation and yield.

| | *g9664* | *g10234* | RS-CDR1 | RS-CDR2 | SS-CDR1 | SS-CDR2 | RS-Yield | SS-Yield |
|---|---|---|---|---|---|---|---|---|
| *g9664* | 1.000 | | | | | | | |
| *g10234* | 0.532 ** | 1.000 | | | | | | |
| RS-CDR1 | 0.053 | 0.061 | 1.000 | | | | | |
| RS-CDR2 | 0.433 * | 0.350 | 0.536 ** | 1.000 | | | | |
| SS-CDR1 | 0.317 | 0.312 | 0.386 * | 0.400 * | 1.000 | | | |
| SS-CDR2 | 0.171 | 0.126 | 0.416 * | 0.488 ** | 0.629 ** | 1.000 | | |
| RS-Yield | 0.317 | 0.274 | 0.573 ** | 0.623 ** | 0.334 | 0.431 * | 1.000 | |
| SS-Yield | 0.200 | 0.151 | 0.526 ** | 0.577 ** | 0.468 ** | 0.660 ** | 0.517 ** | 1.000 |

Note: * Correlation is significant at the 0.05 level (2-tailed). ** Correlation is significant at the 0.01 level (2-tailed).

## 4. Discussion

*A. heimuer* mycelium secretes lignocellulase to the outside world in the process of growth and development, decomposing and using lignocellulose in culture material to obtain the required nutrients. In this process, the level of cellulase activity will directly affect the absorption and utilization of nutrients in culture material by mycelium. However, the degradation amount of lignocellulose in different cultures at different periods could reflect the ability of the *A. heimuer* strain to utilize lignocellulose and the feeding situation of different substrates, to a certain extent [48,49]. Therefore, in this study, we took different *A. heimuer* strains from multiple regions as research objects, to study the relationship between cellulase genes of different strains and their cellulose utilization ability. These strains included cultivated strains and wild strains, and the sample size was abundant. Firstly, we made use of the characteristics of extracellular enzymes secreted by *A. heimuer* mycelia, and preliminarily screened 28 tested strains by Congo red staining on a plate containing only one carbon source (carboxymethyl cellulose Na). The colony diameter was set to be greater than 1.90 cm, and D/d was greater than 0.80. Finally, 13 strains were obtained. Then, the fruiting experiments were carried out on the reed substrate and the sawdust substrate with the 13 tested strains and the hybrid obtained from four parents. The yield of the strains on the two substrates was measured, and the material was extracted at the full-bag period of the mycelium and the fruiting period; the cellulase activity and cellulose degradation amounts were measured, and the correlation analysis of them was conducted. It was found that there were differences between strains in enzyme activity and cellulose degradation amount in the two growth stages, and the differences in the strains had certain effects on cellulase activity and degradation amount, but did not affect the change trend of the two: for most strains, the higher the cellulase activity, the higher the amount of cellulose degradation [50,51]. Compared with the general strains, the strains with strong cellulose utilization ability had stronger extracellular enzyme activity, stronger damage to the cellulose structure of the reed substrate and the sawdust substrate, and had a stronger feeding ability. Therefore, the higher the amount of cellulose degradation of the strain, the higher its yield in general; the correlation analysis among the cellulase activity, degradation amount and yield also proved this point. This was consistent with the fact that the peak of cellulase activity of wood-rotting edible mushrooms, such as *Pleurotus cornucopiae* and *Pleurotus abalonus* [52] and *Pleurotus citrinopileatus* [53], all appeared at the mature period of the fruiting-body, while the enzyme activity was lower at the mycelium period. The amount of degradation was also consistent with the trend found by previous studies [29,54]. In addition, in the two growth stages, most strains had a stronger ability to utilize the cellulose of the sawdust substrate than that of the reed substrate, and their yields were also higher than those of the reed substrate. This conclusion was further confirmed by the research on the relevant data of hybrids in the cross-breeding experiment in this paper, which was inseparable from the fact that *A. heimuer* is a wood-rotting edible mushroom, and breeding of strains better suited to herbaceous substrates also takes time.

The cellulase activity of the *A. heimuer* strain could not be separated from the regulation of related cellulase genes. In general, high levels of gene expression are usually accompanied by high enzyme activity and high substrate degradation. Guan et al. [51] compared and analyzed the expression levels of β-glucosidase coding genes in the mycelia full-length period (30 d) and the mid-color transformation period (60 d) by using real-time quantitative PCR technology, and found that with the increase in enzyme activity, compared with the mycelia full-length period (30 d), the relative expression of the β-glucosidase coding gene was significantly increased during the mid-color transformation period (60 d) ($p < 0.05$). Chen [55] showed that the expression level of UDP glucuronuronyltransferase gene (UGP gene) of *Lentinula edodes* was significantly positively correlated with the activity of the enzyme. In addition, under different carbon- and nitrogen-source fermentation conditions, the expression of the UGP gene and the activity of the UGPase enzyme showed the same trend, indicating that the higher the expression of the UGP gene, the higher the activity of this enzyme. Duan et al. [56] studied the expression level of the LeOAH1 gene in *Lentinula edodes* under different pH conditions, and its close relationship with oxalic acid secretion, and found that the expression level of the LeOAH1 gene was positively correlated with oxalic acid content. The results of our findings were basically consistent with the above studies. The relative expression levels of 4 cellulase genes in 13 tested strains and 2 enzyme genes in 26 hybrids were different among strains, and it was speculated that the strains with high cellulase-gene expression and degradation might have the characteristics of high cellulase-gene expression themselves or be induced by the reed substrate to have such high expression and degradation characteristics [47,57,58]. The expression levels of strains A2, A184, and A12 and hybrids Z6, Z12, Z19 and Z26 in the mycelium period were relatively high, and those of strains A2, A12, A15, A184, A224, A496 and A593 in the fruiting period were relatively high, while the expression levels of strains A125, A127, A134 and A314 were low in the two growth stages, which was basically consistent with the trend of cellulase activity, degradation and yield measured in the previous stage. The correlation analysis between cellulase gene expression and cellulase activity also proved this result; that is, the higher the cellulase-gene expression level of the strain, the higher its cellulase activity. In addition, like the expression levels of *g9664* and *g10234* in strains A15 and A593 were lower in the mycelium period, while their relative expression levels were significantly increased in the fruiting period. It was speculated that cellulase genes in strains A15 and A593 might be more easily induced by the reed substrate during the fruiting period, and thus exhibited higher transcriptional expression levels.

At the same time, we found that some strains might have high gene expression but low enzyme activity and degradation at different stages, or have low gene expression but high enzyme activity and degradation, possibly because the related enzymes could not be properly folded, were improperly modified, or might not function. On the contrary, due to post-transcriptional or post-translational regulatory mechanisms or other factors affecting the efficiency of gene expression, some strains might have high enzyme activity and high degradation amount but low gene expression amount, which was speculated to contain or restrict genes in the translation process [59]. This also explained the reason why the relative expression of enzyme genes and the enzyme activity of some strains in this study, such as A14, A15 and A593, were inconsistent at the two growth stages. In conclusion, the relationship between enzyme gene expression level, enzyme activity and degradation capacity is intricate; a high gene expression level leads to high enzyme activity and degradation capacity, but this is not always the case, and further studies are needed to understand the underlying mechanism of related enzyme function and production. At the same time, the interdependence of the three is of great reference and significance for breeding more suitable strains for reed-substrate cultivation.

In summary, real-time quantitative PCR technology was used to determine the cellulase gene expression of the tested strains on the reed substrate, while the cellulase activity and cellulose degradation ability of the tested strains were measured, and correlation analysis was performed. The results showed that the strains with high cellulase-gene expression also had high enzyme activity and degradation capacity, and there was a positive correlation between them. qRT-PCR technology could be used as an auxiliary breeding method to evaluate the cellulose degradation ability of the strain. At the same time, since the gene expression level did not always correspond to the enzyme activity level, the fruiting test of the target strain could be further conducted on the reed substrate and the relevant agronomic traits could be determined; finally, the suitability of the selected strains for production application could be comprehensively confirmed.

## 5. Conclusions

1. Real-time quantitative PCR technology could be used as a technology to rapidly evaluate the cellulose degradation ability of the *A. heimuer* strain. The use of qRT-PCR technology to evaluate the cellulose degradation ability of the strain in the early stage can reduce the workload of mushroom production in the later stage, improve the work efficiency, shorten the breeding time, and quickly provide a batch of *A. heimuer* strains that are more suitable for growth on the herbaceous substrate, promoting the development of the *A. heimuer* wood-rotting edible-mushroom grass-rotting cultivation. Finally, the feasibility of qRT-PCR as a rapid and auxiliary method to evaluate the cellulose degradation ability of strains was further verified through hybridization experiments.

2. Seven strains (A12, A15, A184, A224, Z6, Z12, and Z18) with high expression of cellulase gene, strong degradation ability and high utilization rate were comprehensively screened. Among them, one hybrid with a higher yield than the parent was Z6 on the reed substrate, and one hybrid with a higher yield than the parent was Z18 on the sawdust substrate.

**Author Contributions:** F.Y. designed the experiments; L.L. and J.L. revised the manuscript; M.F. and X.S. (Xu Sun) guided the experiment; X.S. (Xianqi Shan) prepared the materials for the experiments, analyzed the data and wrote the manuscript. All authors have read and agreed to the published version of the manuscript.

**Funding:** This study was supported by National Key Research and Development Program of China (No. 2023YFD1201604-4).

**Institutional Review Board Statement:** Not applicable.

**Data Availability Statement:** The data presented in this study are available in the article.

**Acknowledgments:** The authors thank the reviewers for their valuable suggestions.

**Conflicts of Interest:** The authors declare no conflicts of interest.

## References


1. Wu, F.; Yuan, Y.; Malysheva, V.F.; Du, P.; Dai, Y.C. Species clarification of the most important and cultivated *Auricularia* mushroom "Heimuer": Evidence from morphological and molecular data. *Phytotaxa* **2014**, *186*, 241–253. [CrossRef]
2. Wu, F.; Dai, Y.C. Notes on the nomenclature of the *Auricularia auricula-judae* complex. *Mycosystema* **2015**, *34*, 604–611. [CrossRef]
3. China Edible Fungi Association. Analysis of the results of the national edible fungi statistical survey in 2022. *Edible Fungus China* **2024**, *43*, 118–126. [CrossRef]
4. Yao, F.J. Research and thinking on the cultivation of wood-rotting edible fungi grass-rotting. In Proceedings of the China Mycological Society 2018 Annual Conference, Tai'an, China, 11 August 2018.
5. Li, Y. Reflections on the development of straw fungus industry in China—Report of 2023 Mushroom Festival in Zhangzhou, Fujian Province. *J. Fungal Res.* **2024**, *22*, 113–117. [CrossRef]
6. Han, X.L.; Fang, Z.H.; Liang, J.Q.; Xiong, L.B.; Xiong, Z.; Lv, H.Y. Development status, model and suggestion of edible fungus industry of reed in Yuanjiang City. *Hunan Agric. Sci.* **2021**, *11*, 96–99. [CrossRef]
7. Santos, F.A.; de Carvalho-Gonçalves, L.C.T.; de Carvalho Cardoso-Simões, A.L.; de Melo Santos, S.F. Evaluation of the production of cellulases by *Penicillium* sp. FSDE15 using corncob and wheat bran as substrates. *Bioresour. Technol. Rep.* **2021**, *14*, 100648. [CrossRef]
8. Liu, Y.; Luo, Y.; Guo, W.; Zhang, X.; Zheng, W.; Chen, X. Study on the Effects of Different Light Supply Modes on the Development and Extracellular Enzyme Activity of *Ganoderma lucidum*. *Agriculture* **2024**, *14*, 835. [CrossRef]
9. Xia, Y.; Wang, J.; Guo, C.; Xu, H.; Wang, W.; Yang, M.; Shen, Q.; Zhang, R.; Miao, Y. Exploring the multi-level regulation of lignocellulases in the filamentous fungus *Trichoderma guizhouense* NJAU4742 from an omics perspective. *Microb. Cell Factories* **2022**, *21*, 144. [CrossRef]
10. Meng, W. Study on Mutagenesis Breeding and Fermentation Technology of Cellulase Producing Bacteria. Master's Thesis, Jilin University, Changchun, China, 2008.
11. Fatani, S.; Saito, Y.; Alarawi, M.; Gojobori, T.; Mineta, K. Genome sequencing and identification of cellulase genes in *Bacillus paralicheniformis* strains from the Red Sea. *BMC Microbiol.* **2021**, *21*, 254. [CrossRef]
12. Bai, X.; Wang, X.; Wang, S.; Ji, X.; Guan, Z.; Zhang, W.; Lu, X. Functional Studies of β-Glucosidases of *Cytophaga hutchinsonii* and Their Effects on Cellulose Degradation. *Front. Microbiol.* **2017**, *8*, 140. [CrossRef]
13. Madan, M.; Bisaria, R. Cellulolytic enzymes from an edible mushroom, *Pleurotus sajor-caju*. *Biotechnol. Lett.* **1983**, *5*, 601–604. [CrossRef]
14. Fan, B.W.; Gong, J.L.; Lin, J.J.; Zhao, Y.; Shi, X.X.; Tang, C.S.; Wang, Z.H.; Li, Z.T.; Yang, K.J.; Zhao, C.J. Effects of carbon and nitrogen sources on mycelium and fruiting body growth and extracellular enzymes of *Hericium erinaceus*. *Jiangsu Agric. Sci.* **2018**, *46*, 122–126. [CrossRef]
15. Lu, X.; Zhao, Y.; Li, F.; Liu, P. Active polysaccharides from *Lentinula edodes* and *Pleurotus ostreatus* by addition of corn straw and xylosma sawdust through solid-state fermentation. *Int. J. Biol. Macromol.* **2023**, *228*, 647–658. [CrossRef]
16. Zhou, Y.; Liu, P.H.; Liu, B.; Su, R.R.; Zhou, J.J. Study on growth and development and extracellular enzyme activity of *Agrocybe cylindracea* with different cultivation formulas. *North. Hortic.* **2023**, *19*, 114–121.
17. Lin, X.J.; Jiang, X.H.; Lin, R.B.; Chen, J.C. Correlations between extracellular enzyme levels and fruit body yields in strains of *Agaricus blazei* Murill. *Acta Edulis Fungi* **2007**, *3*, 24–28. [CrossRef]
18. Al Makishah, N.H.; Elfarash, A.E. Molecular characterization of cellulase genes in *Pseudomonas stutzeri*. *Electron. J. Biotechnol.* **2022**, *59*, 55–61. [CrossRef]
19. Huang, W.B.; Hou, D.; Zhou, C.L.; Li, Y.; Yang, R.H.; Bao, D.P. Analyses of genes related to different mycelial growth rate of *Lentinula edodes* monokaryons. *Mycosystema* **2023**, *42*, 2111–2118. [CrossRef]
20. Li, Z.H.; Han, J.D.; Xie, H.Y.; Hu, Q.X.; Gong, Z.Y.; Zou, Y.J. Expression profiles of laccase gene family in *Flammulina filiformis* cultivated on different agricultural waste media. *Acta Edulis Fungi* **2021**, *28*, 47–54. [CrossRef]
21. Liang, Z.Y.; Liu, F. Research progress on real-time quantitative PCR technology and its application. *Mod. Agric. Sci. Technol.* **2020**, *6*, 1–3+8.
22. Niimi, Y.; Han, D.-S.; Mori, S.; Kobayashi, H. Detection of cucumber mosaic virus, lily symptomless virus and lily mottle virus in *Lilium* species by RT-PCR technique. *Sci. Hortic.* **2003**, *97*, 57–63. [CrossRef]
23. Bai, J.; Chen, M.J.; Tang, L.H.; Du, J.H.; Feng, Z.Y.; Zhang, J.J.; Chen, H. Effects of temperature on antioxidant enzyme activities and gene expression in *Morchella importuna*. *Mycosystema* **2021**, *40*, 3276–3285. [CrossRef]
24. Mao, W.J.; Li, Y.; Zhou, C.L.; Wang, Y.; Zhu, G.; Bao, D.P.; Wang, Y. Relative expression of laccase genes at different stages of *Volvariella volvacea* fruit body development. *Acta Edulis Fungi* **2016**, *23*, 1–6. [CrossRef]
25. Zhang, Y.; Yao, F.J.; Sun, W.J.; Fang, M.; Wu, C.S. Screening of reference genes for qRT-PCR amplification in *Auricularia heimuer*. *Mycosystema* **2020**, *39*, 1510–1519. [CrossRef]
26. Xiang, Q.; Li, J.; Qin, P.; He, M.; Yu, X.; Zhao, K.; Zhang, X.; Ma, M.; Chen, Q.; Chen, X.; et al. Identification and evaluation of reference genes for qRT-PCR studies in *Lentinula edodes*. *PLoS ONE* **2018**, *13*, e0190226. [CrossRef] [PubMed]
27. Liao, F.; Yan, H.; Zhao, S.F.; Li, X.H. Real-time fluorescent quantitative RT-PCR method to determine the antiviral activity of edible mushroom proteins against *Tobacco mosaic virus* (TMV). *Acta Phytopathol. Sin.* **2010**, *40*, 622–627. [CrossRef]

28. Wang, M.L.; Zhao, Y.; Chen, M.J.; Wang, H. Expression analysis of xyn and activity determination of xylanase in different *Volvariella volvacea* strains. *Mycosystema* **2014**, *33*, 1074–1083. [CrossRef]
29. Zhai, Y. Genetic Analysis of Wild Germplasm Resources and Development of New Substrates for Excellent Strains of *Auricularia heimuer*. Master's Thesis, Jilin Agricultural University, Changchun, China, 2017.
30. Wang, R. Evaluation of Cellulose Utilization Capacity of *Auricularia heimuer* Germplasm Resources and Screening of Excellent Strains. Master's Thesis, Jilin Agricultural University, Changchun, China, 2023.
31. Fang, M.; Wang, X.; Chen, Y.; Wang, P.; Lu, L.; Lu, J.; Yao, F.; Zhang, Y. Genome Sequence Analysis of *Auricularia heimuer* Combined with Genetic Linkage Map. *J. Fungi* **2020**, *6*, 37. [CrossRef]
32. Fang, M.; Sun, X.; Yao, F.; Lu, L.; Ma, X.; Shao, K.; Kaimoyo, E. A Combination of Transcriptome and Enzyme Activity Analysis Unveils Key Genes and Patterns of Corncob Lignocellulose Degradation by *Auricularia heimuer* under Cultivation Conditions. *J. Fungi* **2024**, *10*, 545. [CrossRef]
33. Si, J.; Yang, C.; Ma, W.; Chen, Y.; Xie, J.; Qin, X.; Hu, X.; Yu, Q. Screen of high efficiency cellulose degrading strains and effects on tea residues dietary fiber modification: Structural properties and adsorption capacities. *Int. J. Biol. Macromol.* **2022**, *220*, 337–347. [CrossRef]
34. Zhou, L.; Zhuang, W.Y. Screening high cellulase production *Trichoderma* strains and optimization of fermentation conditions. *Mycosystema* **2023**, *42*, 1966–1980. [CrossRef]
35. Sun, J. Mining and Functional Study of Genes Related to Cellulose Degradation in *Auricularia heimuer*. Ph.D. Thesis, Northeast Forestry University, Harbin, China, 2023.
36. Wang, Z.Y.; Wang, R.X.; Zhou, J.S.; Cheng, J.F.; Li, Y.H. An assessment of the genomics, comparative genomics and cellulose degradation potential of *Mucilaginibacter polytrichastri* strain RG4-7. *Bioresour. Technol.* **2020**, *297*, 122389. [CrossRef] [PubMed]
37. Nie, T.; Jiang, Z.; Sun, L.; Chen, Y.; Li, J.; Yang, A.; Wei, Q.; Yin, Z. Reference genes selection for qRT-PCR analysis in various flowering transition events of *Magnolia* × *soulangeana* 'Changchun'. *Sci. Hortic.* **2023**, *316*, 112006. [CrossRef]
38. Zhang, Y. Screening of Reference Genes and Study Expression Levels of Functional Genes in *Auricularia heimuer* for qRT-PCR. Master's Thesis, Jilin Agricultural University, Changchun, China, 2021.
39. Wang, J.N.; Yao, F.J.; Zhang, W.; Liu, L.J. Study on effects of smallhole-fruiting method on physical and chemical nature for compost and growth and development of *Auricularia heimuer*. In Proceedings of the Second National Academic Exchange Meeting of Young and Middle-aged Experts in Edible Fungi, Hangzhou, China, 23 October 2008.
40. Bian, Y.B. *Edible Mushroom Cultivation*, 3rd ed.; Higher Education Press: Beijing, China, 2017; pp. 60–66.
41. Sun, Y.; Tian, Y.Q.; Zhao, L.K. Study on the conditions for determination of CMC enzyme activity of cellulase. *Sci. Technol. Food Ind.* **2013**, *34*, 68–71+74. [CrossRef]
42. Mboowa, D.; Chandra, R.P.; Hu, J.; Saddler, J.N. Substrate Characteristics That Influence the Filter Paper Assay's Ability to Predict the Hydrolytic Potential of Cellulase Mixtures. *ACS Sustain. Chem. Eng.* **2020**, *8*, 10521–10528. [CrossRef]
43. Liu, Q.; Niu, S.; Hu, S.; Cui, X.; Shi, Z.; Wu, J.; Zhang, Y.; Kong, W. Lignocellulose degradation pattern and structural change of the sawdust substrate and enzyme secretion by *Lentinula edodes* during its production. *Wood Sci. Technol.* **2023**, *57*, 389–405. [CrossRef]
44. Jin, M.-Y.; Zhang, T.; Yang, Y.-S.; Ding, Y.; Li, J.-S.; Zhong, G.-R. A simplified and miniaturized glucometer-based assay for the detection of β-glucosidase activity. *J. Zhejiang Univ.-Sci. B* **2019**, *20*, 264–272. [CrossRef]
45. Liu, S.; Zhang, M.; Hong, D.; Fang, Z.; Xiao, Y.; Fang, W.; Zhang, X. Improving the cellobiose hydrolysis activity of glucose-stimulating β-glucosidase Bgl2A. *Enzym. Microb. Technol.* **2023**, *169*, 110289. [CrossRef]
46. Li, F.; Xie, Y.; Gao, X.; Shan, M.; Sun, C.; Niu, Y.D.; Shan, A. Screening of cellulose degradation bacteria from Min pigs and optimization of its cellulase production. *Electron. J. Biotechnol.* **2020**, *48*, 29–35. [CrossRef]
47. Zhang, F.; Zhang, T.; Dai, D.; Zhang, Z.; Zhang, B.; Li, Y. Screening of efficient lignin-degrading fungal strains and their degradation on cornstalk. *Mycosystema* **2021**, *40*, 1869–1880.
48. Atila, F. Compositional changes in lignocellulosic content of some agro-wastes during the production cycle of shiitake mushroom. *Sci. Hortic.* **2019**, *245*, 263–268. [CrossRef]
49. Öztürk, C.; Atila, F. Changes in lignocellulosic fractions of growing substrates during the cultivation of *Hypsizygus ulmarius* mushroom and its effects on mushroom productivity. *Sci. Hortic.* **2021**, *288*, 110403. [CrossRef]
50. Zhai, F.-H.; Han, J.-R. Decomposition of asparagus old stalks by *Pleurotus* spp. under mushroom-growing conditions. *Sci. Hortic.* **2018**, *231*, 11–14. [CrossRef]
51. Guan, W.; Chu, T.; Bao, D.P.; Wang, J.N.; Li, F.H.; Tang, L.H. Detection and analysis of physiological indices during mycelium color conversion of *Lentinula edodes*. *Acta Edulis Fungi* **2021**, *28*, 47–52. [CrossRef]
52. Chu, Y.; Ni, X.J.; Yang, G.W.; Yuan, Y. Comparison of activity of eight extracellular enzymes of *Pleurotus cornucopiae* and *Pleurotus abalonus* in different growth periods. *J. Yantai Univ. (Nat. Sci. Eng. Ed.)* **2008**, *2*, 138–142. [CrossRef]
53. Lei, P.; Wu, Y.Z.; Zhang, W.J.; Ma, J.J.; Jia, J.; Ma, Y. Studies on Changes of Extracellular Enzyme Activities in Different Growth Stages of *Pleurotus citrinopileatus*. *Edible Fungus China* **2020**, *39*, 57–61. [CrossRef]
54. Cong, S. Study on Suitable Variety Screening and Nutritional Physiology of *Auricularia heimuer* Composite Substrate. Master's Thesis, Jilin Agricultural University, Changchun, China, 2014.
55. Chen, X.M.; Wu, H.B.; Xiang, Q.J.; Gu, Y.F.; Zhang, X.P. Transcriptional expression profiles and enzyme activity of UGP from *Letinous edodes* under different carbon and nitrogen sources. *J. Sichuan Univ. (Nat. Sci. Ed.)* **2018**, *55*, 214–220.

56. Duan, Y.C.; Hu, Z.Y.; Yang, F.; Li, J.T.; Wu, X.L.; Zhang, R.Y. Cloning and expression analysis of oxaloacetate hydrolase (LeOAH1) gene from *Lentinula edodes*. *Biotechnol. Bull.* **2020**, *36*, 227–234. [CrossRef]
57. Zhang, Z.Y.; Raza, M.F.; Zheng, Z.; Zhang, X.; Dong, X.; Zhang, H. Complete genome sequence of *Bacillus velezensis* ZY-1-1 reveals the genetic basis for its hemicellulosic/cellulosic substrate-inducible xylanase and cellulase activities. *3 Biotech* **2018**, *8*, 465. [CrossRef]
58. Doria, E.; Altobelli, E.; Girometta, C.; Nielsen, E.; Zhang, T.; Savino, E. Evaluation of lignocellulolytic activities of ten fungal species able to degrade poplar wood. *Int. Biodeterior. Biodegrad.* **2014**, *94*, 160–166. [CrossRef]
59. Wang, M.L. Study on Hemicellulose Related Enzyme Activity and Gene Expression of *Volvariella volvacea*. Master's Thesis, Nanjing Agricultural University, Nanjing, China, 2016.

agriculture

MDPI

Article

# Transcriptome Analysis of *Ganoderma lingzhi* Liquid Fermentation Process Using Corn Straw as Matrix

**Sheng Wang [1,†], Jintao Li [1,†], Qi Fan [1], Shufang Wang [1], Changwei Sun [1,2] and Meixia Yan [1,*]**

[1] Institute of Special Animal and Plant Sciences, Chinese Academy of Agricultural Sciences, Changchun 130112, China; 82101215186@caas.cn (S.W.); lijintao@caas.cn (J.L.); 82101225207@caas.cn (Q.F.); 82101235241@caas.cn (S.W.); sunchangwei@caas.cn (C.S.)

[2] Jilin Provincial Key Laboratory of Traditional Chinese Medicinal Materials Cultivation and Propagation, Changchun 130112, China

* Correspondence: yanmeixia@caas.cn

† These authors contributed equally to this work.

**Abstract:** *Ganoderma lingzhi*, a species of white rot fungus, possesses the highest abundance of lignocellulose-degrading enzymes among these fungi, as well as a relatively high carbon conversion rate. Corn straw, as an important sustainable resource, is used as a substrate for the liquid culture of *G. lingzhi*. However, little is known about the genes encoding the lignocellulose degradation and polysaccharide and triterpenoid biosynthetic pathways involved in this process. This paper employs transcriptomics to uncover the key genes involved in lignocellulose degradation and the synthesis of polysaccharides and triterpenoids during the liquid fermentation of *G. lingzhi* using corn straw as the substrate, as well as their associations. Carbohydrate-Active enzymes analysis of differential genes in the sequencing results was used to analyze the genes related to lignocellulose degradation. Among these, 43 core genes encoding CAZymes were obtained after 0 to 5 days of fermentation, and 25 core genes encoding CAZymes were obtained after 5 to 12 days of fermentation. The differential expression levels of DN3690_c0_g1 (EGL), DN3627_c0_g2 (XYN), DN4778_c0_g1 (XYN), DN2037_c0_g1 (LACC), and DN277_c2_g1 (MnP) were used to identify the key genes. The polysaccharide synthesis metabolic pathway favored mannitol synthesis, and the expression of triterpene precursor-metabolizing enzyme genes revealed higher expression levels of key enzyme genes such as ACAT, HMGS, and MPK. A correlation clustering analysis of genes related to lignocellulose degradation, polysaccharide, and triterpene anabolism during liquid fermentation showed that lignocellulose degradation genes mainly influenced arabinose and mannitol anabolism, as well as the synthesis of triterpene precursors.

**Keywords:** *Ganoderma lingzhi*; corn straw; transcriptome; carbon metabolism

## 1. Introduction

*Ganoderma lingzhi*, as a widely cultivated edible and medicinal fungus, mainly encompasses such active constituents as polysaccharides, triterpenoids, polypeptides, sterols, alkaloids, and so on, which possess anti-tumor [1], immune-boosting [2], antioxidant [3], hypolipidemic [4], liver-protecting [5], and anti-inflammatory [6] effects. Compared to other white rot fungi, *G. lingzhi* has the highest abundance of genes encoding lignocellulose-degrading enzymes [7,8], and it also has a higher carbon conversion rate [9]. *G. lingzhi* breaks down the lignin on the substrate surface by secreting diverse ligninases, uncovering the inner hemicellulose and cellulose. Afterwards, *G. lingzhi* secretes a series of cellulases and hemicellulases to hydrolyze them into oligosaccharides and monosaccharides that can be absorbed and utilized by the mycelium [10].

Corn straw is a multifunctional and renewable biological resource that is abundant in cellulose (41.93%), hemicellulose (24.71%), and lignin (23.68%), along with other nutrients [11]. However, as a type of crop straw resource in China [12], the utilization rate of corn straw is relatively low, resulting in its random burning and stacking, causing serious resource waste and environmental pollution [13]. At the same time, the large molecular cross-linking structure formed by cellulose, hemicellulose, and lignin in corn straw [14,15] raises the difficulty of utilizing lignocellulose in corn straw [16].

*Ganoderma* has potential applications in the utilization of corn straw, but the lack of an in-depth understanding of the mechanisms involved in degrading corn straw lignocellulose significantly restricts the efficient use of corn straw using this fungus [16]. In addition, the effectiveness of using *Ganoderma* to utilize corn straw is not ideal, with a certain level of uncertainty in its utilization. Lignocellulose-degrading enzymes belong to the Carbohydrate-Active EnZymes (CAZy) family in carbohydrate metabolism, playing a crucial role in the growth process of white rot fungi [17]. Lignocellulolytic enzymes are Carbohydrate-Active Enzymes (CAZy) that play an important role in carbohydrate metabolism in organisms [10]. Cellulose-degrading enzymes can be classified into GHs, including endo-β-1,4-glucanase, cellobiohydrolase, and β-glucosidase [17,18]. Hemicellulose-degrading enzymes, mainly endo-β-1,4--xylanase (XYN), β-xylosidase (XYL), and α-glucuronidase (AGU), are also classified as GHs [10,12]. In terms of lignin-degrading enzymes, AAs play a key role, and they are classified as lignin oxidases (LOs) and lignin-degrading auxiliary enzymes (LDAs) [19]. LOs are responsible for the production of highly reactive, nonspecific free radicals that cleave lignin's carbon-carbon and ether-unit interbonds [17]. LOs include laccase (LACC) and class II heme peroxidases (class II PODs, including lignin peroxidase (LiP), manganese peroxidase (MnP), and versatile peroxidase (VP)). LDAs include glucose/methanol/choline oxidases (GMCs) and copper radical oxidases (CROs) [20,21]. GMCs are mainly composed of cellobiose dehydrogenase (CDH), aryl alcohol oxidase, aryl alcohol oxidase (AAO), glucose oxidase (GOX), and alcohol oxidase (AOX), and CROs are mainly composed of galactose oxidase (GAOX) and glyoxal oxidase (GLOX) [16,19]. However, reports on the downstream synthetic pathways involved in *Ganoderma* triterpenes are still scarce, and the relationship between these three metabolic components has rarely been investigated in response to the question of how substrate degradation is related to the synthesis of polysaccharides or triterpenes and how these metabolic pathways are altered during liquid fermentation.

The degradation products of lignocellulose, such as glucose and xylose, participate in carbohydrate catabolic pathways like glycolysis (EMP) and the phosphogluconate pathway (HMP), providing energy and coenzymes for the growth of fungus, and large molecular polysaccharides act as cell wall components and carbon source reserves [22,23]. However, there are fewer reports on whether the expression of lignocellulase in *Ganoderma* is related to the expression of carbohydrate metabolic enzymes or the changes in carbohydrate components. *Ganoderma* triterpenoids are synthesized through the mevalonate pathway [24], and lanosterol undergoes oxidation, reduction, and acylation by various cytochrome oxidases to generate structurally distinct *Ganoderma* triterpenoids. However, studies on the downstream synthetic pathways of triterpenoids in *G. lingzhi* are still very scarce, and further research is needed on how the substrate degradation is associated with polysaccharide or terpene synthesis, as well as how these metabolic pathways are involved in the process of liquid fermentation change, while the relationship between the metabolism of these three components has also seldom been studied.

In addition, studies in fungi such as *G. ingzhi* have shown that terpene metabolism and glucose metabolism are also related to each other. Since terpene precursor synthesis starts with acetyl CoA, which is a core intermediate in carbon metabolism, the synthesis of terpene precursors can be promoted by adding acetyl CoA or regulating the expression of related metabolizing enzymes [25,26]. Meanwhile, various differences in triterpene content and the expression of triterpene precursor-metabolizing enzymes were found after inhibiting MAPKs or Skn7, transcription factors related to cell wall production and spore production in *Ganoderma lucidum* [27,28], suggesting that triterpene metabolism is related to the metabolic processes of butyric acid, polysaccharide synthesis, and spore production, and that there is an association between lignocellulase and sugar metabolism, as well as between terpene metabolism and sugar metabolism. However, the current research results are varied, and there is no clear consensus.

In *G. lingzhi*'s liquid fermentation with corn straw as a matrix, lignocellulose degradation-related genes play a key role, and the process of lignocellulose degradation involves the anabolism of polysaccharides and triterpenes. With cellulose enzymes as the breakthrough point, this study adopts Illumina sequencing technology to identify that cellulase enzyme activity is higher in the fermentation process of samples. Transcriptome sequencing using CAZyme analysis gene sequencing results are used to identify the differences in lignocellulose degradation-related genes and to determine the important genes involved in the process of the liquid fermentation of *G. lingzhi* with a corn stover substrate. Next, we determine the fermentation process of *G. lingzhi* polysaccharides in terpene synthesis involving a series of related genes and perform correlation analysis of *G. lingzhi* with liquid fermentation-related genes during the process of corn straw lignocellulose degradation. The change rules for polysaccharide and terpene synthesis metabolic enzyme gene expression provide new ideas and methods for the subsequent use of *G. lingzhi* in facilitating corn straw resource utilization. These results also provide an important reference for research into the mechanisms of other medicinal fungi in the process of corn straw liquid fermentation.

## 2. Materials and Methods

### 2.1. Fungal Strains and Culture Conditions

The strain LZ-8 of *G. lingzhi* was maintained at 4 °C in potato dextrose agar (PDA) tubes. Three mycelium-covered cakes with a diameter of 5 mm were pre-cultivated by inoculating them with potato dextrose broth (PDB) on PDA plates. Thereafter, the cakes were left undisturbed for 6 to 7 days at 25 °C, were agitated at 150 rpm for 6 days, and were shielded from light. After being transferred to 250 mL triangular flasks containing 90 mL of corn straw liquid medium (CSL, with 30 g/L corn straw powder with a particle size less than 0.2 mm), the 10 mL mycelial suspension in PDB was incubated at 25 °C under light protection and with 150 rpm oscillation. Periodic sampling was conducted with three replicates for each time point.

### 2.2. Enzyme Extraction Solution and Enzymatic Activity Assay

The crude enzyme solution was obtained by filtering the fermentation broth via a 0.45 mm sterile membrane. The activities of cellulase and xylanase were evaluated by adopting the DNS method [29,30], while the activities of laccase and manganese peroxidase were determined by using the ABTS and guaiacol methods, respectively [31,32].

### 2.3. RNA Extraction, Library Construction, and Sequencing

The mycelium cultured in CSL for 0 days was set as the control group CK, that cultured for 5 days was set as the treatment group LM, and that cultured for 12 days was set as the treatment group CH. After filtering the mycelial solution through three layers of gauze, the mycelium was rinsed with sterile water, promptly collected, and then stored in liquid nitrogen.

The total RNA of the mycelium was extracted by using Trizol reagent. Oligo(dT) magnetic beads were employed to enrich the mRNA with a polyA structure in the total RNA, and ion interruption was utilized to cut the RNA into fragments of 300 bp in length. The first strand of cDNA was generated with RNA as the template, 6-base random primers, and reverse transcriptase, while the second strand of cDNA was synthesized with the first strand as the template. After the library was constructed, PCR amplification was applied to enrich the library fragments, and the library was quality tested using an Agilent 2100 Bioanalyzer. Then, the total concentration and the effective concentration of the library were determined. Suzhou Panomic BioPharma Co., Ltd. (Suzhou, China) sequenced the libraries through paired-end (PE) next-generation sequencing (NGS) on the Illumina HiSeq sequencing platform.

### 2.4. Bioinformatics Analysis of RNA-Seq Data

By utilizing the synthesis sequencing (SBS) technology on the Illumina high-throughput platform to sequence the cDNA libraries, large quantities of high-quality raw data were generated. Subsequently, Trinity-v2.14.0 software was employed to assemble the clean reads obtained from the raw control data to yield transcripts, which were then analyzed [33]. After the assembly, the optimal alignment results were taken from the NCBI non-redundant protein sequences database (NR) [34], Evolutionary Genealogy of Genes: Non-supervised Orthologous Groups (eggNOG), Swiss-Prot [35], Pfam, Gene Ontology (GO), and the Kyoto Encyclopedia of Genes and Genomes (KEGG) [36] were used to annotate the Unigene gene functions.

Correlation analysis was carried out on the three repeated experiments to calculate the Pearson correlation coefficient in order to assess the reproducibility between samples. To display the expression levels of each transcript more precisely, normalization was conducted based on the size of the transcripts, and the number of fragments per kilobase of transcript per million mapped reads (FPKM) was used to measure the expression levels of a gene or transcript [37].

### 2.5. Analysis of CAZyme DEGs Related to the Decomposition of Corn Straw Lignocellulose

Differentially expressed genes (DEGs) were recognized with the criteria of a fold change (FC) $\geq$2 and a false discovery rate (FDR) $< 0.05$. GO enrichment analysis was carried out to identify significantly enriched GO functions among the DEGs, disclosing the potential functions of the differentially expressed genes in the samples. KEGG pathway analysis was executed to calculate the number of differentially expressed genes at different levels, determining the main pathways and signaling pathways induced by the DEGs.

The concatenated sets and samples of differential genes from all comparison groups were analyzed by means of bidirectional clustering, based on the expression levels of the same gene in different samples and the expression patterns of different genes in the same samples. The distances were calculated using the Euclidean method and hierarchical clustering was carried out using the method of the longest distance for hierarchical clustering (Complete Linkage). DESeq was utilized for differential gene expression analysis, with the criteria for selecting differentially expressed genes being as follows: $|\log_2\text{FoldChange}| > 1$ and a significant $p$-value $< 0.05$. By comparing the CKvsLM group and the LMvsCH group, the related genes of *G. lingzhi* for degrading corn straw lignocellulose were analyzed.

### 2.6. Analysis of Enzyme Genes Involved in Carbohydrate Metabolism and Triterpenoid Metabolism

In order to explore the expression alterations of enzyme genes in polysaccharide and triterpene anabolic pathways when *G. lingzhi* undergoes liquid fermentation with corn straw as the substrate, the expression of enzymes related to glycolytic metabolic pathways (EMP, HMP, TCA) and glycolytic synthesis pathways (mannitol, arabinitol, alginate, etc.) during the liquid fermentation was subjected to cluster analysis.

The mevalonate pathway, which is involved in triterpene metabolism, was analyzed by clustering gene expression during liquid fermentation.

### 2.7. Analysis of the Co-Regulation Mechanism of Triterpene Metabolism, Matrix Degradation, and Glucose Metabolism Genes

Gene expression data in the transcriptome were subjected to correlation cluster network analysis and correlation clustering analysis so as to analyze the correlation between matrix degradation and the polysaccharide and triterpene anabolism pathways during the liquid fermentation of *G. lingzhi* with corn straw as a substrate. This further resolved the regulatory mechanisms of matrix degradation in connection with the processes of triterpene anabolism and polysaccharide anabolism.

### 2.8. Real-Time Quantitative PCR Experiments

An HiScript III 1st Strand cDNA Synthesis Kit (+gDNA wiper) (Vazyme, Nanjing, China) was utilized to synthesize cDNA from the total RNA in line with the instructions for use in real-time quantitative PCR (RT-qPCR) analysis. The internal reference gene selected was 18S. In order to guarantee the accuracy of our RNA-Seq data, eight genes were picked from the transcriptome of *G. lingzhi* and their relative expression levels were normalized to that of 18S. The sequences of specific primers (Shenggong Biotech, Shanghai, China) that were synthesized are presented in Table 1.

**Table 1.** Primer sequences used in qRT-PCR.

| Gene ID | Description | Forward Primer Sequences (5′-3′) | Reverse Primer Sequences (5′-3′) |
|---|---|---|---|
| DN4778_c0_g1 | Endo-1,4-beta-xylanase 3 | TGAAGAACCTTGCTGCACTC | TCTCAAAGACACAAGTCGAGC |
| DN1646_c0_g1 | Endo-1,4-beta-xylanase A | GGGACTTCACCGACAAGTATTC | TGCCACGATACCGTTGTAC |
| DN5168_c0_g1 | Manganese dependent endoglucanase Eg5A | TGGTTCAGCAACTTCTACGAC | CAGACGCAAACTGGTCATTG |
| DN38_c0_g1 | Exoglucanase 1 | CTGGATGGTGCTGACTACG | AATATCTCGTACTTGGCGTCG |
| DN377_c2_g1 | Manganese transporter pdt1 | TCATCTTCGCACTCGCAC | GCCGATGAGTCTGGTGATAAT |
| DN3627_c0_g2 | Beta-xylanase | TTAACCAGCTCAACGGTCC | CACCAAAAGCGATGCAAGAG |
| DN682_c2_g1 | Laccase 1 | TCGTGGTCAATGGTGTCTTC | GTTCGTGCCCTTTTGGAAG |
| DN2037_c0_g1 | Laccase C | TCACTGCCACATTGACTGG | TGGGATCGTTTGCTATGGAC |

## 3. Results

### 3.1. Lignocellulose-Degrading Enzyme Activities of G. lingzhi

The activities of LACC and MnP reached their peaks on the fifth day of the liquid fermentation of *G. lingzhi* with corn straw as a substrate. Thereafter, they rapidly decreased and remained at a relatively low level (Figure 1C,D), mainly breaking down the surface lignin and uncovering cellulose and hemicellulose. The activity of the enzymes cellulase and xylanase, which mainly decompose cellulose and hemicellulose, increased with the duration of fermentation, peaked on the 13th day, and then declined (Figure 1A,B).

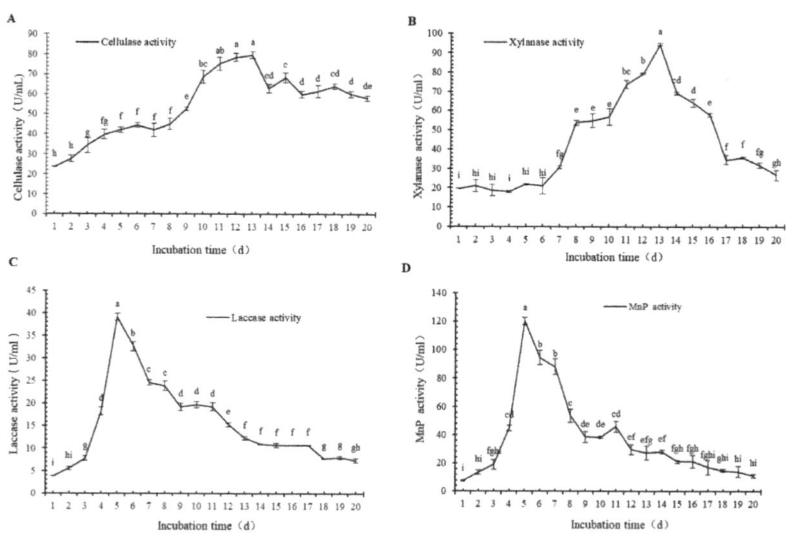


**Figure 1.** Lignocellulolytic enzyme activity of LZ-8. (**A**) Cellulase activity. (**B**) Xylanase activity. (**C**) Laccase activity. (**D**) Manganese peroxidase activity. Note: Each graph point represents the mean of three biological replicates SD ($p \leq 0.05$), the letters represent salience and the error bar represents the standard error.

With the characterization of cellulase's activity and xylanase's activity on cellulose and hemicellulose, the level of gene expression of laccase, and manganese peroxidase's live characterization of lignin, the level of gene expression of the combination of enzyme protein translation lags behind that of gene transcription. We selected the timing of the peak before sampling, choosing 5-day CSL cultures of mycelia as the treatment group LM. For *G. lingzhi* RNA-Seq, the mycelia cultured in CSL for 5 days were selected as the treatment group LM, the mycelia cultured in CSL for 12 days were selected as the treatment group CH, and the mycelia cultured in CSL for 0 days were selected as the control group (CK).

### 3.2. Functional Annotation of Novel Genes

Through screening 48.52, 43.96, and 49.25 million raw reads from the *G. lingzhi* transcriptome samples of the CK, CH, and LM groups, respectively, a total of 46.2, 41.94, and 47.6 million clean reads were obtained. There was no less than 423 Mb of clean reads in each cDNA library, and the percentage of clean reads and the Q30 minimum for the nine cDNA libraries were 94.96% and 94.05%, respectively (Table 2).

**Table 2.** Summary of the sequencing.

| Sample | Read No. | Clean Read No. | Clean Read % | N (%) | Q20 (%) | Q30 (%) |
|---|---|---|---|---|---|---|
| CK_1 | 49,716,670 | 47,483,608 | 95.5 | 0.001485 | 98.32 | 95.15 |
| CK_2 | 44,655,274 | 42,407,348 | 94.96 | 0.001512 | 98.32 | 95.23 |
| CK_3 | 51,196,280 | 48,724,776 | 95.17 | 0.001471 | 98.33 | 95.25 |
| LM_1 | 48,361,614 | 46,107,672 | 95.33 | 0.001457 | 98.3 | 95.15 |
| LM_2 | 44,591,096 | 42,515,008 | 95.34 | 0.001436 | 98.28 | 95.08 |
| LM_3 | 54,795,248 | 52,157,262 | 95.18 | 0.00147 | 98.32 | 95.17 |
| CH_1 | 43,475,624 | 41,499,254 | 95.45 | 0.00157 | 98.09 | 94.67 |
| CH_2 | 44,928,998 | 42,852,550 | 95.37 | 0.001836 | 97.85 | 94.05 |
| CH_3 | 43,479,448 | 41,455,396 | 95.34 | 0.001553 | 98.06 | 94.53 |

Note: Sample: sample name; Read No: Total number of reads; Clean Read No: read number of high-quality sequences; Clean Read %: the percentage of high-quality sequence reads in sequencing reads; N (%): the percentage of ambiguous bases; Q20 (%): the percentage of bases with a base recognition accuracy above 99%; Q30 (%): the percentage of bases with a base recognition accuracy above 99.9%.

From the statistics of the assembly findings, 81,677 overlapping-cluster constructed transcript sequences with an average length of 2044.92 bp and an N50 length of 2800 bp were obtained. To construct single-gene sequences, 18,157 transcript sequences with an average length of 1612.48 bp and an N50 length of 2661 bp were acquired. The single-gene sequences had a GC percentage of 56.65%, an N50 length of 2661 nt, and a fairly good assembly quality (as shown in Table 3).

**Table 3.** Assembly results of the *G. lingzhi* transcriptome.

| | Total Length (bp) | Sequence Number | Max. Length (bp) | Mean Length (bp) | N50 (bp) | N90 (bp) | GC% |
|---|---|---|---|---|---|---|---|
| Transcript | 167,022,854 | 81,677 | 15,429 | 2044.92 | 2800 | 1089 | 57.07 |
| Unigene | 29,277,836 | 18,157 | 15,429 | 1612.48 | 2661 | 654 | 56.65 |

As presented in Figure 2A, the correlation analysis of the three replicated biological experiment samples, which included the CK, LM, and CH groups, indicated that the correlation coefficients among the samples of each group ranged from 0.9 to 1, signifying that the experiment was repeatable. As demonstrated in Figure 2B, the gene expression levels of the nine samples were typically distributed—that is, the density at the two ends was lower and the density in the middle was higher, and the majority of the genes were concentrated in the range of $10^{-2}$ to $10^2$, with little dispersion among the sample groups and samples from the same period presenting the same expression level.

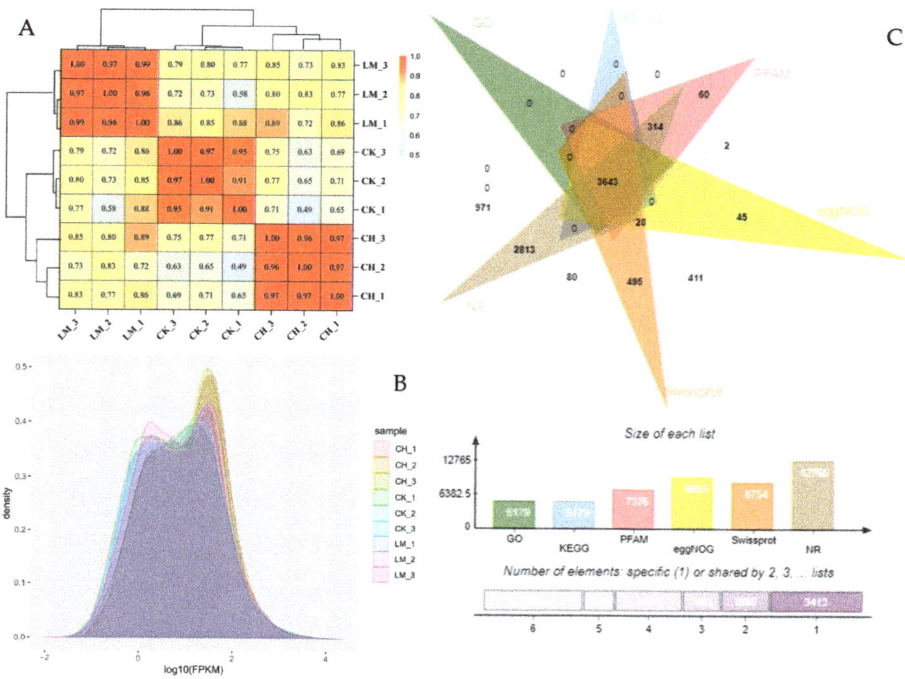


**Figure 2.** Annotation of functions and correlation plots of FPKM for each sample. (**A**) Correlation heatmap between samples. (**B**) FPKM box plot of each sample. (**C**) Number of single-gene interactions annotated by six databases.

Gene function annotation was carried out on Unigene, and a total of 12,765 single genes were annotated from 18,157 single genes in six databases. The numbers of single genes with significant sequence similarity to the GO, KEGG, PFAM, eggNOG, Swiss-Prot, and NR databases were 5179 (40.6%), 5179 (40.6%), 7376 (57.8%), 9655 (75.6%), 8754 (68.6%), and 12,765 (100%), respectively, and the number of interactions shared by single genes with different databases is presented in Figure 2C.

### 3.3. Differential Expression Analysis

To gain an in-depth understanding of the gene expression variations of *G. lingzhi* during the liquid fermentation of corn straw, we compared the DEGs from the *G. lingzhi* transcriptome samples of the CH, CK, and LM groups. In Figure 3A, 1793 genes were differentially expressed ($p < 0.05$) between the CK and LM groups, with 758 genes being up-regulated and 1035 genes being down-regulated. Likewise, in Figure 3B, a total of 1375 genes were differentially expressed ($p < 0.05$) across the LM and CH groups, with 484 being up-regulated and 891 being down-regulated. The number of shared differentially expressed genes between the two groups was computed by utilizing the differential expression values from the samples, as depicted in Figure 3C. There were 1793 and 1375 differentially expressed genes in the CK_vs._LM and LM_vs._CH groups, respectively, amounting to a total of 230 genes between CK_vs._LM and LM_vs._CH.

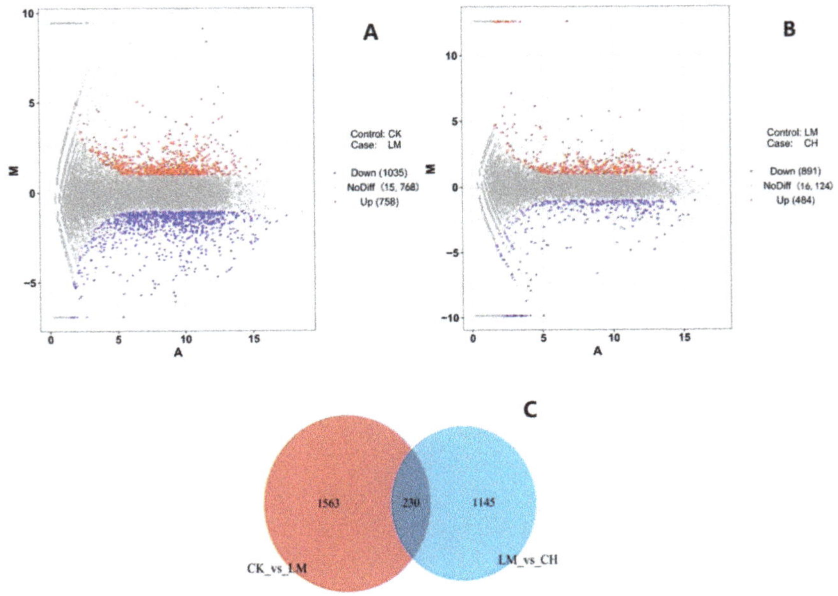


**Figure 3.** Correlation plots of DEGs. (**A**) MA map of differentially expressed genes (CK_vs._LM). (**B**) MA map of differentially expressed genes (LM_vs._CH). (**C**) Shared unique differential genes among groups, Red for CK_vs._LM, blue for LM_vs._CH.

Differentially expressed genes (DEGs) carried out various functions during the liquid fermentation of *G. lingzhi* with corn straw as the substrate, and the DEGs were analyzed through GO enrichment analysis for biological process (BP), cellular composition (CC), and molecular function. GO enrichment analysis divided the CK_vs._LM DEGs into 14 related to cellular components, 106 related to molecular functions, and 44 related to biological processes. Figure 4A presents a histogram of the top ten GO-enriched items with the lowest *p*-value after categorizing the GO items, which were associated with catalytic activity, amino acid synthesis, cellular polysaccharide catabolic processes, oxidoreductase activity, and cellulose metabolism. Additionally, the DEGs of LM_vs._CH were divided into 26 related to cellular components, 95 related to molecular functions, and 101 related to biological processes. Figure 4B presents a histogram of the top ten GO-enriched items with the lowest *p*-values after categorizing the GO items, which were related to ribosomal structural components, structural molecular activities, cytoplasmic translation, and peptide biosynthetic processes. The majority of the molecular functions mentioned above are connected to the process of lignocellulose breakdown.

DEGs were analyzed using the KEGG program. The KEGG database was employed to classify and count the different genes in the three treatment groups (PDR $< 0.05$). Then, the screened DEGs were annotated using KEGG pathways. KEGG enrichment bubble plots were generated using the top pathways with the lowest PDR values, with the vertical coordinates representing the pathways and the horizontal coordinates representing the enrichment factor. The CK_vs._LM comparisons indicated that the main metabolic pathways associated with the differentially expressed genes included starch and sucrose metabolism, glutathione metabolism, cyanoamino acid metabolism, and alanine–aspartic acid and glutamate metabolism (Figure 4C).

The LM_vs._CH analysis revealed that the primary pathways related to the differentially expressed genes were oxidative phosphorylation, glyoxylate and dicarboxylic acid metabolism, the citric acid cycle, methane metabolism, and so on (Figure 4D).

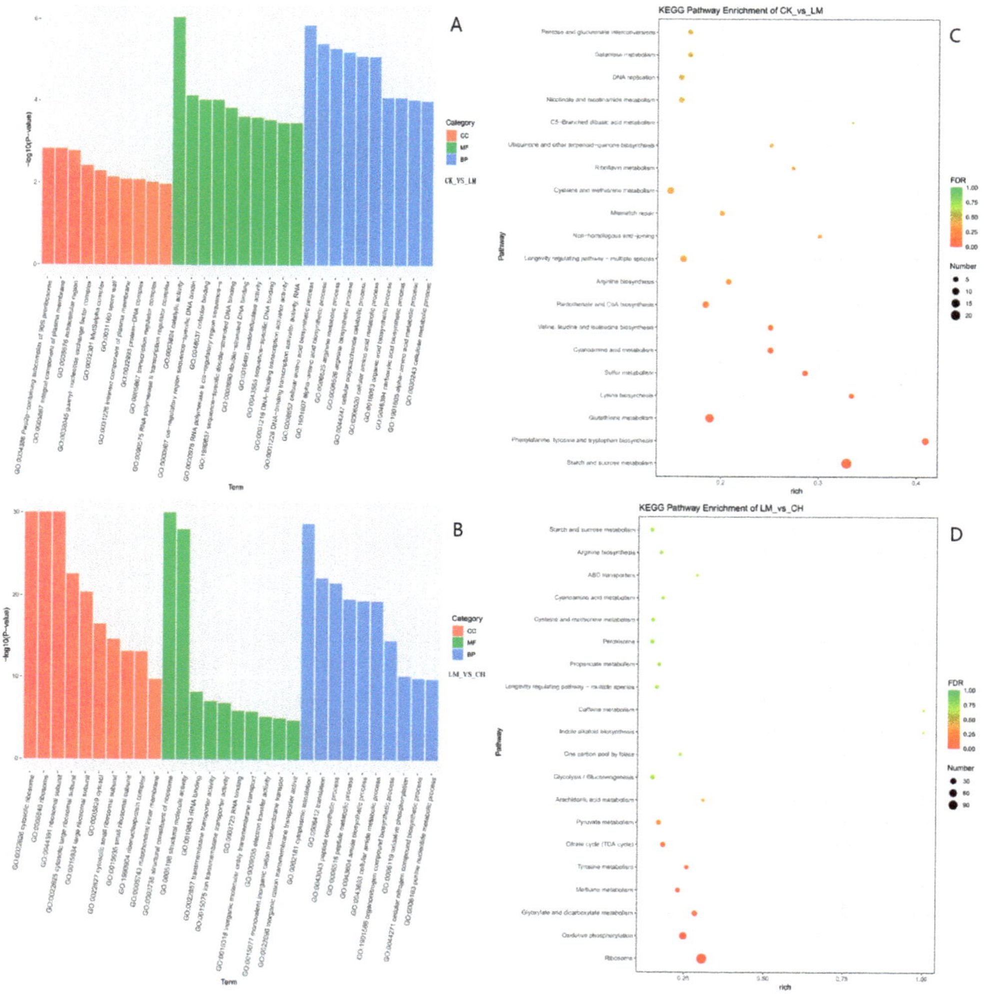


**Figure 4.** GO and KEGG analyses of the differentially expressed genes in the collected hyphae of *G. lingzhi* following different degradation treatments. (**A**,**B**) Quantitative histograms of GO gene enrichment for CK_vs._LM and LM_vs._CH. MF: molecular function; BP: biological process; CC: cell composition. The horizontal axis represents the GO term, and the vertical axis represents −log10 (*p*-value) enriched by the GO term. (**C**,**D**) KEGG enrichment bubble charts. The horizontal axis is the enrichment factor (expressed as the ratio of differentially expressed genes annotated to the pathway to the total number of expressed genes annotated to the pathway). The top pathways with the lowest *p*-values were used to produce the map, where the ordinate represents the pathway; the abscissa represents the enrichment factor (the number of differences in the pathway is divided by all the numbers); and the circle size indicates the number, where the redder the color, the smaller the *p*-value. More differentially expressed genes are enriched in the pathways with redder and bigger bubbles.

### *3.4. Genes Encoding Putative CAZymes Related to the Decomposition of Corn Straw Lignocellulose*

The expression of lignocellulase genes in *G. lingzhi* was analyzed through clustering (Figure 5). Some genes were relatively highly expressed on days 0 and 5 of liquid fermentation, including two BGLs, one XYN, one MnP, one LAC, two GMCs, and one AOX. Some genes were relatively highly expressed on days 5 and 12 of liquid fermentation, including two EGLs, one BGL, one XYL, one AGL, three LACs, one GMC, and one VP. Clustering was utilized to identify the main lignocellulase genes expressed during *G. lingzhi* liquid fermentation with corn straw as the substrate, as well as the lignocellulase genes with significant expression changes at specific stages. Some genes were overexpressed on days 0 and 5 of liquid fermentation, such as two BGLs, one XYN, one MnP, one LAC, two GMCs, and one AOX. On days 5 and 12 of liquid fermentation, certain genes were particularly significantly expressed, including two EGLs, one BGL, one XYL, one AGL, three LACs, one GMC, and one VP. Between days 5 and 12 of liquid fermentation, a larger number of genes, including two EGLs, one BGL, one XYL, one AGL, three LACs, one GMC, and one VP, were relatively significantly expressed. During days 5 and 12 of liquid fermentation, a greater number of genes were expressed. The notable genes expressed included seven BGLs, two LPMOs, two CBHs, four XYNs, two XYLs, one AXE, two GMCs, one LAC, one CDH, and one MnP. On day 0, liquid fermentation was manifested at a fairly high level. On day 5 of liquid fermentation, the genes with the greatest expression included three EGLs, two BGLs, two CBHs, two XYLs, one AGL, and seven LACs. After 12 days of liquid fermentation, the following genes showed greater expression: one EGL, two BGLs, two CBHs, one XYN, two AGLs, seven LACs, two MnPs, two VPs, three AAOs, and one AOX. In conclusion, hemicellulase genes expressed more of the enzymes involved in the degradation of xylan, especially during the first five to twelve days of liquid fermentation. Among the lignin enzymes, laccase expressed more than manganese peroxidase, acting as the predominant enzyme system for lignin degradation in G. *lingzhi*, and the peak expression of these enzymes mostly occurred on day 5. Cellulase expression gradually increased from the beginning of day 0. On day 12, the expression of most of the enzymes reached the highest level. The changes in ligninase and cellulase gene expression followed the trend of the corresponding changes in enzyme activity.

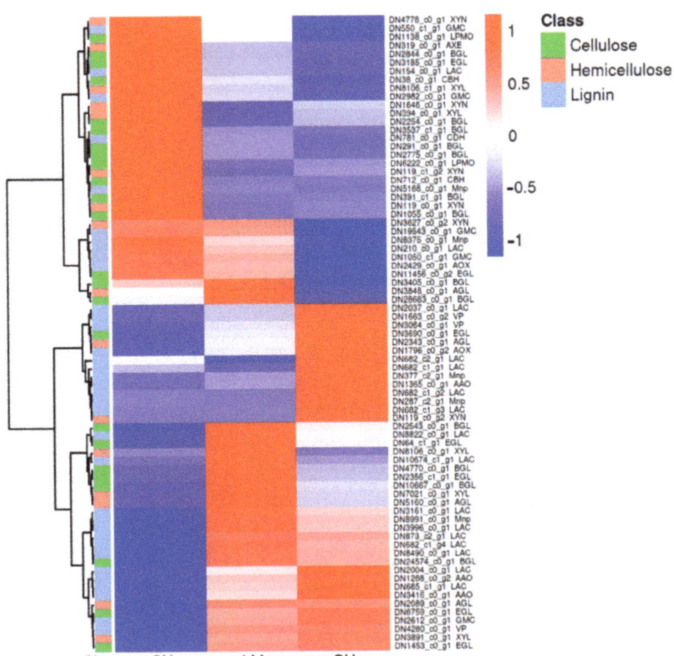


**Figure 5.** Differential expression of genes (DEGs) encoding CAZymes related to the decomposition of corn straw lignocellulose biomass in the *G. lingzhi* trancriptome.

Some example genes are provided to illustrate the specific expression patterns. For instance, two BGL genes were highly expressed on days 0 and 5, while three EGL genes had significant expression on day 5. This shows the dynamic nature of gene expression during different stages of the fermentation process. Additionally, the consideration of the predominant enzyme systems like laccase and manganese peroxidase helps us to understand the specific roles of these enzymes in lignin degradation. The description of the gradual increase in cellulase expression over time provides insights into the progression of cellulose breakdown during the liquid fermentation.

This study utilized the carbohydrate-active enzymes gene database (www.cazy.org, accessed on 11 June 2023) to analyze the differentially expressed genes (DEGs) among *G. lingzhi* transcriptome samples from the CK, LM, and CH groups in order to mine genes related to the degradation of corn straw by *G. lingzhi*. Through significant difference analysis of the 1793 DEGs obtained from the CK_vs._LM group, a total of 432 genes associated with the liquid fermentation of corn straw in *G. lingzhi* were obtained, and from the 1375 DEGs of the LM_vs._CH group, 412 such genes were identified.

In the *G. lingzhi* transcriptome, CK_vs._LM identified a total of seventeen CAZymes related to cellulose biodegradation, including four endoglucanases (EGLs), two cellobiohydrolases (CBHs), ten β-glucosidases (BGLs), and two lytic polysaccharide monooxygenases (LPMOs). LM_vs._CH identified seven such CAZymes, including three EGLs, one CBH, two BGLs, and one LPMO (Figure 6), indicating that *G. lingzhi* can produce multiple cellulases that work together to degrade corn straw cellulose. Some genes were down-regulated during liquid fermentation from 0 to 5 days (CK_vs._LM group treatment), but their expression was not obvious during liquid fermentation from 5 to 12 days (LM_vs._CH treatment group). Among them, DN10667_c0_g1 (GH2) was only expressed in the LM_vs._CH treatment group compared to other genes encoding BGLs, such as GH3 and GH1. DN3690_c0_g1 (GH5_15) was up-regulated in both phases, DN38_c0_g1 (GH7) was up-regulated first and then down-regulated, DN6222_c0_g1 (AA9) was down-regulated first and then up-regulated, and several other genes that were significantly expressed in both phases were down-regulated. These results suggest that EGL and LPMO play more important roles in the catabolism of corn straw cellulose than CBH and BGL, with the GH5_15 family of EGLs and the AA9 family of LPMOs being particularly significant.

For example, in other studies on microbial cellulose degradation, similar findings have been reported where certain families of enzymes like the EGLs and LPMOs have shown enhanced activity and importance [38,39]. This study provides valuable insights into the specific enzymes and their roles in the degradation process of *G. lingzhi* on corn straw, which can guide further studies and potential applications in the field of biomass utilization.

In CK_vs._LM, a total of nine CAZymes related to hemicellulose biodegradation were identified, encompassing four xylanases (XYNs), three xylosidases (XYLs), and two arabinogalactan lyases (AGLs). In LMvsCH, four CAZymes associated with cellulose biodegradation were detected, including three XYNs and one XYL (Figure 6). Among these genes, the single genes DN3627_c0_g2 and DN4778_c0_g1 encoding an XYN (GH10) showed up-regulated expression at both stages. The single gene DN1646_c0_g1 encoding an XYN (GH11) was expressed first in a down-regulated manner and then was up-regulated at both stages, while the single gene DN8106_c1_g1 encoding an XYL (GH5_22) had down-regulated expression at both stages, and these CAZymes might play a more prominent role in the catabolism of corn straw hemicellulose.

A total of seventeen CAZymes were found in CK_vs._LM, including thirteen ligninolytic oxidases (LOs) and four lignin degradation-associated proteins (LDAs) for degrading lignin (Figure 6). The thirteen LOs comprised ten laccase copper oxidases (LACCs) and three manganese peroxidases (MnPs). The four LDAs included two aryl-alcohol oxidases (AAOs) and one cellobiose dehydrogenase (CDH). In LM_vs._CH, fourteen CAZymes were discovered, including twelve LOs and two LDAs for degrading lignin. The thirteen LOs included six LACCs, five MnPs, and one versatile peroxidase (VP). The four LDAs consisted of two AAOs.

| Log2 Ratio LM/CK | Log2 Ratio CH/LM | Gene ID | CAZy family | Function annotation | EC No. | Substrates |
|---|---|---|---|---|---|---|
| −1.7 | −1.86 | TRINITY_DN3185_c0_g1 | GH5_5 | EGL | EC3.2.1.4 | Cellulose |
| 2.06 | 1.37 | TRINITY_DN3690_c0_g1 | GH5_15 | EGL | | |
| −1.34 | −1.17 | TRINITY_DN64_c1_g1 | GH12 | EGL | | |
| −1.61 | | TRINITY_DN2356_c1_g1 | GH5_5 | EGL | | |
| 1.92 | −1.42 | TRINITY_DN38_c0_g1 | GH7 | CBH | EC3.2.1.91 | Cellulose |
| −4.41 | | TRINITY_DN712_c0_g1 | GH7 | CBH | | |
| −1.95 | −2.25 | TRINITY_DN2844_c0_g1 | GH3 | BGL | EC3.2.1.21 | Cellulose |
| −1.21 | | TRINITY_DN291_c0_g1 | GH1 | BGL | | |
| −2.72 | | TRINITY_DN3537_c1_g1 | GH1 | BGL | | |
| 1.3 | | TRINITY_DN2543_c0_g1 | GH1 | BGL | | |
| −1.41 | | TRINITY_DN1055_c0_g1 | GH3 | BGL | | |
| −1.11 | | TRINITY_DN391_c1_g1 | GH3 | BGL | | |
| −1.53 | | TRINITY_DN2254_c0_g1 | GH3 | BGL | | |
| −2.14 | | TRINITY_DN2775_c0_g1 | GH3 | BGL | | |
| 1.25 | | TRINITY_DN28683_c0_g1 | CE3 | BGL | | |
| | −1.58 | TRINITY_DN10667_c0_g1 | GH2 | BGL | | |
| −4.63 | 1.44 | TRINITY_DN6222_c0_g1 | AA9 | LPMO | | Cellulose |
| −1.05 | | TRINITY_DN1138_c0_g1 | AA9 | LPMO | | |
| 2.96 | 2.08 | TRINITY_DN3627_c0_g2 | GH10 | XYN | EC3.2.1.8 | Hemicellulose |
| 2.62 | 2.46 | TRINITY_DN4778_c0_g1 | GH10 | XYN | | |
| −3.48 | 1.84 | TRINITY_DN1646_c0_g1 | GH11 | XYN | | |
| −1.12 | | TRINITY_DN119_c0_g1 | GH10 | XYN | | |
| −1.15 | −1.13 | TRINITY_DN8106_c1_g1 | GH5_22 | XYL | EC3.2.1.37 | Hemicellulose |
| 1.2 | | TRINITY_DN3891_c0_g1 | GH3 | XYL | | |
| 1.14 | | TRINITY_DN7021_c0_g1 | GH3 | XYL | | |
| 2.27 | | TRINITY_DN2089_c0_g1 | GH27 | AGL | EC3.2.1.22 | Hemicellulose |
| 2.49 | | TRINITY_DN5160_c0_g1 | GH27 | AGL | | |
| 3.42 | 3.7 | TRINITY_DN2037_c0_g1 | AA1_1 | LACC | EC1.10.3.2 | Lignin |
| −1.09 | 2.53 | TRINITY_DN682_c1_g1 | AA1_1 | LACC | | |
| −2.17 | 2.84 | TRINITY_DN682_c2_g1 | AA1_1 | LACC | | |
| −1.15 | | TRINITY_DN154_c0_g1 | AA1_1 | LACC | | |
| 1.34 | | TRINITY_DN3996_c0_g1 | AA1_1 | LACC | | |
| 1.3 | | TRINITY_DN682_c1_g4 | AA1_1 | LACC | | |
| 1.22 | | TRINITY_DN8490_c0_g1 | AA1_1 | LACC | | |
| 1.87 | | TRINITY_DN8822_c0_g1 | AA1_1 | LACC | | |
| 1.4 | | TRINITY_DN3161_c0_g1 | GH35 | LACC | | |
| 1.59 | | TRINITY_DN873_c2_g1 | GH35 | LACC | | |
| | −3.29 | TRINITY_DN10674_c1_g1 | AA1_1 | LACC | | |
| | 5.27 | TRINITY_DN682_c1_g2 | AA1_1 | LACC | | |
| | 1.96 | TRINITY_DN682_c1_g3 | AA1 | LACC | | |
| | 5.82 | TRINITY_DN287_c2_g1 | AA2_3 | VP | EC1.11.1.16 | Lignin |
| −1.56 | | TRINITY_DN781_c0_g1 | AA3_1 | CDH | EC1.1.99.18 | Lignin |
| 1.25 | 1.08 | TRINITY_DN1268_c0_g2 | AA3_2 | AAO | EC1.1.3.7 | Lignin |
| 1.43 | | TRINITY_DN3416_c0_g1 | AA3_2 | AAO | | |
| | 1.54 | TRINITY_DN1365_c0_g1 | AA3_2 | AAO | | |
| 1.47 | | TRINITY_DN1796_c0_g2 | AA3_3 | AOX | EC1.1.3.13 | Lignin |
| 3.09 | 3.5 | TRINITY_DN377_c2_g1 | AA2-2 | MnP | EC1.11.1.13 | Lignin |
| −2.63 | 2.67 | TRINITY_DN5168_c0_g1 | AA2-2 | MnP | | |
| 0.61 | −0.22 | TRINITY_DN8991_c0_g1 | AA2 | MnP | | |
| | 5.82 | TRINITY_DN287_c2_g1 | AA2 | MnP | | |
| | −1.34 | TRINITY_DN8375_c0_g1 | AA2 | MnP | | |

**Figure 6.** Expression of putative CAZyme DEGs related to the decomposition of corn straw lignocellulose in the *G. lingzhi* transcriptome.

Most of the encoded genes, with the exception of some genes encoding CDH, were up-regulated in the LM transcriptome samples compared to the CH transcriptome samples. The results suggest that the breakdown of lignin with LOs is still crucial for utilizing corn straw polysaccharide biomass. Some of the genes exhibited a high fold change, such as DN682_c1_g2 encoding LACs, DN287_c2_g1 encoding MnP gene, and DN287_c2_g1, a single gene encoding VPs.

Important related genes, including DN3690_c0_g1 (EGL), DN3627_c0_g2 (XYN), DN4778_c0_g1 (XYN), DN2037_c0_g1 (LACC), and DN277_c2_g1 (MnP), were identified based on the differential expression levels.

### 3.5. Expression Cluster Analysis of Enzyme Genes Involved in Carbohydrate and Triterpene Metabolism

We conducted a cluster analysis of the expression of enzyme genes involved in the fundamental anabolic pathways in order to comprehensively examine the expression alterations of these genes in the polysaccharide and triterpene anabolic pathways during the liquid fermentation of *G. lingzhi* with corn straw. According to the findings of these studies, there were fluctuations in the expression levels of enzyme genes related to glycolytic metabolism, including mannitol, galactose, rhamnose, alginose, and arabinose, as well as in the triterpene anabolic pathway during the fermentation process (Figure 7A).

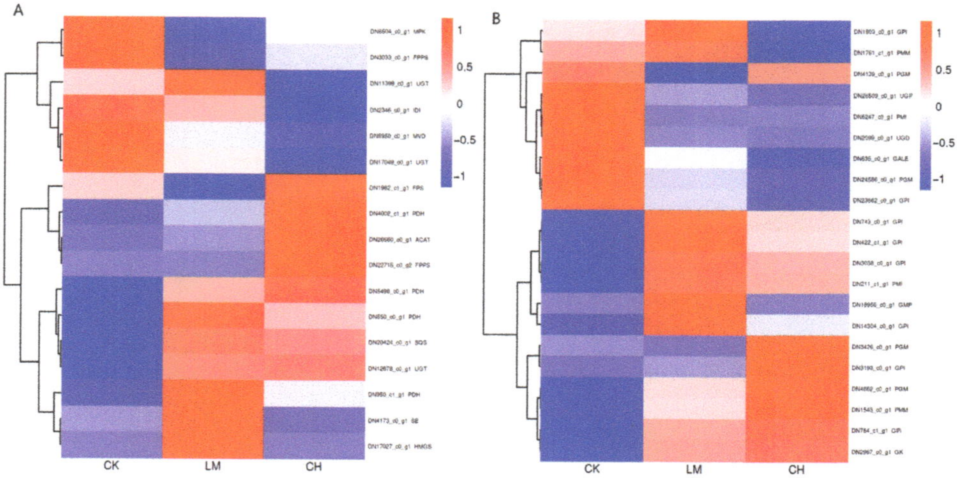


**Figure 7.** Cluster analysis of carbohydrate metabolism and triterpene metabolism gene expression in liquid fermented mycelium of *G. lingzhi*. (**A**) Polysaccharides. (**B**) Triterpenes.

Firstly, during the fermentation process, a large amount of glucose-6-phosphatic was produced via GK (DN2967_c0_g1), which is a necessity for the synthesis of numerous sugars. Secondly, eight genes with similar overall expression trends controlled GPI, an enzyme crucial for the synthesis of mannitol and alginate precursors. Additionally, the mannitol anabolism-related genes exhibited high overall expression levels that persisted throughout the development stage. Notably, one GALE gene (DN18956_c0_g1) peaked in expression on the fifth day of fermentation. The anabolism of arabinose, galactose, and rhamnose, as well as the transformation of glucose-6-phosphatic acid into glucose-1-phosphate, was regulated by a group of four PGM genes. Among the enzyme genes involved in arabinose anabolism, one UGD (DN2099_c0_g1) and one UGP (DN26509_c0_g1) showed a down-regulated expression. The arabinose-related genes and the enzyme genes involved in the galactose anabolic pathway presented similar expression change tendencies, but the expression of GALE 1 (DN635_c0_g1) was down-regulated. Furthermore, during liquid fermentation, the genes encoding the rhamnose anabolism-related enzymes RFFH and RHS, as well as the alginose anabolism-related enzymes GMDH and GFS, showed decreased expression, though the differences were not very significant. In conclusion, due to sugar catabolism during the liquid fermentation of *G. lingzhi* using corn straw as the substrate, there may be a bias towards mannitol anabolism.

Both on the fifth day and the twelfth day of fermentation, there was an increase in the expression of enzyme genes in the *G. lingzhi* mycelium involved in the metabolism of triterpene precursor substances (Figure 7B). Among these genes, HMGS, MPK, MVD, IDI, and ACAT (DN26660_c0_g1) had higher expression levels. The three genes ACAT (DN26660_c0_g1), MPK (DN6504_c0_g1), and MVD (DN5950_c0_g1) are involved in the conversion of acetyl CoA to acetoacetyl CoA, phosphomesylate to mevalonate, and isopentenyl pyrophosphate to mevalonate pyrophosphate. These genes may be the key enzyme genes involved in the synthesis of triterpenoid compound precursors.

### *3.6. Co-Regulation of Triterpene Metabolism and Matrix Degradation and Glucose Metabolism Genes*

We found that the red, blue, and green modules encompassed the majority of the genes associated with matrix degradation, sugar metabolism, and triterpene alterations. (Figure 8A,B). The production of acetoacetyl CoA, the breakdown of lignocellulose, and the generation of the polysaccharide precursors glucose-6-phosphate and mannose-6-phosphate were all related to changes in gene expression within the red module. The acetoacetyl CoA synthesis gene PDH (DN4002_c1_g1) was positively correlated with one EGL (DN3690_c0_g1), one VP (DN287_c2_g1), and one AOX (DN1796_c0_g2) among the genes with highly significant levels ($p \leq 0.001$) (Figure 8C,D). In contrast, the other triterpene gene UGT (DN17049_c0_g1) had a correlation that was opposite to that mentioned above for the lignocellulosic genes. This indicates that the genes involved in lignocellulose degradation are crucial for the synthesis of acetoacetyl CoA, and that interactions with genes such as AOX, VP, and EGL might be part of the regulatory mechanism. The hemicellulose degradation-associated gene AGL (DN5160_c0_g1) and the lignin degradation-associated gene LAC (DN10674_c1_g1) had a positive correlation with each other, as did the polysaccharide synthesis precursor gene GPI (DN14304_c0_g1) with GMP (DN18956_c0_g1).

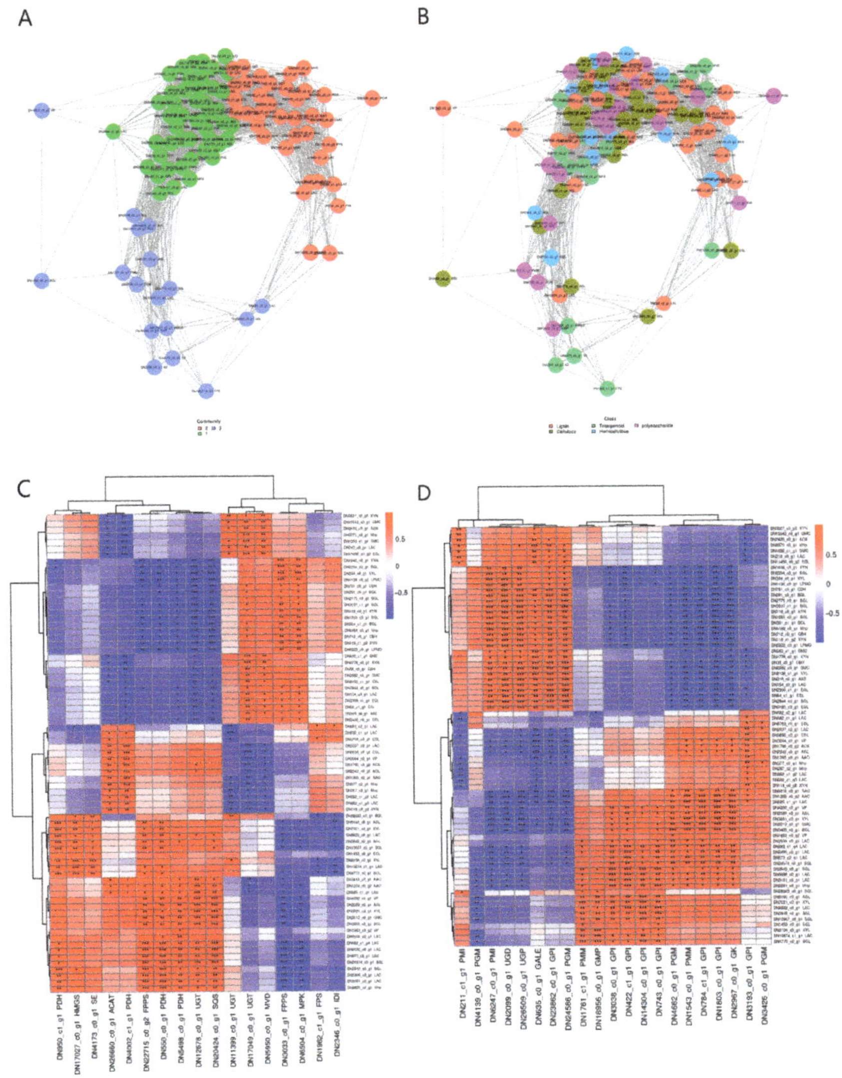

Figure 8. Correlation between genes. (A) Correlation cluster network (module). (B) Correlation cluster network (classification). (C) Heat map of association between matrix degradation and triterpene metabolism. (D) Heat map of association between matrix degradation and polysaccharide metabolism. $p$ significance level: *, <0.05; **, <0.01; ***, <0.001.

The degradation of lignocellulose, the formation of arabinose, and the conversion of acetyl CoA and a certain pyrophosphate to squalene were all associated with changes in gene expression in the green module. The hemifiber degradation genes LPMO (DN1138_c0_g1) and BGL (DN2254_c0_g1) were positively correlated with FPPS (DN3033_c0_g1); the genes SQS (DN20424_c0_g1) and PDH (DN550_c0_g1) involved in the conversion of acetyl CoA and that of pyrophosphate to squalene were positively correlated with two BGL and three LAC genes. The arabinose synthesis genes UGD (DN2099_c0_g1) and UGP (DN26509_c0_g1) were positively correlated with cellulose degradation-related genes, including five BGLs, two EGLs, two LPMOs, and one CBH (DN712_c0_g1); GK (DN2967_c0_g1) was positively correlated with lignin degradation genes, namely an AAO (DN1268_ c0_g2) and a VP (DN287_c2_g); and PGM (DN4662_c0_g1) was positively correlated with a cellulose degradation gene, a BGL (DN3405_c0_g1).

The gene expression alterations in the blue module were related to lignocellulose degradation, mannitol synthesis, and the conversion of acetoacetyl CoA to 2,3-oxidosqualene production. LAC (DN10674_c1_g1) had a favorable correlation with HMGS (DN17027_c0_g1), which converts acetoacetyl CoA to hydroxymethylglutaryl CoA. Moreover, the genes encoding GPI, GMP, and PMM in the mannitol synthesis process were favorably correlated with four BGL (DN2543_c0_g1), two AGL, and one LAC gene involved in lignocellulose breakdown.

### 3.7. Validating RNA-Seq Results via RT-qPCR

To validate the *G. lingzhi* transcriptome in our investigation, RT-qPCR analysis was performed on eight DEGs (Table 1). The gene expression information gleaned from the *G. lingzhi* transcriptome and the RT-qPCR results agreed rather well (Figure 9). This suggests that the RNA-Seq dataset is a trustworthy source.

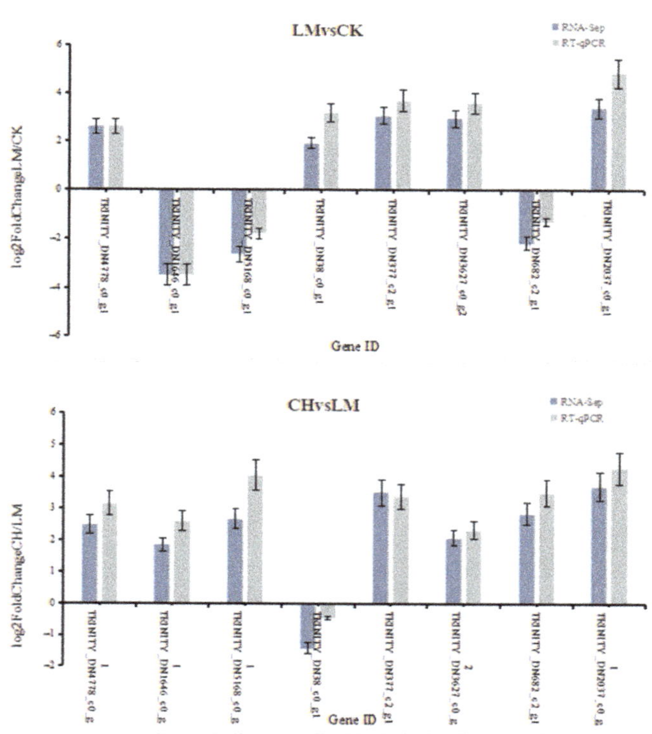


**Figure 9.** Validation of the RNA−Seq of 8 selected DEGs in the *G. lingzhi* transcriptome by RT−qPCR.

## 4. Discussion

The current investigations of *G. lingzhi* liquid fermentation mainly aimed to optimize fermentation conditions and the selection of productive strains, with a high-yield of biologically active metabolites (e.g., laccases, polysaccharides, triterpenes, etc.) through liquid fermentation [40–42]. Cui [43] discovered that malt wort could substitute the commonly used glucose as the sole carbon source. Studies related to liquid fermentation with corn straw as the substrate have also been carried out using different fungi in terms of enzyme activity and biofermentation related to lignocellulose degradation, such as the study by Wang Mengmeng [44] on the optimization of conditions for the production of xylanase through the liquid fermentation of striga siderophores and that by Chen Qingsen [45] on the improvement of cellulose utilization in a fermentation broth by using corn straw as a raw material for multi-strain mixed fermentation. In this study, we identified the transcript samples through the changes in lignocellulases during liquid culture, and then identified the related genes in *G. lingzhi* liquid fermentation with corn straw as the substrate through the transcriptional changes, and found that the process of corn straw lignocellulose degradation involves the cooperation of a variety of enzyme genes [36], providing information on *G. lingzhi* liquid fermentation in the utilization of corn straw at the molecular level.

Eleven and ten CAZymes were found in the transcriptomes of CK_vs._LM and LM_vs._CH, respectively, which are crucial for the exploitation of corn straw. Some genes exhibited high fold changes, such as DN3690_c0_g1 (EGL), DN3627_c0_g2 (XYN), DN4778_c0_g1 (XYN), DN2037_c0_g1 (LACC), and DN277_c2_g1 (MnP), which may play more significant roles in the synergistic decomposition of corn straw lignocellulose. This is similar to the CAZymes gene expression results of *Pleurotus ostreatus* on corn straw [6]. In summary, the results of this study show that *G. lingzhi* degraded maize stover, expressing degradation genes related to cellulose, hemicellulose, and lignin, similarly to the action of white rot fungus siderophores on maize stover [38], with the family of GHs still being the mainstay. Meanwhile, similar results were obtained from a study of 22 transcriptome datasets from 10 *basidiomycete* species [46].

LPMOs can break down components other than cellulose in plant cells by oxidizing the carbon of the sugar ring, resulting in chain scission, which enhances the activity of the classical GHs by introducing new chain ends [47] and is a key factor in the decomposition of lignocellulose [39,48]. VPs are relatively rarer [21], with dual LiP and MnP activities that can degrade phenolic and non-phenolic substrates of lignin [49,50]. In this experiment, DN6222_c0_g1 (LPMO) had a remarkable expression during fermentation, and the single gene encoding VPs, DN287_c2_g1, showed a high fold change, and LPMOs and VPs may act as very important factors in the degradation of the lignocellulose decomposition of corn straw by *G. lingzhi*.

Mannitol can be produced by fungi through the utilization of carbohydrates. Smiley disclosed that *Aspergillus candidus* can synthesize mannitol from glucose, and a further study revealed that other carbon sources can also be combined with glucose to generate mannitol. The majority of fungi have the ability to produce mannitol, but their volumetric yield is relatively low when glucose is used as a substrate, and they also consume the generated mannitol [3,4]. The high expression level of GALE (DN18956_c0_g1) and other alterations in the expression levels of genes related to polysaccharide anabolism in this study were more conducive to mannitol anabolism, and further research on these genes may be crucial for enhancing substrate utilization and producing more mannitol. It is generally acknowledged that the MVA pathway is the mechanism by which the common precursor of terpenoids, lanosterol, is synthesized during the upstream metabolism of terpenoids in *Ganoderma* triterpenoids. ACAT (DN26660_c0_g1), which participates in the conversion of acetyl-CoA to acetoacetyl-CoA, and MPK (DN6504_c0_g1), which participates in the conversion of mevalonate to phosphomelane, were found to have high expression among the genes of individual metabolism enzymes involved in this metabolic pathway according to the results of the current study. Higher expression levels were observed for DN6504_c0_g1 and

MVD (DN5950_c0_g1), which are involved in the conversion of mevalonate pyrophosphate to isopentenyl pyrophosphate [51–53].

Higher fungi growth is induced by expressing interrelated genes for lignocellulose-degrading and polysaccharide-metabolizing enzymes. These gene expressions can be controlled to modulate the expression of intracellular sugar transporters, glycolytic metabolizing enzymes, and chitinases [54]. Moreover, research on fungi such as *G. lingzhi* has demonstrated a connection between the metabolism of glucose and terpenes. Since acetyl CoA is the starting point for terpene precursor synthesis and is a key intermediate in carbon metabolism, terpene precursor synthesis can be accelerated by increasing acetyl CoA or controlling the expression of related metabolic enzymes [25,26]. Alternatively, terpene precursor synthesis can be accelerated by inhibiting MAPKs, transcription factors related to the formation of cell walls and the production of spores in *G. lingzhi*. It was discovered that the production of chitin and β-glucan in the mycelium was reduced along with the triterpene content and the expression of triterpene precursor-metabolizing enzymes, along with transcription factors such as MAPKs and Skn7 in *G. lucidum* that are associated with cell wall generation and spore production [27,28]. However, these findings of previous investigations are fragmented, making it difficult to correlate metabolic processes, such as matrix breakdown, sugars, and terpenes. In this study, we focused on the analysis of the key genes of three carbon metabolism pathways during liquid fermentation and found that cellulose-degrading enzymes EGLs and lignin-degrading enzymes VPs and AOXs were positively correlated with the acetyl CoA convertase gene PDH (DN4002_c1_g1) and negatively correlated with UGT (DN17049_c0_g1), and that BGLs and LACs were involved in the conversion of acetyl CoA and that of pyrophosphate to squalene. Meanwhile, the arabinose synthesis genes UGD (DN2099_c0_g1), UGP (DN26509_c0_g1), PGM (DN4662_c0_g1), and GK (DN2967_c0_g1) were positively correlated with hemicellulose degradation-related genes, and laccases had an influence on HMGS expression and mannitol synthesis. The polysaccharide synthesis precursor gene GPI (DN14304_c0_g1) was positively correlated with the hemicellulose degradation-related gene AGL and the lignin degradation-related LAC gene, FPPS was correlated with hemicellulose expression. One way to assess the relationships between lignocellulases and sugar metabolism and terpene metabolism and sugar metabolism is to perform a comprehensive examination of metabolic processes such as matrix degradation, sugar metabolism, and terpene metabolism.

**Author Contributions:** Conceptualization, C.S., M.Y. and S.W. (Sheng Wang); methodology, S.W. (Sheng Wang); validation, S.W. (Sheng Wang), J.L., S.W. (Shufang Wang) and Q.F.; formal analysis, S.W. (Sheng Wang); investigation, S.W. (Sheng Wang) and J.L.; resources, S.W. (Sheng Wang); data curation, S.W. (Sheng Wang); writing—original draft preparation, S.W. (Sheng Wang) and J.L.; writing—review and editing, M.Y., S.W. (Sheng Wang) and J.L.; visualization, S.W. (Sheng Wang); supervision, M.Y. and C.S.; project administration, M.Y.; funding acquisition, M.Y. All authors have read and agreed to the published version of the manuscript.

**Funding:** This research was funded by the Agricultural Science and Technology Innovation Program (CAAS-ASTIP-2021-ISAPS) and Demonstration and Promotion of Key Agricultural Core Technologies in Jilin Province (Industrial Technology System 202400601).

**Institutional Review Board Statement:** Not applicable.

**Data Availability Statement:** The data presented in this study are openly available in NCBI at BioProject: PRJNA1107921, Biosample: SAMN41216530, SAMN41216531, SAMN41216532, SAMN41216533, SAMN41216534, SAMN41216535, SAMN41216536, SAMN41216537 and SAMN41216538.

**Conflicts of Interest:** The authors declare no conflicts of interest.

## References


1. Sohretoglu, D.; Huang, S.L. *Ganoderma lucidum* Polysaccharides as an Anti-cancer Agent. *Anti-Cancer Agents Med. Chem.* **2018**, *18*, 667–674. [CrossRef]
2. Zhu, N.; Lv, X.C.; Wang, Y.Y.; Li, J.L.; Liu, Y.M.; Lu, W.F.; Yang, L.C.; Zhao, J.; Wang, F.J.; Zhang, L.S.W. Comparison of immunoregulatory effects of polysaccharides from three natural herbs and cellular uptake in dendritic cells. *Int. J. Biol. Macromol.* **2016**, *93*, 940–951. [CrossRef] [PubMed]
3. Chiu, H.F.; Fu, H.Y.; Lu, Y.Y.; Han, Y.C.; Shen, Y.C.; Venkatakrishnan, K.; Golovinskaia, O.; Wang, C.K. Triterpenoids and polysaccharide peptides-enriched *Ganoderma lucidum*: A randomized, double-blind placebo-controlled crossover study of its antioxidation and hepatoprotective efficacy in healthy volunteers. *Pharm. Biol.* **2017**, *55*, 1041–1046. [CrossRef] [PubMed]
4. Xiao, C.; Wu, Q.P.; Xie, Y.Z.; Tan, J.B.; Ding, Y.R.; Bai, L.J. Hypoglycemic mechanisms of *Ganoderma lucidum* polysaccharides F31 in db/db mice via RNA-seq and iTRAQ. *Food Funct.* **2018**, *9*, 6496–6508. [CrossRef] [PubMed]
5. Zhang, J.; Liu, M.; Yang, Y.; Lin, L.; Xu, N.; Zhao, H.; Jia, L. Purification, characterization and hepatoprotective activities of mycelia zinc polysaccharides by *Pleurotus djamor*. *Carbohydr. Polym.* **2016**, *136*, 588–597. [CrossRef] [PubMed]
6. Li, Y.F.; Li, M.M.; Wang, R.; Wang, B.Y.; Athari, S.S.; Wang, J.L. *Ganoderma* modulates allergic asthma pathologic features via anti-inflammatory effects. *Respir. Physiol. Neurobiol.* **2022**, *299*, 103843. [CrossRef] [PubMed]
7. Liu, D.; Sun, X.; Diao, W.; Qi, X.; Bai, Y.; Yu, X.; Li, L.; Fang, H.; Chen, Z.; Liu, Q.; et al. Comparative transcriptome analysis revealed candidate genes involved in fruiting body development and sporulation in *Ganoderma lucidum*. *Arch. Microbiol.* **2022**, *204*, 514. [CrossRef] [PubMed]
8. Sun, J.; Peng, R.H.; Xiong, A.S.; Tian, Y.; Zhao, W.; Xu, H.; Liu, D.T.; Chen, J.M.; Yao, Q.H. Secretory expression and characterization of a soluble laccase from the *Ganoderma lucidum* strain 7071-9 in Pichia pastoris. *Mol. Biol. Rep.* **2012**, *39*, 3807–3814. [CrossRef] [PubMed]
9. Liu, L.Y.; Huang, Z.X.; Xing, S.H.; Wang, B.Q.; Luo, X.H.; Liu, P.H. Carbon transformation and $CO_2$ emission in cultures during growth of *Ganoderma lucidum*. *J. Hortic.* **2019**, *46*, 2047–2054.
10. Alfaro, M.; Castanera, R.; Lavín, J.L.; Grigoriev, I.V.; Oguiza, J.A.; Ramírez, L.; Pisabarro, A.G. Comparative and transcriptional analysis of the predicted secretome in the lignocellulose-degrading basidiomycete fungus *Pleurotus ostreatus*. *Environ. Microbiol.* **2016**, *18*, 4710–4726. [CrossRef]
11. Liu, X. *Construction of Composite Bacterial Colony for Efficient Degradation of Corn Stover and Research on Its Degradation Effect*; Northeast Agricultural University: Harbin, China, 2019.
12. Zhang, X.Q.; Wang, Z.F.; Sen, M.Y.; Bai, H.H.; Ta, N. Analysis of crop straw production and comprehensive utilisation in China. *J. China Agric. Univ.* **2021**, *26*, 30–41.
13. Liu, J.M.; Ju, W.; Wu, B.; Liu, L.; Zhan, M.; Wu, P.; Wang, Y.; Liu, S.T. Lignocellulolytic Enzyme Production in Solid-State Fermentation of Corn Stalk with Ammoniation Pretreatment by *Lentinus edodes* L-8. *Bioresources* **2014**, *9*, 1430–1444. [CrossRef]
14. Adebayo, E.A.; Martinez-Carrera, D. Oyster mushrooms (*Pleurotus*) are useful for utilizing lignocellulosic biomass. *Afr. J. Biotechnol.* **2015**, *14*, 52–67.
15. Hamelinck, C.N.; Hooijdonk, G.V.; Faaij, A.P.C. Ethanol from lignocellulosic biomass: Techno-economic performance in short-middle-and long-term. *Biomass Bioeng.* **2005**, *28*, 384–410. [CrossRef]
16. Marinovic, M.M.V.; Aguilar-Pontes, M.; Zhou, O.; Miettinen, R.P.V.; Makela, M.R.; Hilden, K. Temporal transcriptome analysis of the white-rot fungus Obba rivulosa shows expression of a constitutive set of plant cell wall degradation targeted genes during growth on solid spruce wood. *Fungal Genet. Biol.* **2018**, *112*, 47–54. [CrossRef]
17. Rytioja, J.; Hildén, K.; Hatakka, A.; Mäkelä, M.R. Transcriptional analysis of selected cellulose-acting enzymes encoding genes of the white-rot fungus *Dichomitus squalens* on spruce wood and microcrystalline cellulose. *Fungal Genet. Biol.* **2014**, *72*, 91–98. [CrossRef] [PubMed]
18. Baldrian, P.; Valaskova, V. Degradation of cellulose by basidiomycetous fungi. *Fems Microbiol. Rev.* **2008**, *32*, 501–521. [CrossRef]
19. Lombard, V.; Ramulu, H.G.; Drula, E.; Coutinho, P.M.; Henrissat, B. The carbohydrate-active enzymes database (CAZy) in 2013. *Nucleic Acids Res.* **2014**, *42*, D490–D495. [CrossRef] [PubMed]
20. Bourbonnais, R.; Paice, M.G. Oxidation of non-phenolic substrates. An expanded role for laccase in lignin biodegradation. *FEBS Lett.* **1990**, *267*, 99–102. [CrossRef]
21. Knop, D.; Yarden, O.; Hadar, Y. The ligninolytic peroxidases in the genus *Pleurotus*: Divergence in activities, expression, and potential applications. *Appl. Microbiol. Biotechnol.* **2015**, *99*, 1025–1038. [CrossRef]
22. Xie, C.L.; Yan, S.W.; Zhang, Z.M.; Gong, W.B.; Zhu, Z.H.; Zhou, Y.J.; Yan, L.; Hu, Z.X.; Ai, L.Z.; Peng, Y.D. Mapping the metabolic signatures of fermentation broth, mycelium, fruiting body and spores powder from *Ganoderma lucidum* by untargeted metabolomics. *Lwt-Food Sci. Technol.* **2020**, *129*, 109494. [CrossRef]
23. Ma, Z.B.; Ye, C.; Deng, W.W.; Xu, M.M.; Wang, Q.; Liu, G.Q.; Wang, F.; Liu, L.M.; Xu, Z.H.; Shi, G.Y.; et al. Reconstruction and Analysis of a Genome-Scale Metabolic Model of *Ganoderma lucidum* for Improved Extracellular Polysaccharide Production. *Front. Microbiol.* **2018**, *9*, 3076. [CrossRef] [PubMed]
24. Shiao, M.S. Triterpenoid natural-products in the fungus *Ganderma lucidum*. *J. Chin. Chem. Soc.* **1992**, *39*, 669–674. [CrossRef]
25. Huang, Y.Y.; Jian, X.X.; Lv, Y.B.; Nian, K.Q.; Gao, Q.; Chen, J.; Wei, L.J.; Hua, Q. Enhanced squalene biosynthesis in *Yarrowia lipolytica* based on metabolically engineered acetyl-CoA metabolism. *J. Biotechnol.* **2018**, *281*, 106–114. [CrossRef] [PubMed]

26. Wei, L.J.; Kwak, S.; Liu, J.J.; Lane, S.; Hua, Q.; Kweon, D.H.; Jin, Y.S. Improved squalene production through increasing lipid contents in *Saccharomyces cerevisiae*. *Biotechnol. Bioeng.* **2018**, *115*, 1793–1800. [CrossRef] [PubMed]
27. Zhang, G.; Sun, Z.H.; Ren, A.; Shi, L.; Shi, D.K.; Li, X.B.; Zhao, M.W. The mitogen-activated protein kinase GlSlt2 regulates fungal growth, fruiting body development, cell wall integrity, oxidative stress and ganoderic acid biosynthesis in *Ganoderma lucidum*. *Fungal Genet. Biol.* **2017**, *104*, 6–15. [CrossRef] [PubMed]
28. Wang, S.L.; Shi, L.; Hu, Y.R.; Liu, R.; Ren, A.; Zhao, M.W. Roles of the Skn7 response regulator in stress resistance, cell wall integrity and GA biosynthesis in *Ganoderma lucidum*. *Fungal Genet. Biol.* **2018**, *114*, 12–23. [CrossRef] [PubMed]
29. Xiao, Z.Z.; Storms, R.; Tsang, A. Microplate-based filter paper assay to measure total cellulase activity. *Biotechnol. Bioeng.* **2004**, *88*, 832–837. [CrossRef] [PubMed]
30. Bailey, M.J.; Biely, P.; Poutanen, K. Interlaboratory testing of methods for assay of xylanase activity. *J. Biotechnol.* **1992**, *23*, 257–270. [CrossRef]
31. Coconi-Linares, N.; Magana-Ortiz, D.; Guzman-Ortiz, D.A.; Fernandez, F.; Loske, A.M.; Gomez-Lim, M.A. High-yield production of manganese peroxidase, lignin peroxidase, and versatile peroxidase in *Phanerochaete chrysosporium*. *Appl. Microbiol. Biotechnol.* **2014**, *98*, 9283–9294. [CrossRef]
32. Ruiz-Duenas, F.J.; Guillén, F.; Camarero, S.; Pérez-Boada, M.; Martínez, M.J.; Martínez, A.T. Regulation of peroxidase transcript levels in liquid cultures of the ligninolytic fungus *Pleurotus eryngii*. *Appl. Environ. Microbiol.* **1999**, *65*, 4458–4463. [CrossRef] [PubMed]
33. Grabherr, M.G.; Haas, B.J.; Yassour, M.; Levin, J.Z.; Thompson, D.A.; Amit, I.; Adiconis, X.; Fan, L.; Raychowdhury, R.; Zeng, Q.; et al. Full-length transcriptome assembly from RNA-Seq data without a reference genome. *Nat. Biotechnol.* **2011**, *29*, 644–652. [CrossRef]
34. Yang, Y.D.; Jian, Q.L.I.; Song, F.W.U.; Yun, P.Z.H.U.; Yao, W.C.; Fu, H.E. Integrated nr database in Protein Annotation System and Its Localization. *Comput. Eng.* **2006**, *32*, 71–73,76.
35. Renaux, A.; UniProt, C. UniProt: The universal protein knowledgebase (vol 45, pg D158, 2017). *Nucleic Acids Res.* **2018**, *46*, 2699.
36. Kanehisa, M.; Goto, S.; Kawashima, S.; Okuno, Y.; Hattori, M. The KEGG resource for deciphering the genome. *Nucleic Acids Res.* **2004**, *32*, D277–D280. [CrossRef]
37. Mortazavi, A.; Williams, B.A.; McCue, K.; Schaeffer, L.; Wold, B. Mapping and quantifying mammalian transcriptomes by RNA-Seq. *Nat. Methods* **2008**, *5*, 621–628. [CrossRef] [PubMed]
38. Chen, Q.S.; Liu, J.H.; Li, Y.T.; Yan, L.Y.; Pang, G.C. Establishment of multi-strain co-fermentation system and bioconversion of corn stover. *Guangzhou Chem. Ind.* **2000**, *4*, 69–73+27.
39. Monclaro, A.V.; Ferreira Filho, E.X. Fungal lytic polysaccharide monooxygenases from family AA9: Recent developments and application in lignocelullose breakdown. *Int. J. Biol. Macromol.* **2017**, *102*, 771–778. [CrossRef]
40. Lei, T.C.; Long, J.X.; Tian, C.E. Screening of laccase-producing straw mushroom strains. *Ind. Microbiol.* **2008**, *3*, 51–55.
41. Chen, Q.H.; Zhou, Y.P.; Zhou, Y.P.; Bi, F.S.; Cheng, H.Z.; Tian, C.E. Screening of high yielding strains of fungal laccase. *J. Guangzhou Univ. (Nat. Sci. Ed.)* **2009**, *8*, 53–57.
42. Liu, X.D.; Liu, X.D.; Wang, J.L.; Wang, X. Selection of laccase high-yielding strains by compound mutagenesis of tree tongue *Ganoderma lucidum*. *North. Hortic.* **2019**, *14*, 124–129.
43. Cui, M.L.; Yang, H.Y.; He, G.Q. Submerged fermentation production and characterization of intracellular triterpenoids from *Ganoderma lucidum* using HPLC-ESI-MS. *J. Zhejiang Univ. Sci. B* **2015**, *16*, 998–1010. [CrossRef] [PubMed]
44. Wang, M.M. *Effects of Maize Straw Cultivation of Pleurotus ostreatus and Its Residue on the Growth of Maize Seedling*; Tianjin Agricultural College: Tianjin, China, 2018.
45. Li, Y.L.; Liu, J.H.; Wang, G.; Yang, M.Y.; Yang, X.; Li, T.B.; Chen, G. De novo transcriptome analysis of *Pleurotus djamor* to identify genes encoding CAZymes related to the decomposition of corn stalk lignocellulose. *J. Biosci. Bioeng.* **2019**, *128*, 529–536. [CrossRef] [PubMed]
46. Peng, M.; Aguilar-Pontes, M.V.; Hainaut, M.; Henrissat, B.; Hilden, K.; Makela, M.R.; de Vries, R.P. Comparative analysis of *basidiomycete* transcriptomes reveals a core set of expressed genes encoding plant biomass degrading enzymes. *Fungal Genet. Biol.* **2017**, *112*, 40–46. [CrossRef] [PubMed]
47. Hemsworth, G.R.; Davies, G.J.; Walton, P.H. Recent insights into copper-containing *lytic polysaccharide mono-oxygenases*. *Curr. Opin. Struct. Biol.* **2013**, *23*, 660–668. [CrossRef] [PubMed]
48. Hemsworth, G.R.; Johnston, E.M.; Davies, G.J.; Walton, P.H. Lytic Polysaccharide Monooxygenases in Biomass Conversion. *Trends Biotechnol.* **2015**, *33*, 747–761. [CrossRef] [PubMed]
49. Salame, T.M.; Knop, D.; Levinson, D.; Mabjeesh, S.J.; Yarden, O.; Hadar, Y. Release of *Pleurotus ostreatus* Versatile-Peroxidase from $Mn^{2+}$ Repression Enhances Anthropogenic and Natural Substrate Degradation. *PLoS ONE* **2012**, *7*, e52446. [CrossRef] [PubMed]
50. Pérez-Boada, M.; Ruiz-Dueñas, F.J.; Pogni, R.; Basosi, R.; Choinowski, T.; Martínez, M.J.; Piontek, K.; Martínez, A.T. Versatile peroxidase oxidation of high redox potential aromatic compounds: Site-directed mutagenesis, spectroscopic and crystallographic investigation of three long-range electron transfer pathways. *J. Mol. Biol.* **2005**, *354*, 385–402. [CrossRef] [PubMed]
51. Zhang, D.H.; Li, N.; Yu, X.Y.; Zhao, P.; Li, T.; Xu, J.W. Overexpression of the homologous lanosterol synthase gene in ganoderic acid biosynthesis in *Ganoderma lingzhi*. *Phytochemistry* **2017**, *134*, 46–53. [CrossRef]

52. Feng, Z.R.; Feng, Z.R.; Li, H.J.; Xu, J.W. Ganoderic Acid Accumulation and Biosynthetic Gene Expression during Fruiting Body Development in *Ganoderma lucidum*. In Proceedings of the 2015 Asia-Pacific Energy Equipment Engineering Research Conference (ap3er 2015), Zhuhai, China, 13–14 June 2015; pp. 354–358.
53. Zhou, J.S.; Ji, S.L.; Ren, M.F.; He, Y.L.; Jing, X.R.; Xu, J.W. Enhanced accumulation of individual ganoderic acids in a submerged culture of *Ganoderma lucidum* by the overexpression of squalene synthase gene. *Biochem. Eng. J.* **2014**, *90*, 178–183. [CrossRef]
54. Ma, L.; Chen, L.; Zhang, L.; Zou, G.; Liu, R.; Jiang, Y.P.; Zhou, Z.H. RNA Sequencing Reveals Xyr1 as a Transcription Factor Regulating Gene Expression beyond Carbohydrate Metabolism. *Biomed Res. Int.* **2016**, *2016*, 4841756. [CrossRef] [PubMed]

agriculture

MDPI

*Article*

# Study on the Effects of Different Light Supply Modes on the Development and Extracellular Enzyme Activity of *Ganoderma lucidum*

**Yihan Liu [1,2], Yuan Luo [1], Wenzhong Guo [1], Xin Zhang [1], Wengang Zheng [3,*] and Xiaoli Chen [1,*]**

1 Intelligent Equipment Research Center, Beijing Academy of Agriculture and Forestry Sciences, Beijing 100097, China
2 College of Horticultural and Landscape Architecture, Tianjin Agricultural University, Tianjin 300384, China
3 Nongxin Technology (Beijing) Co., Ltd., Beijing 10097, China
* Correspondence: zhengwg@nercita.org.cn (W.Z.); kindrabbit@126.com (X.C.)

**Abstract:** Edible fungi have certain photo-sensitivity during the mushroom emergence stage, but there have been few relevant studies on the responses of *Ganoderma lucidum* to different light irradiation conditions. *Ganoderma lucidum* were planted in an environmentally controllable mushroom room with different light supply modes that were, respectively, continuous white light (CK), red light (R), green light (G), blue light (B), and intermittent red light (R-), green light (G-), and blue light (B-), with a total light intensity of 15 $\mu mol \cdot m^{-2} \cdot s^{-1}$ and a light/dark (L/D) period of 12 h/12 h for each treatment. The interval in intermittent light treatments was 30 min. The optimal light supply mode suitable for the growth of *Ganoderma lucidum* was explored by analyzing the characteristics, nutritional quality, and extracellular enzyme activity in mushrooms exposed to different light treatments. The results showed that red light (whether in continuous or intermittent supply modes) inhibited the fruiting body differentiation of *Ganoderma lucidum*, showing delayed differentiation or complete undifferentiation. The highest stipe length and pileus diameter of fruiting bodies were detected under G- treatment, which were, respectively, increased by 71.3% and 3.2% relative to the control. The highest weight of fruiting bodies was detected under G treatment, which was significantly increased by 21.4% compared to the control ($p < 0.05$). Intermittent light mode seemed to be more conducive to the size development of the fruiting body, while continuous light mode was beneficial for increasing the weight. The highest contents of crude protein and total triterpenes in pileus were detected under G treatment (significantly 14.9% and 28.1% higher than the control, respectively), while that of the crude polysaccharide was detected under G- treatment (significantly 35.7% higher than the control) ($p < 0.05$). The highest activities of extracellular enzymes such as cellulase, hemicellulase, laccase, lignin peroxidase, and amylase were detected in fruiting bodies subjected to G treatment, which were significantly increased by 11.9%~30.7% in the pileus and 9.5%~44.5% in the stipe. Green light might increase the weight and nutrient accumulation in the pileus of *Ganoderma lucidum* via up-regulating the extracellular enzyme activities. This study provides an effective light supply strategy for regulating the light environment in the industrial production of *Ganoderma lucidum*.

**Keywords:** edible fungi; *Ganoderma lucidum*; light supply mode; nutritional quality; extracellular enzyme

## 1. Introduction

*Ganoderma lucidum* is an important edible and medicinal fungus belonging to the family Polyporaceae and genus *Ganoderma*. *Ganoderma lucidum* has high nutritional value and is rich in various components such as polysaccharides, total triterpenes, proteins, amino acids, alkaloids, adenine, etc. [1]. Among them, *Ganoderma lucidum* polysaccharides and triterpenes are the main bioactive substances used to evaluate the value of *Ganoderma lucidum*. At the same time, *Ganoderma lucidum* also has medicinal functions such as the prevention and treatment of hepatitis, hypertension, and stomach cancer, which is why it is also known as the "magic herb" [2]. Therefore, it is very important to study the growth and quality of *Ganoderma lucidum*.

In addition to green plants, many non-photosynthetic organisms, such as edible fungi, also exhibit photo-sensitivity. The growth of edible mushrooms includes two stages: nutritional growth and reproductive growth. Light plays an important role in the transition between these two stages. At present, research on photo-sensitivity in fungi is mainly focused on the model organism *Neurospora crassa* [3]. However, different edible fungi have different light requirements. Therefore, the response of model fungi to light is not applicable to other varieties, such as *Ganoderma lucidum*. In recent years, researchers have mainly studied the effects of light intensity, light quality, and light cycle on the growth and development of edible fungi [4]. Some studies have shown that certain mushrooms, such as *Agaricus bitoquis*, *Agaricus bisporus*, and underground *Poria cocos*, could produce fruiting bodies in total darkness. However, light is indispensable during the reproductive growth stage of most edible mushrooms; meanwhile, the light requirements of edible fungi are related to the variety and growth stage [5–7]. Previous studies have reported that light quality can affect mycelium activity, as well as the size and weight of the fruiting bodies, and edible fungi exhibit different fruiting body morphologies under different light environments. Research has shown that blue light could regulate the production of asexual spores and the development of fruiting bodies in *Coprinopsis atramentaria* [8]. Compared with white light, red and yellow light promoted the mycelial growth of *Pleurotus eryngii* in solid culture [9]. Dong et al. showed that pink light increased the dry matter content of *Cordyceps militaris* fruiting bodies compared to blue light [10]. Wu et al. showed that the exopolysaccharides (EPS) content of *Pleurotus eryngii* was highest under blue light conditions [9]. Light also has a significant impact on the synthesis, accumulation, and gene expression of secondary metabolites in fungi. Jang and Lee showed that the yield and ergot content of *Pleurotus ostreatus* were higher under mixed blue and white light relative to the white light [11]. Palanivel et al. studied the effect of light quality on five filamentous fungi and found that red, blue, white, green, and yellow light all inhibited the accumulation of fungal pigments [12]. The mRNA level of the gene *madA*, which is homologous to the *WC-1* photoreceptor in *Phycomyces*, decreased under blue light irradiation [13,14].

The nutrients required for the growth and development of edible fungi are mainly provided by cultivation materials that are rich in a large amount of cellulose, hemicellulose, lignin, etc. However, edible fungi cannot directly utilize these substances as their energy source and must decompose these substances into micromolecular substances by secreting extracellular enzymes. These small molecule substances, which are easily absorbed and transformed by the mycelium and fruiting bodies, can provide nutrients for hypha growth, primordial formation, and fruiting body growth. The main extracellular enzymes during the growth of edible fungi include the cellulase system, hemicellulase system, lignin-degrading enzyme system, and amylase system [15–17]. Cellulases include endo-1,4-β-D-glucanohydrolase (E.C.3.2.1.4), exo-1,4-β-D-glucannase (E.C.3.2.1.91) and β-1,4- glucosidase (E.C.3.2.1.21), etc. Hemicellulases include endo-1,4-β-xylanase (E.C.3.2.1.8) and exo-1,4-β-xylosidase (E.C.3.2.1.37), etc. Lignin-degrading enzymes include lignin peroxidase (E.C.1.11.1.14), manganese peroxidase (E.C.1.11.1.13) and laccase (E.C.1.10.3.2). Amylase includes α-amylase (E.C.3.2.1.1), β-amylase (E.C.3.2.1.2), glucoamylase (E.C.3.2.1.3.), and isoamylase (E.C.3.2.1.68) [18]. The synthesis of extracellular enzymes in edible fungi is influenced not only by genetic factors and cultivation substrates but also by environmental factors. Studies have shown that green light can increase the activities of total cellulase, endo-1,4-β-D-glucanohydrolase, and xylanase in *Pleurotus ostreatus*, but reduce the activity of laccase [19]. The activity changes of extracellular enzymes have a significant impact on the growth of mycelium. On the one hand, it directly affects the growth rate and momentum of mycelium, and on the other hand, it indirectly affects the formation time and biological efficiency of fruiting bodies. Therefore, it is of great practical significance to study the response of extracellular enzymes in edible fungi to different light qualities.

*Ganoderma lucidum* has been proven to have phototropism; the pileus edge always grows toward the light source. It was also indicated that the biomass of *Ganoderma lucidum* was higher under blue light than that under red light [20,21]. It can be seen that a possible way to regulate the growth and nutritional quality of *Ganoderma lucidum* is through light control in the environment. However, there are few studies on the responses of the pileus (the main edible part of *Ganoderma lucidum*) to different light environments. Moreover, the mechanism of how light quality affects the development and nutrient accumulation of *Ganoderma lucidum* still remains unclear. Therefore, the current study evaluated the growth characteristics and the nutrient content as well as the extracellular enzyme activities in the pileus and stipe of *Ganoderma lucidum* exposed to different light irradiation supplied by light-emitting diodes (LEDs) to reveal the function mechanism and explore the optimal light condition for *Ganoderma lucidum* cultivation.

## 2. Materials and Methods

### *2.1. Experimental Design*

This experiment was conducted in an environmentally controllable mushroom factory of Beijing Academy of Agriculture and Forestry Sciences, using an LED system that could set any light formula. The *Ganoderma lucidum* mushroom spawn bags were treated with different light qualities from the day when the mycelium was full.

Seven light treatments were set up in the experiment, which were, respectively, continuous white light (CK), continuous red light (R), continuous green light (G), continuous blue light (B), intermittent red light (R-), intermittent green light (G-), and intermittent blue light (B-), with a total light intensity of 15 $\mu mol \cdot m^{-2} \cdot s^{-1}$ and a light/dark (L/D) period of 12 h/12 h for each treatment. The interval in intermittent light treatments was 30 min (30 min of light treatment and 30 min of dark treatment). The light intensity and spectrum were all measured approximately 10 cm below the light source using an LI-180 spectrometer; the peak wavelengths of red, blue, and green light were, respectively, 660 nm, 450 nm, and 520 nm (Table 1 and Figure 1). The temperature, $CO_2$ concentration, and the air relative humidity during the entire growth period of *Ganoderma lucidum* were monitored by sensors and controlled by an intelligent control system, with values of 28 ± 1 °C, 500 ± 20 $\mu mol \cdot mol^{-1}$ and (90 ± 1)%, respectively. During the growth period, purified water was sprayed twice a day at 8 am and 8 pm for 1 min each time.

**Table 1.** Light supply modes in each treatment.

| Treatment | | Supplementary Lighting | Light Intensity ($\mu mol \cdot m^{-2} \cdot s^{-1}$) | | | |
|---|---|---|---|---|---|---|
| | | | **Red Light** | **Green Light** | **Blue Light** | **White Light** |
| Continuous light | R | Continuous red light | 15 | 0 | 0 | 0 |
| | G | Continuous green light | 0 | 15 | 0 | 0 |
| | B | Continuous blue light | 0 | 0 | 15 | 0 |
| Intermittent light | R- | Intermittent red light, interval of 30 min | 15 | 0 | 0 | 0 |
| | G- | Intermittent green light, interval of 30 min | 0 | 15 | 0 | 0 |
| | B- | Intermittent blue light, interval of 30 min | 0 | 0 | 15 | 0 |
| | CK | Continuous white light | 0 | 0 | 0 | 15 |

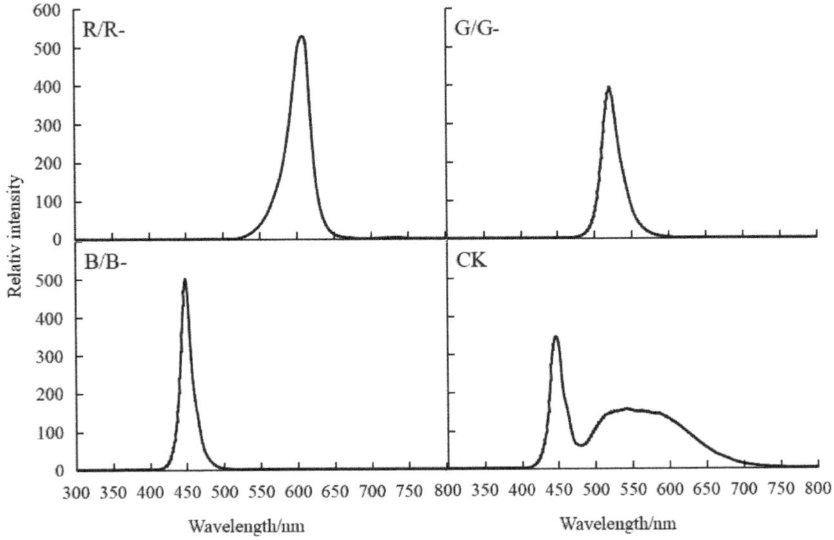


**Figure 1.** LED spectral distribution in each treatment.

### 2.2. Sampling and Index Determination

The length and diameter of mushroom stipes, the diameter and thickness of mushroom pilei, as well as the number of pilei, were measured dynamically at 26, 33, 40, 47, and 51 days after treatment (DAT). The weight of fruiting bodies was measured at harvest (51 DAT). The stipe length, stipe diameter, pileus diameter, and pileus thickness of *Ganoderma lucidum* were measured with a vernier caliper. The weight of fruiting bodies was measured using an electronic balance. Eight fruiting bodies randomly taken from each treatment were regarded as a repetition, and there were three repetitions in each treatment.

Mixed 0.1 g of mushroom tissue (ground in liquid nitrogen) with 0.9 mL of PBS buffer (pH = 7.4), centrifuged at 4 °C and 8000 r/min for 30 min, and then the supernatant was collected and stored at 4 °C for use. The nutritional indicators and extracellular enzyme activities of *Ganoderma lucidum* were determined using an Elisa assay kit purchased from Shanghai C-reactive Biotechnology Co. Ltd. The content of crude polysaccharides [22], crude proteins [23], and total triterpenes [24] was determined according to the instructions of the biochemical analysis kit. The activities of extracellular enzymes, including cellulase [25], hemicellulase [26], laccase [27], manganese peroxidase [28], lignin peroxidase [28], and amylase [29], were measured according to the instructions of the enzyme-linked immunosorbent assay kit.

### 2.3. Statistical Analysis

The relative spectral curve was extracted using Avasoft 8, and data were organized and plotted using Excel 2016 and SPSS Statistics 22. Cluster analysis was performed using Hiplot, and correlation analysis was performed using Origin 2021. The data are presented as mean ± errors.

## 3. Results

### 3.1. Effects of Different Light Supply Modes on the Characteristics of Ganoderma lucidum

As shown in Figure 2, compared with the other treatments, red light (no matter whether in continuous or intermittent supply modes) inhibited the fruiting body differentiation of *Ganoderma lucidum*, showing delayed differentiation or complete undifferentiation. This indicated that red light was not conducive to the production of *Ganoderma lucidum*.

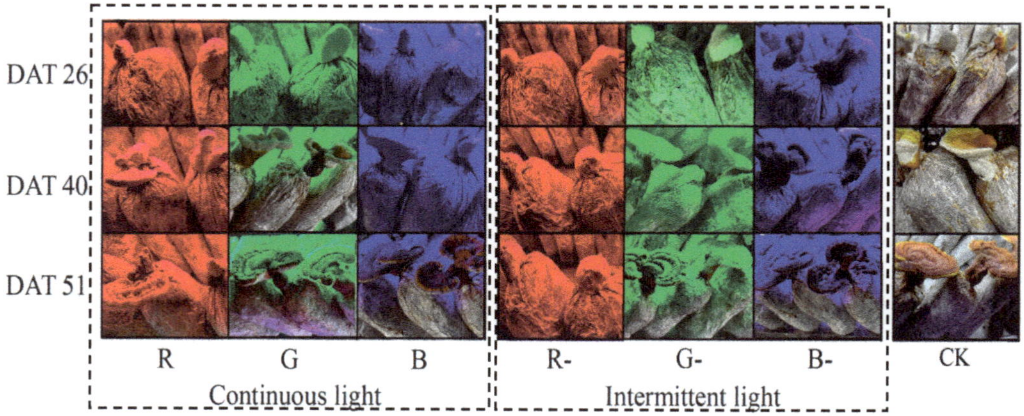


**Figure 2.** Photos of *Ganoderma lucidum* growth under different treatments.

At DAT 51, the stipe length and pileus thickness were increased by all treatments compared with the white light (with an increase of 16.0–91.3% and 4.9–21.4%, respectively) (Figure 3a,d). On the contrary, all treatments were not as conducive as the white light regarding the increase in *Ganoderma lucidum* stipe diameter (with a decrease of 5.2–42.5%) (Figure 3b). Among them, the longest mushroom stipe was detected under G- treatment, which was significantly increased by 71.3% relative to the control ($p < 0.05$). As shown in Figure 3c, the largest pileus diameter of mushrooms was also detected under G-, which was increased by 3.2%, while those exposed to the other treatment displayed a decrease of 2.6–46.4% compared to the control. The thickest pileus and the maximum number of pilei were all detected under the B- treatment, which were significantly increased by 31.2% and 37.5% ($p < 0.05$) (Figure 3e). As shown in Figure 3f, the heaviest weight of the fruiting body was detected under G treatment, which was increased by 21.4% relative to the control. In addition, when comparing the two light supply modes of the same light quality, it was found that continuous light mode was more conducive to an increase in the weight, while intermittent light mode was more conducive to an increase in the size of the *Ganoderma lucidum* fruiting body.

On the whole, continuous green light promoted an increase in fruiting body weight and intermittent green light was conducive to an increase in fruiting body size, while intermittent blue light was positive to an increase in pileus thickness and the number of pilei.

### 3.2. *Effects of Different Light Supply Modes on the Nutritional Quality of* Ganoderma lucidum

Due to the pileus is the main edible and medicinal part of *Ganoderma lucidum*, only the mushroom pileus was analyzed for its nutrient content. As shown in Figure 4a,c, G and G- treatments both raised the contents of crude protein and total triterpenes in *Ganoderma lucidum* pileus compared with the control. The highest contents of crude protein (16.56%) and total triterpenes (429.46 mg/g) in the pileus were detected under G treatment; they increased by 14.9% and 28.1% relative to the control. As shown in Figure 4b, all treatments raised the content of the crude polysaccharide in the fruiting bodies to various degrees compared with the control. The highest crude polysaccharide content in the pileus (78.99 mg/g) was detected under G- treatment, which was significantly increased by 35.7% relative to the control ($p < 0.05$). In addition, the results also indicated that although both green light modes could increase the contents of crude protein and crude polysaccharides in the fruiting body, it was found that continuous green light was more conducive to the accumulation of crude protein while intermittent green light was more conducive to that of crude polysaccharides.

On the whole, green light was beneficial for the synthesis and accumulation of crude proteins, crude polysaccharides, and total triterpenes in the pileus of *Ganoderma lucidum*.

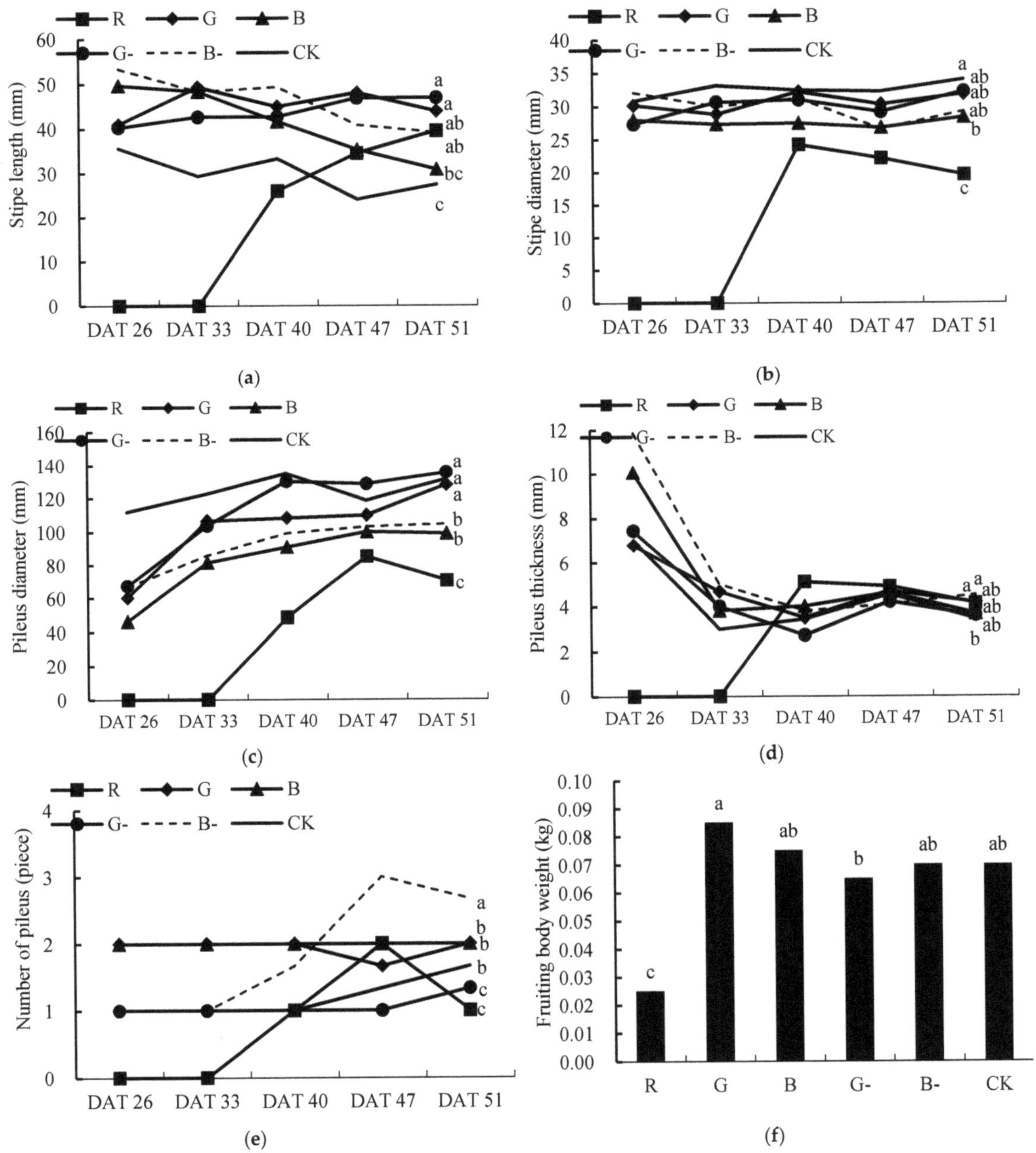


**Figure 3.** Effects of different light supply modes on the stipe length (**a**), stipe diameter (**b**), pileus diameter (**c**), pileus thickness (**d**), the number of pilei (**e**), and the weight (**f**) of *Ganoderma lucidum* fruiting body. Note: Different lowercase letters indicate significant differences between groups ($p < 0.05$).

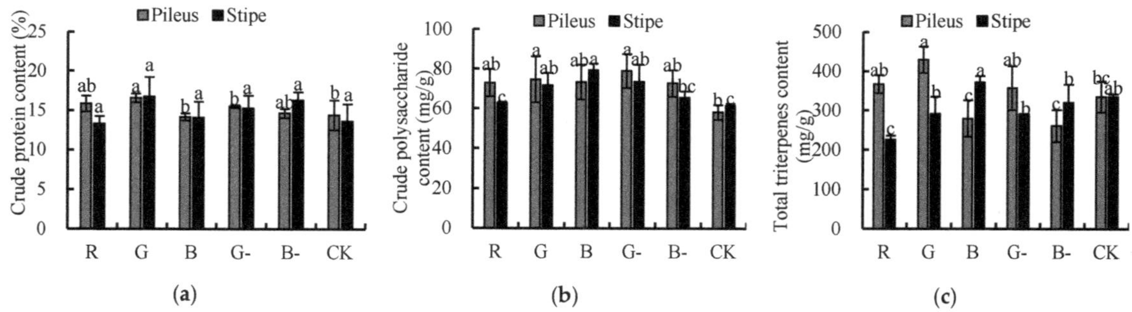


**Figure 4.** Effects of different light supply modes on the nutritional quality of crude proteins (**a**), crude polysaccharides (**b**), and total triterpenes (**c**) of *Ganoderma lucidum*. Note: Different lowercase letters indicate significant differences between groups ($p < 0.05$).

### 3.3. *Effects of Different Light Supply Modes on the Extracellular Enzyme Activity of Ganoderma lucidum*

As shown in Figure 5, G, G-, and B treatments increased the activities of cellulase, hemicellulase, laccase, manganese peroxidase, lignin peroxidase, and amylase in the pileus and stipe of *Ganoderma lucidum* compared with the control. Among them, the highest activities of cellulase, hemicellulase, laccase, lignin peroxidase, and amylase in the pileus and stipe were all observed in *Ganoderma lucidum* subjected to G treatment. G treatment significantly increased the activity of the above-mentioned extracellular enzymes in the pileus by 13.6%, 16.5%, 18.5%, 30.7%, and 11.9%, respectively, relative to the control, and significantly increased those in the stipe by 44.5%, 12.8%, 14.3%, 9.5%, and 12.9% ($p < 0.05$). Furthermore, when comparing the two different light modes of green light or blue light, it was found that continuous light was more conducive to an improvement in the extracellular enzyme activity in *Ganoderma lucidum*.

The clustering heatmap (Figure 6) intuitively shows that the activities of six extracellular enzymes were obviously up-regulated under G treatment, which might explain the reason for the height weight of fruiting bodies detected in G treatment (Figure 3f).

### 3.4. *Correlation Analysis between Different Characteristics, Nutritional Quality, and Extracellular Enzymes Activity of Ganoderma lucidum*

As shown in Figure 7, a significant positive relationship was observed between the stipe length and the activities of hemicellulase in the stipe ($p < 0.05$). The diameter of the stipe and pileus was positively correlated with the activities of manganese peroxidase and lignin peroxidase in the fruiting bodies ($p < 0.05$). This might indicate that hemicellulase mainly promoted the vertical growth of *Ganoderma lucidum*, while manganese peroxidase and lignin peroxidase mainly promoted its horizontal growth. The number of pilei was positively correlated with the activities of cellulase, laccase, manganese peroxidase, and lignin peroxidase in the fruiting bodies. The weight of the fruiting body was significantly ($p < 0.05$) or extremely significantly ($p < 0.01$) positively correlated with the activities of these six enzymes in the fruiting bodies of *Ganoderma lucidum*.

As for the correlation between the nutrients and extracellular enzyme activity in the fruiting bodies, it was observed that the contents of crude protein and crude polysaccharide in the stipe and pileus were positively correlated with the activities of all the six extracellular enzymes in the corresponding parts of the fruiting body. Additionally, the total triterpenes content in the stipe and pileus was positively correlated with the activities of cellulase, laccase, manganese peroxidase, lignin peroxidase, and amylase in the corresponding parts.

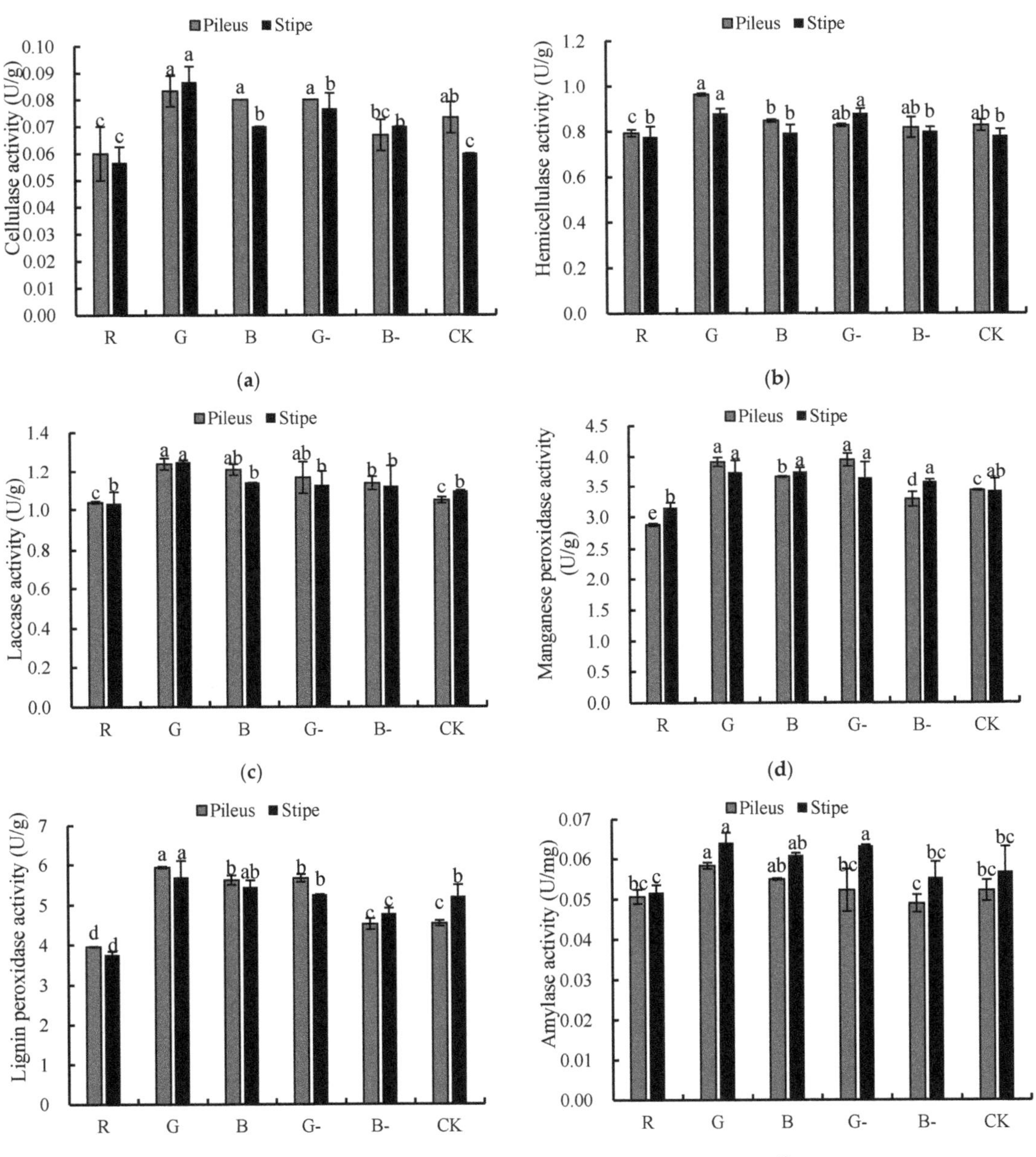


**Figure 5.** Effects of different light supply modes on the activities of cellulase (**a**), hemicellulase (**b**), laccase (**c**), manganese peroxidase (**d**), lignin peroxidase (**e**), and amylase (**f**) of *Ganoderma lucidum*. Note: Different lowercase letters indicate significant differences between groups ($p < 0.05$).

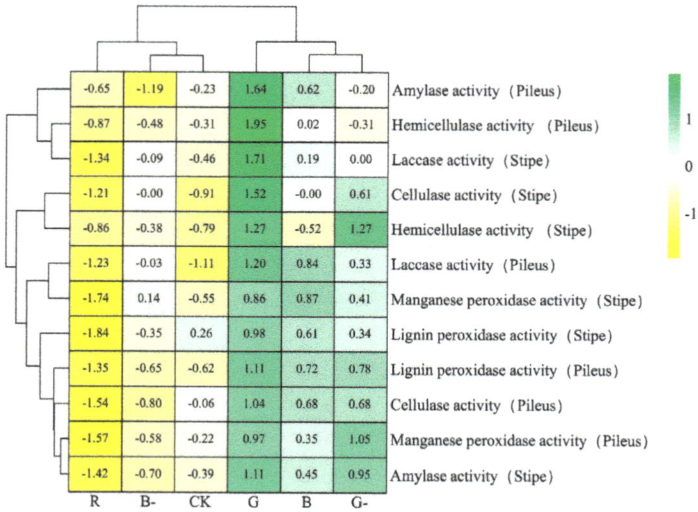


**Figure 6.** Cluster analysis of extracellular enzyme activity in *Ganoderma lucidum* under different treatments.

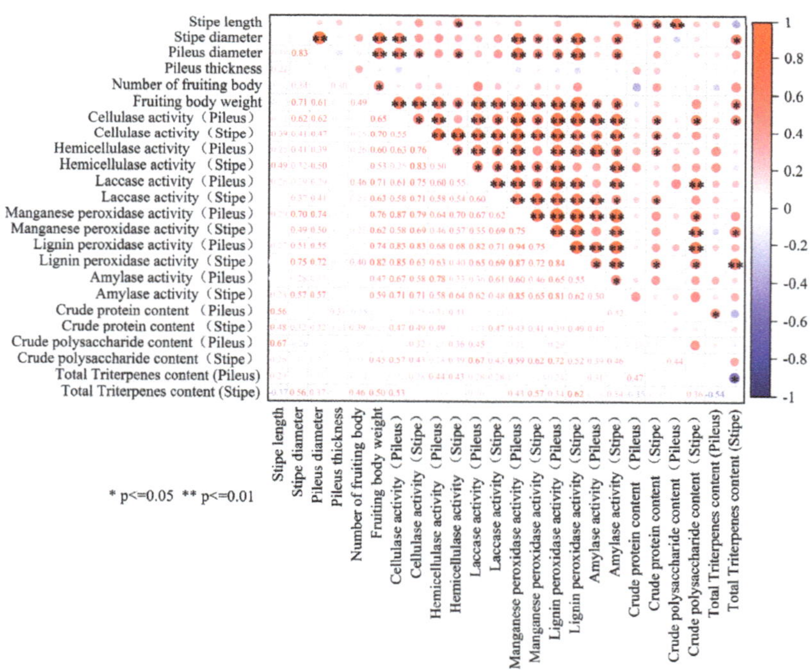


**Figure 7.** Correlation analysis between different characteristics, nutritional quality, and extracellular enzyme activity of *Ganoderma lucidum*.

## 4. Discussion

The fruiting stage of edible fungi is a light-sensitive stage, and different growth and development stages have different requirements for the light environment. Light can either stimulate or inhibit fungal development, so light is an indispensable factor in the growth and development of edible fungi. Arjona et al. showed that blue light was the triggering signal for the formation of *Pleurotus ostreatus* primordia, and the formation of primordia could not be achieved in a light environment lacking blue light [30]. Ellis et al. found that blue light with wavelengths of 440–470 nm was the most favorable for the formation of *Coprinus comatus* primordia [31]. Halabura et al. reported that the *Lentinus crusitus* primordia could form under green light irradiation, while it could not form under red light [32]. These studies indicated that blue and green light with shorter wavelengths had the greatest impact on the development of edible mushroom primordia in different light qualities. Therefore, the inhibition of fruiting body differentiation by red light treatment (R and R-) in our study might be due to the longer wavelength of red light, which was not easily perceived and responded to by photoreceptors. This indicated that light quality had an important effect on the differentiation of *Ganoderma lucidum* fruiting bodies, and red light was not conducive to their differentiation and even caused death. In actual production, it is advisable to avoid using red light to produce *Ganoderma lucidum* as much as possible. In addition, studies also have shown that several important biosynthetic pathways in mushrooms, such as the membrane transport protein synthesis and the amino acid biosynthetic were inactive under monochromatic red light irradiation. The function of membrane transporters is to perceive external stimuli and transmit signals to cells, maintaining the activity of mycelial [5]. Therefore, the inhibition of fruiting body differentiation by red light treatment (R and R-) in this study might be due to the lack of blue and green light irradiation or a decreased expression level of membrane transporter protein in *Ganoderma lucidum* caused by monochromatic red light.

Our study found that the stipe length and pileus thickness were increased by all the treatments compared with the control. On the contrary, all the monochromatic light treatments were not as conducive as the white light as regards the increase in *Ganoderma lucidum* stipe diameter. The highest stipe length and pileus diameter of fruiting bodies were detected under G-, which were increased by 71.3% and 3.2% relative to the control. The highest weight of fruiting bodies was detected under G, which was significantly increased by 21.4%, respectively, compared to the control ($p < 0.05$). The thickness and number of *Ganoderma lucidum* pileus exposed to B- treatment were significantly increased by 31.2% and 37.5% compared with the control ($p < 0.05$). On the whole, based on the comparison of various characteristics and two light supply modes of *Ganoderma lucidum*, our study suggests that intermittent green light was more conducive to the growth of the pileus and stipe, while continuous green light was more conducive to an increase in *Ganoderma lucidum* yield. It is worth noting that although blue light had a crucial role in the primordium formation of *Ganoderma lucidum*, monochromatic blue light was not the optimum light quality for the size or weight of the fruiting body of *Ganoderma lucidum*. Correlation analysis showed that the stipe length was significantly positively correlated with the activity of hemicellulase in the stipe. The weight of the fruiting body was positively correlated with the activities of six enzymes, including cellulase, hemicellulase, laccase, manganese peroxidase, lignin peroxidase, and amylase. The cluster analysis in Figure 6 shows that the activities of these enzymes were significantly up-regulated under G treatment, which indicates that the extracellular enzyme could effectively promote an increase in stipe length, stipe and pileus diameter, and fruiting body weight of *Ganoderma lucidum* by degrading culture materials. The green light might promote the degradation of the cultivation materials and absorption of nutrients by *Ganoderma lucidum* via increasing the activities of the extracellular enzymes, which might account for the best growth and weight of the fruiting bodies detected under G treatment.

As an important environmental factor, light quality not only affected the morphology formation of edible fungi but also acted on the synthesis and accumulation of nutrient substances in the fruiting body. Tang et al. used transcriptomics to study the photo response mechanism of *Lentinus edodes* and found that light would affect the transportation and metabolism of carbohydrates [33]. Our study found that the highest contents of crude protein and total triterpenes in the pileus were detected under G treatment, which was increased by 14.9% and 28.1% relative to the control. Therefore, our study suggests that green light was a favorable light quality for the synthesis and accumulation of nutrients in *Ganoderma lucidum*. The possible reason is that organic metabolism-related genes such as hydrophobin genes (*SC1* and *SC3*), lignin-modifying genes (*LAC1*, *LCC2*, and *LCC3*), and tyrosinase-encoding genes (*TYR1* and *MELC2*) were up-regulated or the expression of enzymes related to the synthesis of crude polysaccharides, crude proteins, and total triterpenes in *Ganoderma lucidum* were increased by green light irradiation [34,35]. On the contrary, the highest content of crude polysaccharides in the pileus was detected under G-treatment, which might be due to the correlation between polysaccharide synthesis and the circadian rhythm of *Ganoderma lucidum*. Studies have shown that the blue light receptor *WC-1* directly binds to the negative feedback factor FRQ promoter to mediate light participation in the circadian rhythm cycle. Thus, the intermittent light mode might increase the expression level of *Ganoderma lucidum WC-1*, thereby promoting an increase in polysaccharide synthesis-related enzyme activity [36]. Therefore, the continuous/intermittent green light mode could be dynamically adjusted to achieve the target quality requirements in production.

Extracellular enzymes are involved in almost every process of edible mushroom growth and development. Extracellular enzyme activity reflects the ability of mushrooms to absorb and utilize small molecule nutrients, indirectly affecting the yield and nutrient quality of mushrooms. Correlation analysis showed that the crude protein content in the pileus and stipe was positively correlated with the activities of cellulase, hemicellulase, laccase, lignin peroxidase, and amylase in the corresponding parts. The total triterpenes content in the pileus was positively correlated with the activities of cellulase, hemicellulase, laccase, lignin peroxidase, and amylase in the pileus. The results confirmed the positive correlations between the extracellular enzyme and the organic metabolism. In addition, our study also found that the highest activities of cellulase, hemicellulase, laccase, lignin peroxidase, and amylase in the pileus and stipe were all observed in *Ganoderma lucidum* subjected to G treatment. Thus, the increased activity of cellulase, hemicellulase, laccase, lignin peroxidase, and amylase in *Ganoderma lucidum* treated with G may also account for the higher contents of organic substances observed in the light treatment. Xie et al. investigated the effects of blue light on the activity of manganese peroxidase in *Pleurotus eryngii* and found that blue light inhibited the activity of manganese peroxidase [37]. Ramírez et al. showed that blue light significantly reduced the activity of lignin peroxidase in *Phanerochaete chrysosporium Burds* [38]. The above-mentioned results were different from the current study, in which the continuous blue light treatment (B) increased the activities of manganese peroxidase and lignin peroxidase in the pileus and stipe of *Ganoderma lucidum*. This might indicate that the effects of light quality on extracellular enzymes in edible fungi are variety-dependent. Gan et al. found that blue light enhanced the activity of fungal amylase, which was consistent with the present results that continuous blue light treatment (B) increased the activity of amylase in *Ganoderma lucidum* compared with the control [39]. However, our study also found that intermittent blue light treatment (B-) reduced the activity of amylase in the *Ganoderma lucidum* pileus and stipe, indicating that not all blue light could increase the activity of amylase in *Ganoderma lucidum*, possibly due to darkness during the alternating process.

Red light should be avoided in the factory cultivation of *Ganoderma lucidum*. Combining the characteristics and nutritional quality of *Ganoderma lucidum*, continuous green light should be preferred for the production of *Ganoderma lucidum*, while intermittent green light can also be chosen under limited economic conditions. There are differences in the response of different types of edible mushrooms to light quality. It is necessary to further determine the expression of light receptor-related genes in *Ganoderma lucidum* exposed to different light formula conditions to clarify their light response mechanism.

## 5. Conclusions

Red light (whether in continuous or intermittent supply modes) was found to inhibit the fruiting body differentiation of *Ganoderma lucidum*, showing delayed differentiation or thorough undifferentiation. Continuous green light was beneficial for an increase in the weight, extracellular enzyme activities, as well as the contents of crude protein and total triterpenoid in the pileus of *Ganoderma lucidum*. Intermittent green light was conducive to an increase in fruiting body size and crude polysaccharide content. On the whole, green light might enhance growth and nutrient synthesis by up-regulating the activity of extracellular enzymes in *Ganoderma lucidum*.

**Author Contributions:** Y.L. (Yihan Liu) and Y.L. (Yuan Luo) designed the project, performed statistical data analyses, and wrote the main manuscript. W.G. and X.Z. conducted the measurements. W.Z. and X.C. guided the experiment and reviewed the manuscript. All authors have read and agreed to the published version of the manuscript.

**Funding:** This work was financially supported by the National Edible Fungi Industry Technology System (CARS-20) and the Beijing Edible Fungi Innovation Team (BAIC03-2024).

**Institutional Review Board Statement:** Not applicable.

**Data Availability Statement:** The data presented in this study are available in the article.

**Conflicts of Interest:** Author Wengang Zheng was employed by the company Nongxin Technology (Beijing) Co., Ltd. The remaining authors declare that the research was conducted in the absence of any commercial or financial relationships that could be construed as a potential conflict of interest.

## References


1. Xu, J.W.; Zhao, W.; Zhong, J.J. Biotechnological production and application of ganoderic acids. *Appl. Microbiol. Biotechnol.* **2010**, *87*, 457–466. [CrossRef] [PubMed]
2. Liu, G.Q.; Zhang, K.C. Mechanisms of the Anticancer Action of *Ganoderma lucidum* (Leyss. ex Fr.) Karst.: A New Understanding. *J. Integr. Plant Biol.* **2005**, *47*, 129–135. [CrossRef]
3. Schwerdtfeger, C.; Linden, H. *VIVID* is a flavoprotein and serves as a fungal blue light photoreceptor for photoadaptation. *EMBO J.* **2003**, *22*, 4846–4855. [CrossRef] [PubMed]
4. Tisch, D.; Schmoll, M. Light regulation of metabolic pathways in fungi. *Appl. Microbiol. Biotechnol.* **2010**, *85*, 1259–1277. [CrossRef] [PubMed]
5. Ye, D.; Du, F.; Hu, Q.; Zou, Y.; Bai, X. Transcriptome Analysis Reveals Candidate Genes Involved in Light-Induced Primordium Differentiation in Pleurotus eryngii. *Int. J. Mol. Sci.* **2022**, *23*, 435. [CrossRef] [PubMed]
6. Janusz, G.; Sulej, J.; Jaszek, M.; Osińska-Jaroszuk, M. Effect of different wavelengths of light on laccase, cellobiose dehydrogenase, and proteases produced by *Cerrena unicolor*, *Pycnoporus sanguineus* and *Phlebia lindtneri*. *Acta Biochim. Pol.* **2016**, *63*, 223–228. [CrossRef] [PubMed]
7. Kamada, T.; Sano, H.; Nakazawa, T.; Nakahori, K. Regulation of fruiting body photomorphogenesis in *Coprinopsis cinerea*. *Fungal Genet. Biol.* **2010**, *47*, 917–921. [CrossRef] [PubMed]
8. Kuratani, M.; Tanaka, K.; Terashima, K.; Muraguchi, H.; Nakazawa, T.; Nakahori, K.; Kamada, T. The *dst2* gene essential for photomorphogenesis of *Coprinopsis cinerea* encodes a protein with a putative FAD-binding-4 domain. *Fungal Genet. Biol.* **2010**, *47*, 152–158. [CrossRef] [PubMed]
9. Wu, J.Y.; Chen, H.B.; Chen, M.J.; Kan, S.C.; Shieh, C.J.; Liu, Y.C. Quantitative analysis of LED effects on edible mushroom *Pleurotus eryngiiin* solid and submerged cultures. *J. Chem. Technol. Biotechnol.* **2013**, *88*, 1841–1846. [CrossRef]
10. Dong, J.Z.; Lei, C.; Zheng, X.J.; Ai, X.L.; Wang, Y.; Wang, Q. Light Wavelengths regulate growth and active components of *Cordyceps militaris* fruit bodies. *J. Food Biochem.* **2013**, *37*, 578–584. [CrossRef]
11. Jang, M.; Lee, Y. The suitable mixed LED and light intensity for cultivation of oyster mushroom. *J. Mushrooms* **2014**, *12*, 258–262. [CrossRef]
12. Miyake, T.; Mori, A.; Kii, T.; Okuno, T.; Usui, Y.; Sato, F.; Sammoto, H.; Watanabe, A.; Kariyama, M. Light effects on cell development and secondary metabolism in *Monascus*. *J. Ind. Microbiol. Biotechnol.* **2005**, *32*, 103–108. [CrossRef] [PubMed]
13. Rodríguez-Romero, J.; Corrochano, L.M. Regulation by blue light and heat shock of gene transcription in the fungus Phycomyces: Proteins required for photoinduction and mechanism for adaptation to light. *Mol. Microbiol.* **2006**, *61*, 1049–1059. [CrossRef] [PubMed]
14. Velmurugan, P.; Lee, Y.H.; Venil, C.K.; Lakshmanaperumalsamy, P.; Chae, J.C.; Oh, B.T. Effect of light on growth, intracellular and extracellular pigment production by five pigment-producing filamentous fungi in synthetic medium. *J. Biosci. Bioeng.* **2010**, *109*, 346–350. [CrossRef] [PubMed]
15. Gupta, V.K.; Kubicek, C.P.; Berrin, J.G.; Wilson, D.W.; Couturier, M.; Berlin, A.; Filho, E.X.F.; Ezeji, T. Fungal Enzymes for Bio-Products from Sustainable and Waste Biomass. *Trends Biochem. Sci.* **2016**, *41*, 633–645. [CrossRef] [PubMed]
16. Paramjeet, S.; Manasa, P.; Korrapati, N. Biofuels: Production of fungal-mediated ligninolytic enzymes and the modes of bioprocesses utilizing agro-based residues. *Biocatal. Agric. Biotechnol.* **2018**, *14*, 57–71. [CrossRef] [PubMed]

17. Rabemanolontsoa, H.; Saka, S. Various pretreatments of lignocellulosics. *Bioresour. Technol.* **2016**, *199*, 83–91. [CrossRef] [PubMed]
18. Shon, Y.H.; Nam, K.S. Antimutagenicity and induction of anticarcinogenic phase II enzymes by basidiomycetes. *J. Ethnopharmacol.* **2001**, *77*, 103–109. [CrossRef]
19. Araújo, N.L.; Avelino, K.V.; Halabura, M.I.W.; Marim, R.A.; Kassem, A.S.S.; Linde, G.A.; Colauto, N.B.; do Valle, J.S. Use of green light to improve the production of lignocellulose-decay enzymes by Pleurotus spp. in liquid cultivation. *Enzym. Microb. Technol.* **2021**, *149*, 109860. [CrossRef]
20. Zhang, W.; Tang, Y.J. A novel three-stage light irradiation strategy in the submerged fermentation of medicinal mushroom *Ganoderma lucidum* for the efficient production of ganoderic acid and Ganoderma polysaccharides. *Biotechnol. Progr.* **2008**, *24*, 1249–1261. [CrossRef]
21. Tang, Y.J.; Zhong, J.J. Exopolysaccharide biosynthesis and related enzyme activities of the medicinal fungus, *Ganoderma lucidum*, grown on lactose in a bioreactor. *Biotechnol. Lett.* **2002**, *24*, 1023–1026. [CrossRef]
22. Ye, L.Y.; Liu, S.R.; Xie, F.; Zhao, L.L.; Wu, X.P. Enhanced production of polysaccharides and triterpenoids in *Ganoderma lucidum* fruit bodies on induction with signal transduction during the fruiting stage. *PLoS ONE* **2018**, *13*, e0196287. [CrossRef] [PubMed]
23. Löptien, J.; Vesting, S.; Dobler, S.; Mohammadi, S. Evaluating the efficacy of protein quantification methods on membrane proteins. *bioRxiv* **2024**. *preprint*. [CrossRef]
24. Zhu, P.L.; Zhang, J.; Li, S.Y.; Cui, X.G.; Yuan, S.J.; Wang, W.; Peng, C.; Zhou, J. Purification of total triterpenoids from fruiting body and spore powder of *Ganoderma lucidum* by Macroporous adsorption resins. *Food Res. Dev.* **2023**, *44*, 79–85.
25. Chai, S.M.; Zhang, X.L.; Jia, Z.Y.; Xu, X.F.; Zhang, Y.F.; Wang, S.C.; Feng, Z.Y. Identification and characterization of a novel bifunctional cellulase/hemicellulase from a soil metagenomic library. *Appl. Microbiol. Biotechnol.* **2020**, *104*, 7563–7572. [CrossRef] [PubMed]
26. Silva, E.M.; Milagres, A.M.F. Production of Extracellular Enzymes by *Lentinula edodes* Strains in Solid-State Fermentation on Lignocellulosic Biomass Sterilized by Physical and Chemical Methods. *Curr. Microbiol.* **2023**, *80*, 395. [CrossRef]
27. Valle, J.S.; Vandenberghe, L.P.; Oliveira, A.C.; Tavares, M.F.; Linde, G.A.; Colauto, N.B.; Soccol, C.R. Effect of different compounds on the induction of laccase production by *Agaricus blazei*. *Genet. Mol. Res.* **2015**, *14*, 15882–15891. [CrossRef]
28. Jin, J.; Kang, W.L.; Sheng, J.P.; Cheng, F.S.; Wang, Q.S.; Zhang, Y.X.; Zhang, G.P.; Sheng, L. Enzymological characteristics of lignin peroxidase (LiP) from *Coriolus versicolor*. *Food Sci.* **2010**, *31*, 224–227.
29. Peng, H.; Li, R.; Li, F.L.; Zhai, L.; Zhang, X.H.; Xiao, Y.Z.; Gao, Y. Extensive hydrolysis of raw rice starch by a chimeric α-amylase engineered with α-amylase (AmyP) and a starch-binding domain from *Cryptococcus* sp. S-2. *Appl. Microbiol. Biotechnol.* **2018**, *102*, 743–750. [CrossRef]
30. Arjona, D.; Aragón, C.; Aguilera, J.A.; Ramírez, L.; Pisabarro, A.G. Reproducible and controllable light induction of in vitro fruiting of the white-rot basidiomycete *Pleurotus ostreatus*. *Mycol. Res.* **2009**, *113*, 552–558. [CrossRef]
31. Ellis, R.J.; Bragdon, G.A.; Schlosser, B.J. Properties of the blue light requirements for primordia initiation and basidiocarp maturation in Coprinus stercorarius. *Mycol. Res.* **1999**, *103*, 779–784. [CrossRef]
32. Halabura, M.I.W.; Avelino, K.V.; Araújo, N.L.; Kassem, A.S.S.; Seixas, F.A.V.; Barros, L.; Fernandes, A.; Liberal, A.; Ivanov, M.; Soković, M.; et al. Light conditions affect the growth, chemical composition, antioxidant and antimicrobial activities of the white-rot fungus *Lentinus crinitus* mycelial biomass. *Photochem. Photobiol. Sci.* **2023**, *22*, 669–686. [CrossRef] [PubMed]
33. Tang, L.H.; Jian, H.H.; Song, C.Y.; Bao, D.P.; Shang, X.D.; Wu, D.Q.; Tan, Q.; Zhang, X.H. Transcriptome analysis of candidate genes and signaling pathways associated with light-induced brown film formation in *Lentinula edodes*. *Appl. Microbio. Biotechnol.* **2013**, *97*, 4977–4989. [CrossRef] [PubMed]
34. Kim, J.Y.; Kim, D.Y.; Park, Y.J.; Jang, M.J. Transcriptome analysis of the edible mushroom Lentinula edodes in response to blue light. *PLoS ONE* **2020**, *15*, e0230680. [CrossRef]
35. Deshpande, N.; Wilkins, M.R.; Packer, N.; Nevalainen, H. Protein glycosylation pathways in filamentous fungi. *Glycobiology* **2008**, *18*, 626–637. [CrossRef] [PubMed]
36. Zhang, X.; Dong, X.; Song, X.; Wang, F.; Dong, C. Photoperiodic Responses and Characterization of the *Cmvvd* Gene Encoding a Blue Light Photoreceptor from the Medicinal Caterpillar Fungus *Cordyceps militaris* (*Ascomycetes*). *Int. J. Med. Mushrooms* **2017**, *19*, 163–172. [CrossRef] [PubMed]
37. Xie, C.L.; Gong, W.B.; Zhu, Z.H.; Yan, L.; Hu, Z.X.; Peng, Y.D. Comparative transcriptomics of *Pleurotus eryngii* reveals blue-light regulation of carbohydrate-active enzymes (CAZymes) expression at primordium differentiated into fruiting body stage. *Genomic.* **2018**, *110*, 201–209. [CrossRef]
38. Ramírez, D.A.; Muñoz, S.V.; Atehortua, L.; Michel, F.C., Jr. Effects of different wavelengths of light on lignin peroxidase production by the white-rot fungi Phanerochaete chrysosporium grown in submerged cultures. *Bioresour. Technol.* **2010**, *101*, 9213–9220. [CrossRef]
39. Gan, P.T.; Lim, Y.Y.; Ting, A.S.Y. Influence of light regulation on growth and enzyme production in rare endolichenic fungi. *Folia Microbiol.* **2023**, *68*, 741–755. [CrossRef]

agriculture

MDPI

Article

# ISSR-Assisted Breeding of Excellent New Strains of *Ganoderma lingzhi* through Single-Spore Selfing

**Jintao Li †, Sheng Wang †, Qi Fan, Linling Liu, Yanliang Gao, Changwei Sun and Meixia Yan ***

Institute of Special Animal and Plant Sciences, Chinese Academy of Agricultural Sciences, Changchun 130112, China; lijintao@caas.cn (J.L.); 82101215186@caas.cn (S.W.); 82101225207@caas.cn (Q.F.); liulinling@caas.cn (L.L.); gaoyanliang@caas.cn (Y.G.); sunchangwei@caas.cn (C.S.)
* Correspondence: yanmeixia@caas.cn
† These authors contributed equally to this work.

**Abstract:** To improve our understanding of the selfing of *G. lingzhi* basidiospore monokaryons and increase the efficiency of breeding excellent strains, 52 basidiospore monokaryons were isolated from a commercial *G. lingzhi* strain (laboratory number P). A severe partial segregation was observed using the chi-square test, the growth rate of the monokaryotic strains was normally distributed, and colonies exhibited 5 forms. The genetic diversity of the monokaryotic strains was further demonstrated by intersimple sequence repeat (ISSR) analysis, and the similarity coefficient was in the range of 0.49–1, which was consistent with the genotype classification results. In total, 14 AxBx monokaryotic strains were randomly selected for selfing with the 1 AyBy strain when the similarity coefficient was 0.76, and a total of 14 offspring were obtained via selfing, all of which were incompatible with their parents. The traits of the selfing progenies were diverse. The mycelial growth rate, fruiting body yield, and polysaccharide, triterpene, and sterol contents were the main indices. According to the membership function value, 71.43% of the selfing progeny were super parent, and the A88 strain with the best comprehensive traits was selected. These findings prove that ISSR molecular marker-assisted breeding reduces blindness, greatly reduces workload, and improves work efficiency.

**Keywords:** *Ganoderma lingzhi*; basidiospore monokaryons; selfing; ISSR; genetic diversity

## 1. Introduction

*Ganoderma lingzhi* is a well-known medicinal fungus with important economic value that has a history of more than 2000 years in China [1]. *Ganoderma lingzhi* reportedly contains hundreds of bioactive secondary metabolites, consisting mainly of triterpenoids, polysaccharides, fatty acids, and sterols [2,3], and has a substantial effect on a variety of diseases. For example, pharmacological experiments showed that the *G. lingzhi* triterpenoids GL-4 and GL-8 had moderate cytotoxic effects on different cancer cell lines, inhibited nitric oxide production induced by lipopolysaccharide in RAW264.7 macrophage cells, and had dual anti-cancer and anti-inflammatory effects [4]. In addition, *G. lingzhi* also exhibits pharmacological effects such as immune regulation, antioxidant, germicidal, and hypoglycaemic properties [3,5–7]. Today, *G. lingzhi* products are greatly welcomed by the market with the increase in people's health awareness. At present, China's *G. lingzhi* industry is developing rapidly. The China Edible Fungi Association has reported that the output of *G. lingzhi* exceeded 250,000 tons in 2022, an increase of 21.63% from the output of 2021, and the output value of about 21.5 billion yuan in 2022.

Excellent strains are the basis of and the key to the healthy and rapid development of the *G. lingzhi* industry. It is highly important to select high-yield and quality strains, because continuous asexual reproduction in commercial production usually leads to strain quality degradation [8,9]. At present, the breeding methods commonly used for *G. lingzhi* include mutagenesis [10], crossbreeding [9], protoplast fusion [11], and molecular breeding [12]. Article 36 of the Standard for the Chinese Standard for the Production Quality Management of Chinese Medicinal Materials requires that polyploid or haploid cultivars, interspecific hybrids, and transgenic varieties produced by artificial intervention are prohibited in the breeding of Chinese medicinal materials. From this point of view, crossbreeding is the best way to select *G. lingzhi* for medicinal fungi. Crossbreeding, which can be divided into sexual and asexual, is one of the most commonly used methods to obtain new high-yield quality cultivars of edible fungi. Mushrooms produce spores through sexual reproduction in nature. After spore germination, sexual monokaryotic strains are produced, and fertile mating occurs between compatible monokaryotic strains [13]. Selfing is a kind of crossbreeding. Because sexual spores undergo meiosis and are the product of genetic recombination, they have high genetic diversity, resulting in differing traits in offspring [14,15], simultaneous selfing aggregation, and homozygous genes [13]. The F1 selfing progeny of *Agrocybe salicacola* were comprehensively analysed based on their qualitative traits (colour, deformity, and growth characteristics) and quantitative traits (shape, yield, size, and the number of fruiting bodies). It was found that 18% of the selfing progenies had better agronomic traits than their parents, and the dominant lines could be selected from the selfing progenies groups according to their different breeding purposes [16]. Strains with excellent agronomic properties were also cultivated in *Lentinula edodes* through selfing [17]. However, the traditional method of selfing breeding randomly selected basidiospore monokaryons for selfing and selected excellent new strains in a large number of progenies, which was blind and had a large workload.

The evaluation of genetic diversity will increase the effective use of genetic variation in breeding [18]. The genetic diversity analysis of strains can be performed in a variety of ways in edible fungi [19], for example, simple sequence repeat (SSR) [20], intersimple sequence repeat (ISSR) [21], random amplified polymorphic DNA (RAPD) [22], and sequence-related amplified polymorphism (SRAP) marker methods [23]. Different molecular markers have different limitations. ISSR markers can overcome these limitations and are widely used because of their high polymorphism, strong repeatability, simplicity, speed, and economy [24]. For example, ISSR molecular marker technology was conditioned to analyse the genetic diversity of 22 wild strains and 2 cultivars of *Auricularia heimuer* in 13 ecological regions in China, and it was found that the genetic differences within the region or between neighbouring strains were minimal, and that the genetic differences between the wild strains and the cultivated strains were the greatest [25]. In addition, ISSR molecular markers have also been conditioned to determine the genetic diversity of macrofungal monokaryotic strains, which is conducive to the breeding of excellent strains [26].

The mating compatibility that allows for heterokaryosis and fruiting body formation exists only between the mycelium carrying genes of different mating types and is the basis of crossbreeding [27]. *Ganoderma lingzhi* exhibits a tetrapolar heterozygotic mating system that produces a dikaryon capable of developing into a fruiting body through the fusion of two monokaryotic strains with different mating alleles at the A and B mating type sites [28]. After fruiting body maturation, the four types of haploid basidiospores are formed in large quantities in basidia via meiosis [9]. The genetic diversity of *G. lingzhi* basidiospore monokaryons was analysed, and the strains were classified according to the genetic distance in this study. Then, according to the classification results, monokaryotic strains were selected for selfing. Finally, the mycelial growth speed, fruiting body yield, and polysaccharide, triterpene, and sterol contents were used as the main indices, and the strains with an excellent overall performance were selected from among the hybrid strains, according to their membership functions.

## 2. Materials and Methods

### 2.1. Fungal Strains

The strains used in this paper are shown in Table 1. The parent strain analysed in this study was a commercial *G. lingzhi* strain, laboratory number strain P, which was preserved by the Edible and Medicinal Fungus Team at the Institute of Special Animal and Plant Sciences, Chinese Academy of Agricultural Sciences. The strain P was maintained on potato dextrose agar (PDA) slants containing 200 g potato (immersed), 20 g of glucose, 1.5 g of $KH_2PO_4$, 1.5 g of $K_2HPO_4$, 2 g of $MgSO_4$, 20 g of agar, and water up to 1 L at a natural pH of 4 °C.

**Table 1.** Sample data of *G. lingzhi* strains.

| Code | Strain No. | Type | Code | Strain No. | Type | Code | Strain No. | Type |
|---|---|---|---|---|---|---|---|---|
| 1 | P | parental strain | 24 | 43 | AyBy | 47 | 306 | AxBx |
| 2 | 20 | AyBy | 25 | 44 | AxBx | 48 | 337 | AxBx |
| 3 | 21 | AxBx | 26 | 45 | AyBy | 49 | 355 | AxBx |
| 4 | 22 | AyBy | 27 | 46 | AyBy | 50 | 382 | AxBx |
| 5 | 23 | AyBy | 28 | 48 | AxBx | 51 | 413 | AxBx |
| 6 | 24 | AyBy | 29 | 49 | AyBy | 52 | 418 | AxBx |
| 7 | 25 | AyBy | 30 | 50 | AxBx | 53 | 426 | AxBx |
| 8 | 26 | AxBx | 31 | 52 | AxBx | 54 | A21 | 21 × 31 |
| 9 | 27 | AyBy | 32 | 53 | AxBx | 55 | A28 | 28 × 31 |
| 10 | 28 | AxBx | 33 | 54 | AxBx | 56 | A53 | 53 × 31 |
| 11 | 29 | AyBy | 34 | 56 | AxBx | 57 | A54 | 54 × 31 |
| 12 | 31 | AyBy | 35 | 63 | AxBx | 58 | A88 | 88 × 31 |
| 13 | 32 | AxBx | 36 | 78 | AxBx | 59 | A97 | 97 × 31 |
| 14 | 33 | AyBy | 37 | 83 | AxBx | 60 | A203 | 203 × 31 |
| 15 | 34 | AyBy | 38 | 88 | AxBx | 61 | A217 | 217 × 31 |
| 16 | 35 | AyBy | 39 | 92 | AxBx | 62 | A226 | 226 × 31 |
| 17 | 36 | AyBy | 40 | 97 | AxBx | 63 | A337 | 337 × 31 |
| 18 | 37 | AyBy | 41 | 159 | AxBx | 64 | A382 | 382 × 31 |
| 19 | 38 | AyBy | 42 | 203 | AxBx | 65 | A413 | 413 × 31 |
| 20 | 39 | AyBy | 43 | 216 | AxBx | 66 | A418 | 418 × 31 |
| 21 | 40 | AyBy | 44 | 217 | AxBx | 67 | A426 | 426 × 31 |
| 22 | 41 | AyBy | 45 | 226 | AxBx | | | |
| 23 | 42 | AyBy | 46 | 300 | AxBx | | | |

× stands for selfing.

### 2.2. Collection of the Basidiospores and Isolation of the Monokaryotic Strains

To harvest the *G. lingzhi* spores, paper was wrapped around fruiting bodies during the maturation period, and sterile sulphate paper was placed on pileus. The spores collected on the sulphate paper were scraped into a centrifuge tube under sterile conditions, and sterile water was added to resuspend the spores on the second day. After the concentration was adjusted to approximately $1 \times 10^6$ spores/mL, 0.1 mL of the spore suspension was collected and uniformly coated on PDA (CM123, Land Bridge, Beijing, China) media and incubated at 25 °C.

After spore germination, a single colony was carefully selected on PDA (CM123A, Land bridge, Beijing, China) culture media with an inoculation needle, and a monokaryotic strain without clamp connections was observed under a 400× optical microscope (NLCD500, Jiangnan Yongxin Optical Co., Ltd., Nanjing, China).

### 2.3. Spore Mating Type Analysis and Chi-Square Test

For mating type determination, the methods of Ke et al. [29] were used. The chi-square test was used to detect 4 types of basidiospore monokaryons to verify whether the proportion of monokaryotic strains of the different mating types exhibited free separation.

### 2.4. ISSR Profiling

Genomic DNA was extracted using an Ezup Column Fungi Genomic DNA Purification Kit (B518259, Sangon Biotech, Shanghai, China). The ISSR primer sequences and amplification reaction conditions followed those described by Li et al. [11].

### 2.5. Mycelial Growth Rate and Fructification

The activated mycelia were inoculated in the centre of the PDA media and cultured in a constant temperature incubator at 25 °C. The colony radius was measured on the 5th day after inoculation, and the mycelial growth rates were calculated (mm/day).

The activated mycelia were inoculated in a test tube containing sawdust media (30 mm × 200 mm). The sawdust media contained 76% sawdust, 20% wheat bran, 2% gypsum, 2% glucose, and 60% water content. After inoculation, the mycelia were cultured vertically in a constant temperature incubator at 25 °C. The mycelia height was measured on the 12th and 17th days after inoculation, and the mycelial growth rate was calculated (mm/d).

The composition of the experimental mushroom cultivation material was the same as that described above. Each mushroom bag weighed 450 g and was incubated in a constant temperature incubator at 25 °C after inoculation. After complete colonization, the cultures were moved to the greenhouse for the fruiting experiment, and the white edge of the pileus disappeared as the fruiting bodies matured.

### 2.6. Determination of Total Polysaccharides, Triterpenes, and Sterols in Fruiting Bodies

The fruiting bodies were placed in an electric blast drying oven (GZX-9140, Boxun, Shanghai, China) at 60 °C and then crushed after drying, after maturation. The crushed fruiting bodies were used for the determination of total polysaccharides, triterpenes, and sterols. Specific methods referred to the method of the Pharmacopoeia of the People's Republic of China improved by Li et al. [11].

### 2.7. Statistical Analysis

Based on the ISSR amplification electrophoretic map, a 0/1 matrix was established, the genetic similarity coefficient was calculated by NTSYS-pc 2.10 software based on the DICE coefficient, and cluster analysis was performed by the unweighted pair group method with an arithmetic mean (UPGMA) [25]. All experiments included three or more biological and technical replicates. IBM SPSS Statistics 23 was used for the statistical analysis of the experimental data, and GraphPad Prism 8.0 was used to construct the diagrams. The letters in the diagrams indicate significant differences at the $p < 0.05$ level. The membership function method was used to evaluate the hybrid strains comprehensively, and the calculation formula was as follows:

$$\mu(Xi) = (Xi - Xmin)/(Xmax - Xmin) \quad (1)$$

where $\mu(Xi)$ is the membership function value, $i$ is the indicator, $Xi$ is the measured value of an indicator, and $Xmax$ and $Xmin$ are the maximum and minimum measured values of an indicator, respectively [30]. After calculating each index of fruiting bodies of the same strain, the comprehensive quality measurement value of the strain was obtained by arithmetic averaging, and the strains were ranked.

## 3. Results

### 3.1. Characteristics of the Monokaryotic Strains

A total of 52 basidiospore monokaryons were selected via microscopic examination (Figure S1), among which AxBx:AyBy:AxBy:AyBx = 15:11:0:0 (Figure 1A), $\chi^2 = 54.46$, and $\chi^2_{0.05} = 7.81 < 54.46$ when n = 3 degrees of freedom, which did not meet the 1:1:1:1 free separation ratio, exhibited partial separation. Figure 1B shows the mycelial growth rate frequency of the monokaryotic strains on PDA media. The growth rate of monokaryotic strains presented a normal distribution with a continuous change trend, exhibiting an average value of 3.7. According to the Kolmogorov–Smirnoff (K–S) test, $p = 0.2$, which was greater than 0.05 and conforms to a normal distribution, indicating that the mycelial growth rate was a quantitative trait controlled by polygenes. The lowest mycelial growth rate was 2.41 ± 0.19 mm/d, and the highest was 6.16 ± 0.17 mm/d, with a coefficient of variation of 0.23, exhibiting a large difference and a significant character separation (Figure 1C). There were five types of mycelia overall, namely, compact, fluffy, central compact, grey, and fading, with most being of the central compact type and three being fading types (Figure 1D).

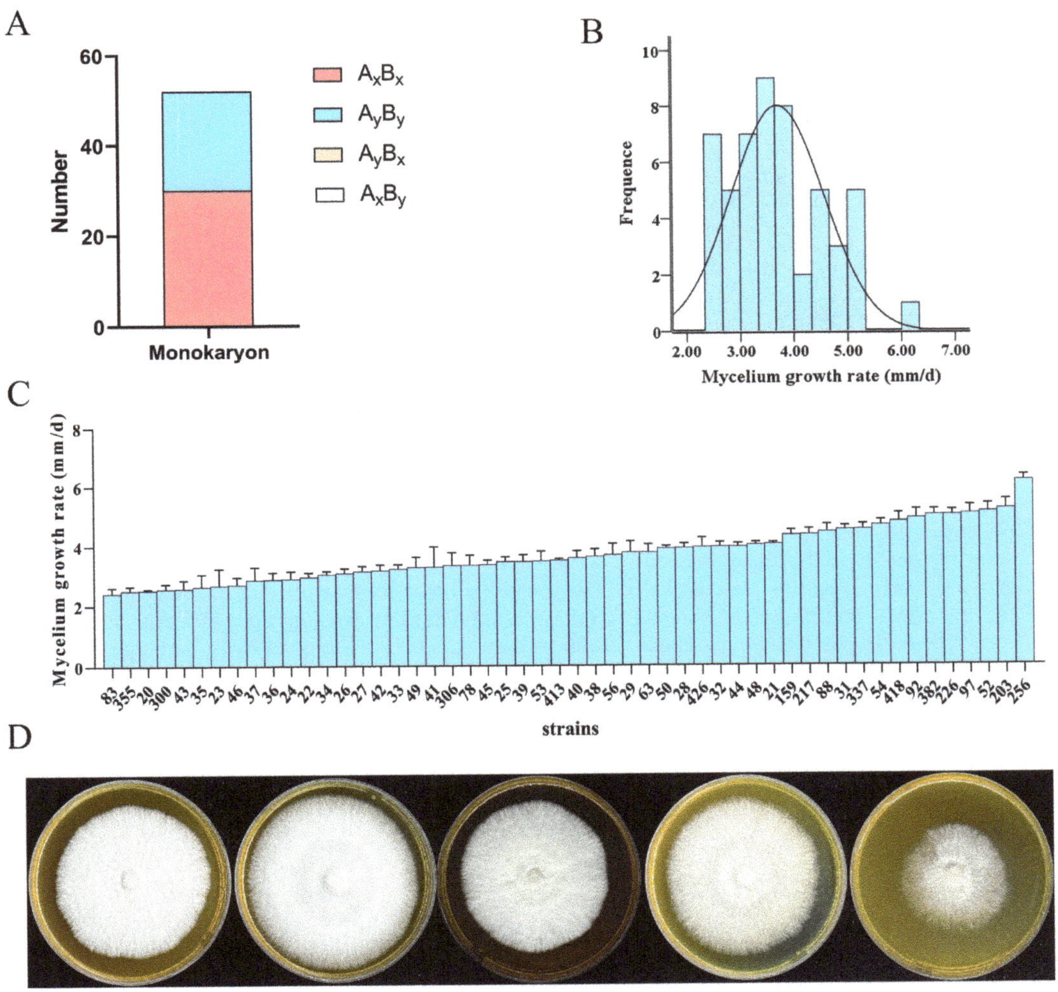


**Figure 1.** Characteristics of the monokaryotic strains. (**A**) The number of monokaryotic strains of the four genotypes. (**B**) Mycelial growth rate frequency distribution. (**C**) Mycelial growth rate. (**D**) Colony morphology of the monokaryotic strains. The morphology of the colony from left to right are named as compact, fluffy, central compact, grey, and fading.

### 3.2. Genetic Diversity Analysis of the Monokaryotic Strains

ISSR-PCR was performed on 52 monokaryotic strains and their parents using 12 primers (Table 2). The different primers were amplified between 1 and 10 bands, and a total of 57 bands were amplified, with an average of 4.75 bands per primer. The numbers of polymorphic bands amplified by each primer and the polymorphism ratio were statistically analysed. The average polymorphic band ratio was 98.32%, and the polymorphism ratio of the bands amplified by the 11 primers was 100%, while the polymorphism ratio of the bands amplified by ISSR4 was 0%. The ISSR4 primer may facilitate the identification of the parent and its basidiospore monokaryons. In addition, the bands amplified with ISSR1, ISSR6, ISSR8, ISSR9, ISSR10, ISSR11, and ISSR16 exhibited one genotype, and the other genotype was absent, which could be used as the basis for genotype identification. In conclusion, the frequency of genetic polymorphisms among the monokaryotic strains was high.

**Table 2.** ISSR primer amplification results and polymorphism analysis.

| Primer | Sequence (5′-3′) | Total Bands | Polymorphic Bands | Percentage of Polymorphic Bands (%) |
|---|---|---|---|---|
| ISSR-1 | CACCACACACACACACA | 5 | 5 | 100 |
| ISSR-3 | GAGAGAGAGAGAGAGACC | 3 | 3 | 100 |
| ISSR-4 | AGCAGCAGCAGCAGCAGCG | 1 | 0 | 0 |
| ISSR-5 | TGCACACACACACAC | 8 | 8 | 100 |
| ISSR-6 | GAGAGAGAGAGAGAGAT | 4 | 4 | 100 |
| ISSR-7 | AGAGAGAGAGAGAGAGC | 2 | 2 | 100 |
| ISSR-8 | CACACACACACACACAT | 4 | 4 | 100 |
| ISSR-9 | GAGAGAGAGAGAGAGACT | 5 | 5 | 100 |
| ISSR-10 | TTCCCTTCCCTTCCC | 4 | 4 | 100 |
| ISSR-11 | GTGACACACACACAC | 10 | 10 | 100 |
| ISSR-12 | AGTGTGTGTGTGTGT | 5 | 5 | 100 |
| ISSR-16 | GGATGCAACACACACACAC | 6 | 6 | 100 |
| | All number | 57 | 56 | |
| | Average value | 4.75 | 4.67 | 98.32 |

The results of the cluster analysis are shown in Figure 2. In total, 53 strains were divided into 2 categories at a similarity coefficient of 0.49, which was consistent with the genotype classification results. The similarity of strains 48, 32, and 50, in terms of the AxBx genotype was 100%. The similarity of all monokaryotic strains of the AyBy genotype was 100%. The parental and AxBx genotype strains were grouped into class 1. The AxBx genotype strains exhibited a greater genetic diversity, and the AyBy genotype strains exhibited a greater similarity.

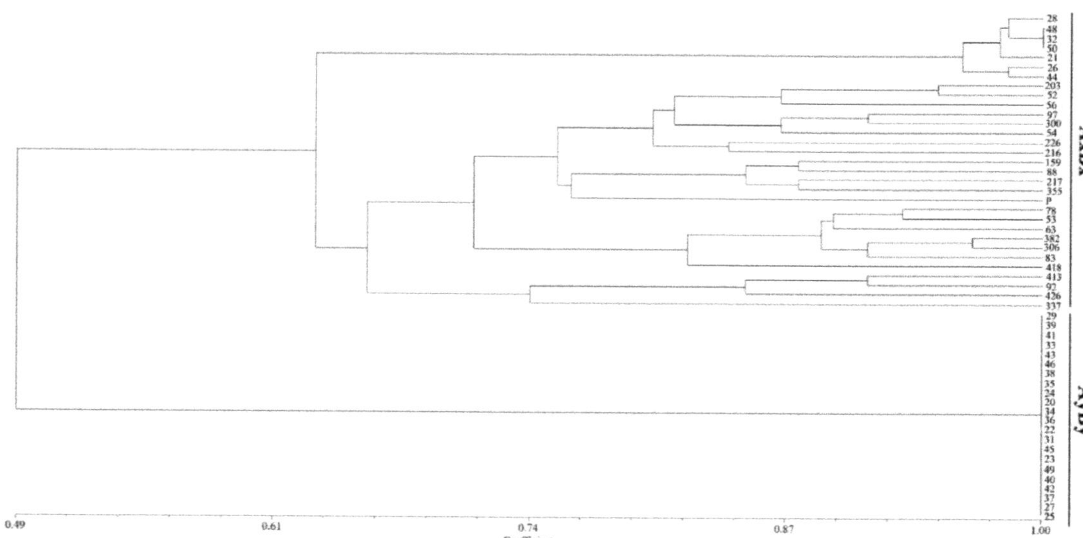


**Figure 2.** ISSR affinity cluster analysis of the basidiospore monokaryons.

The traditional breeding method randomly selects the mating basidiospore monokaryons for selfing, which is blind and usually forms dozens of selfing combinations. According to the results of the cluster map, 53 strains were divided into 8 categories at a similarity coefficient of 0.76, and 14 AxBx strains (28, 21, 203, 226, 97, 54, 88, 217, 53, 382, 418, 413, 426, and 337) were randomly selected for crossing with the AyBy genotype 31 in this study. A total of 14 selfing progenies were formed.

### 3.3. Identification of the Hybrid Strains

The 14 hybrids of the randomly selected AxBx strain and AyBy genotype 31 produced clamp connections (Figure S2), which indicated that the hybridisation was successful. As shown in Figure 3A, the mycelia of the monokaryotic strains were whiter and thicker, while the mycelia of the strain after hybridisation were sparser compared to those of the strain before hybridisation. Another obvious feature was that the AxBx monokaryotic strain had a greater effect on the morphology of the hybrid strains, such as strains 21 and A21 and 28 and A28, which exhibited a flocculent colony morphology. In addition, the colony morphology of the obtained hybrid strains was similar to that of the parent P. ISSR electrophoretic results. (Figure 3B) showed that the electrophoretic bands of hybrids were a combination of the monokaryotic strains bands. According to the antagonism test (Table S1) and the ISSR results (Figure 3C), 14 hybrids were obtained via hybridisation, and they were all different from parent P. The results for hybrids A28 and A21 showed the same strain. The calculated genetic similarity coefficient was 0.98, which indicates that A28 and A21 are very close to each other.

### 3.4. Agronomic Traits and Comprehensive Evaluation

To compare the correlation between the growth rate of the monokaryotic strains and the growth rate of the hybrid progeny mycelia, and between the growth rate of mycelia on different media, the growth rates of mycelia on the PDA media and sawdust media were measured. The growth rate of monokaryotic strain 413 on the PDA (5.40 ± 0.85 mm/d) was significantly greater than that of the other strains, and the growth rate of strain 337 was the lowest (3.64 ± 0.29 mm/d) (Figure 4A). The growth rate of the hybrid strains on the PDA media was greater than that of the monokaryotic strains, and the growth rate of the hybrid strains ranged from 6.95 ± 0.23 mm/d–8.68 ± 0.33 mm/d, with an average growth rate of 7.58 mm/d. The K–S test showed that $p = 0.102$, which was greater than 0.05 and was consistent with a normal distribution. Strain P may have degenerated, and its growth rate (3.69 ± 1.16 mm/d) was significantly lower than that of the hybrid strains (Figure 4B). The growth rate of the monokaryotic strains was relatively concentrated on the sawdust media, and the growth rates of these 10 strains were significantly greater than that of the remaining 5 strains. The growth rate of strain 97 was the lowest, at 3.15 ± 0.21 mm/d, and the growth rates of the monokaryotic strains were normally distributed in the sawdust media (Figure 4C). The growth rate of P was relatively fast (5.75 ± 0.37 mm/d) on the sawdust media, which was greater than the average growth rate of the offspring (5.52 mm/d); the growth rate of A337 was the greatest (6.89 ± 0.43 mm/d), and that of A21 was the lowest (4.77 ± 0.91 mm/d). The median growth rate of the hybrid strains was 5.46 mm/d, which was lower than the average (Figure 4D). By calculating the Pearson correlation coefficient, the growth rate correlation between the AxBx strain and the hybrid offspring on the PDA ($r^2 = -0.299$, $p = 0.299$) and the growth rate correlation on the sawdust media ($r^2 = 0.34$, $p = 0.235$) were poor. There was a negative, but not significant, correlation between the growth rate of the monokaryotic strains on the PDA and sawdust media ($r^2 = -0.367$, $p = 0.197$) and between the growth rate of the hybrid progeny on the PDA and sawdust media ($r^2 = -0.465$, $p = 0.094$).

The fresh weight and total contents of polysaccharides, triterpenes, and sterols in fruiting bodies of *G. lingzhi* were normally distributed ($p$ values were 0.2, 0.17, and 0.2, respectively). A21 and A28 had weak anti-bacterial abilities and failed to produce fruiting bodies, the phenotype of the other strains was normal (Figure S3). The fruiting body formation rate was 85.71%. The fresh weight of the parent strains was 20.94 ± 4.96 g, and the fresh weight of A413 in the offspring was the highest, reaching 33.64 ± 1.97 g, which was 1.6 times that of the parents. The consistency in the fresh weight of A54 was poor (mass range 18.45–38.84 g). There was no significant difference between A418 and its parents, and A426 was the only strain with a lower mean fresh weight than its parents (Figure 4E). The polysaccharide content of P was high (2.47 ± 0.02%), the polysaccharide content of A382 was not significantly different from that of P, and the polysaccharide content of the other progeny was significantly lower than that of P, but the polysaccharide content of all strains exceeded the pharmacopoeia standard (Figure 4F). The triterpene and sterol contents of A382 were the lowest (0.72 ± 0.01%), and the triterpene and sterol contents of P were 1.03 ± 0.02%, which were not significantly different from those of A203 and A426, while those of A54, A88, and A337 were greater than those of their parents. The highest triterpene and sterol contents of A88 was 1.10 ± 0.07, which was not significantly different from that of A337 (Figure 4G).

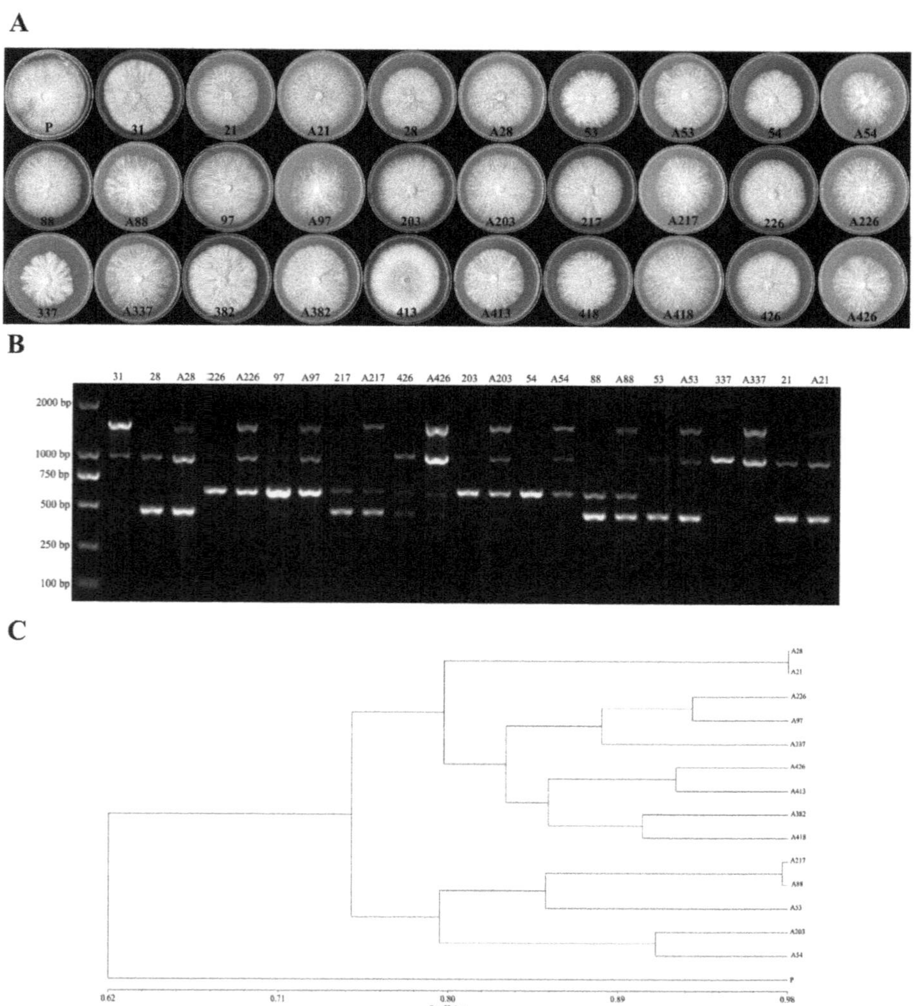


**Figure 3.** Identification of the hybrid strains. (**A**) Colony morphology. (**B**) ISSR electrophoretic map of the monokaryotic strains and their hybrids. (**C**) Phylogenetic cluster analysis of the hybrid strains.

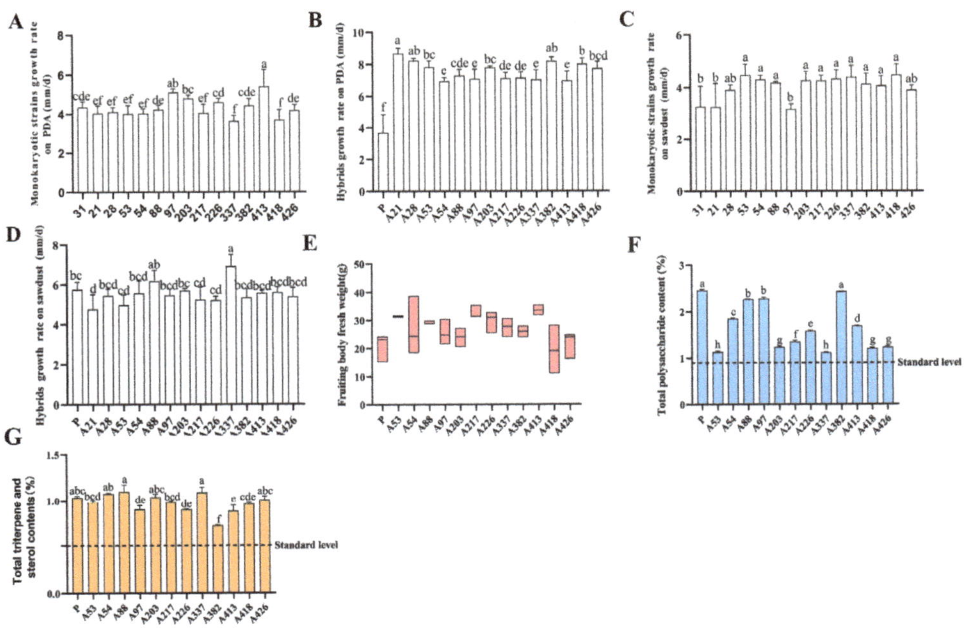

**Figure 4.** Agronomic traits. (**A**) Growth rate of the monokaryotic strains on the PDA media; (**B**) Growth rate of the hybrid strains on the PDA media; (**C**) Growth rate of the monokaryotic strains on the sawdust media; (**D**) Growth rate of the hybrid strains on the sawdust media; (**E**) Fruiting body yield; (**F**) Total polysaccharide content of the fruiting body; (**G**) Triterpenoid and sterol contents in the fruiting bodies. The values are the means ± standard deviations of three or more independent experiments. Different letters indicate significant differences in sample comparisons ($p < 0.05$ according to Duncan's test).

The mycelial growth rate on the PDA media and sawdust media, fruiting body yield, and total polysaccharide, triterpene, and sterol contents were compared, based on the membership functions (Table 3). A88 had the best comprehensive traits, and A418 had the worst. P ranked 11th due to its low growth rate on the PDA and low yield, and 10 strains had more comprehensive traits than P, with a super parent rate of 71.43%. Only a few selfing combinations are needed to obtain a new strain that is superior to the parent, compared to the traditional breeding method.

**Table 3.** Comprehensive evaluation of the hybrid strains.

| Strain | μ (x1) | μ (x2) | μ (x3) | μ (x4) | μ (x5) | Average Value | Ranking |
|---|---|---|---|---|---|---|---|
| P | 0.10 | 1.00 | 0.83 | 0.40 | 0.00 | 0.47 | 11 |
| A53 | 0.84 | 0.01 | 0.69 | 0.00 | 0.92 | 0.49 | 9 |
| A54 | 0.55 | 0.54 | 0.94 | 0.30 | 0.72 | 0.61 | 3 |
| A88 | 0.21 | 0.86 | 1.00 | 0.63 | 0.80 | 0.70 | 1 |
| A97 | 0.40 | 0.86 | 0.47 | 0.26 | 0.75 | 0.55 | 5 |
| A203 | 0.32 | 0.08 | 0.84 | 0.37 | 0.91 | 0.51 | 8 |
| A217 | 0.93 | 0.17 | 0.69 | 0.14 | 0.76 | 0.54 | 6 |
| A226 | 0.75 | 0.34 | 0.47 | 0.13 | 0.77 | 0.49 | 10 |
| A337 | 0.57 | 0.00 | 0.97 | 1.00 | 0.74 | 0.66 | 2 |
| A382 | 0.46 | 0.98 | 0.00 | 0.19 | 1.00 | 0.53 | 7 |
| A413 | 1.00 | 0.42 | 0.43 | 0.31 | 0.73 | 0.58 | 4 |
| A418 | 0.00 | 0.06 | 0.64 | 0.33 | 0.96 | 0.40 | 13 |
| A426 | 0.15 | 0.09 | 0.75 | 0.22 | 0.90 | 0.42 | 12 |

## 4. Discussion

Basidiospore monokaryons are the product of gene recombination during meiosis, and they exhibit high genetic diversity [31]. The screening of the monokaryotic strains with good traits is an important prerequisite for improving the efficiency of new mushroom variety development [17]. Theoretically, *G. lingzhi* can produce spores of four genotypes in the same proportion [32]. In this study, only two spore genotypes were isolated, $\chi^2 = 54.46$, and $\chi^2_{0.05} = 7.81 < 54.46$ when n = 3 degrees of freedom, which did not meet the 1:1:1:1 free separation ratio but exhibited partial separation, resulting in severe partial segregation, which has been reported in a variety of edible fungi [33]. It has been reported that there are differences in the germination ability or the survival ability of spores carrying different mating types, leading to the occurrence of partial segregation in sexual monokaryotic strains isolation [34]. There were differences in the growth rates of monokaryotic strains. In this study, the growth rates of the monokaryotic strains showed a normal distribution with a continuous change trend, indicating that the mycelial growth rate was a quantitative trait controlled by polygenes. Currently, it is believed that the mating type A site is related to the mycelial growth rate, and QTL mapping has revealed the main genomic areas responsible for growth rate regulation and control. The primary effect site that controls the mycelial growth rate is near the mating type A site [35,36]. For example, the overexpression of *LeHD1* gene can increase the mycelial growth rate of mushrooms by 8.6–18.5%, and the silencing of *LeHD1* can reduce the mycelial growth rate by 8.5–26.8% [37]. The diversity of basidiospore monokaryons is also reflected in colony morphology. In this study, the basidiospore monokaryons exhibited five colony morphologies, and one colony morphology showed obvious weakness. Molecular markers were important for analysing genetic diversity, and most previous studies have focused on the dikaryotic cultivars and wild strains [20,38]. Few studies have analysed basidiospore monokaryons. Using ISSR molecular markers, we found that at a similarity coefficient of 0.49, these 53 basidiospore monokaryons of *G. lingzhi* were divided into two categories, which was consistent with genotype classification results. The AxBx strains had high genetic diversity, while the AyBy genotype had a similarity coefficient of 100%. Based on these results, strains can be selected for hybridisation, which is a scientific and reasonable approach for reducing the workload. The genetic diversity of the basidiospore monokaryons of edible fungi, such as *Auricularia heimuer*, *Lentinula edodes*, and *Pleurotus ostreatus*, was also proven by the molecular markers, indicating that it is feasible to screen for Mono–mono mating via molecular markers in edible and medicinal fungi. However, the results showed that hybrid strains were not strictly classified according to the mating type, which may be related to the characteristics of the parental strains. This method can be used as an auxiliary means for identifying the mating type [26,39,40].

ISSR-assisted monokaryotic strains selfing is one of the most promising breeding methods for *G. lingzhi*. On the one hand, the breeding methods for Chinese medicinal materials are restricted by the Chinese Standard for the Production Quality Management of Chinese Medicinal Materials. On the other hand, according to the classification results of ISSR genetic diversity, the mycelial growth rate, yield, and polysaccharide, triterpene, and sterol contents of the 14 hybrids were diverse. The growth rate is an important index for breeding new strains, and the rapid growth of strains can reduce microbial contamination and shorten the production cycle [41]. We found that there was little correlation between the growth rate of monokaryotic strains and the growth rate of the mycelia of hybrids, and the growth rate of hybrids was greater than that of the monokaryotic strains, which was also reported in macrofungi [42], possibly due to the complementary coordination mechanism between the two cell nucleus in somatic cells [34,43]. Transcriptome analysis of the hybrid dikaryon and the two parental monokaryotic strains of *Pleurotus ostreatus* showed that 7953 and 7787 genes were upregulated more than 3 times, and that 8421 and 7425 genes were downregulated, respectively, compared to their parental monokaryotic strains. These genes are related to macromolecule utilization, cell material synthesis, stress resistance and signal transduction. This also proved that these two nuclei were co-ordinated [44]. There may be various reasons for the low growth correlation of mycelia in the PDA media and sawdust media. The composition of the PDA media is simple, and mycelia can directly absorb and utilize substances such as glucose, while the sawdust media is composed mainly of wood chips, and the composition is complex and diverse. Extracellular enzymes, such as lignocellulosic degrading enzymes, need to be secreted to degrade the matrix before it can be absorbed and utilized [45]. It is also necessary to secrete organic acids to regulate environmental pH to create the optimal conditions for growth and development [46]. Polysaccharides and triterpenes are the main active substances of *G. lingzhi.*, which reflects the strain quality [11]. Fruiting body yield is an important agronomic trait, and the traits of selfing offspring are diverse, with yields per packet ranging from 19.49–33.64 g and polysaccharide contents ranging from 1.12–2.44%. The contents of triterpenes and sterols ranged from 0.72 to 1.1%, and the highest polysaccharide, triterpene and sterol contents were observed in A413, A382, and A88, respectively, which could meet different market demands. The membership function value is a comprehensive value obtained by using the basic theory of fuzzy mathematics and the membership function method from the perspective of the membership degree [47]. In combination with the five important traits, the growth rate on PDA and sawdust media, the yield, the polysaccharide content, and the triterpene and sterol contents, the membership function value is sorted according to the membership function value. The results showed that the percentage of selfing progeny that performed better than the parental strains was 71.43%, and A88 was the best selfing progeny, which may be a good strain for commercial cultivation.

In this study, we confirmed the rich genetic diversity of the *G. lingzhi* basidiospore monokaryons by ISSR, which can also be used as an auxiliary means for identifying the mating type. According to the ISSR clustering results, monokaryotic strains can be screened for selfing, the best comprehensive characteristics of the strains can be obtained, and the workload can be greatly reduced.

## 5. Conclusions

In this study, 52 basidiospore monokaryons were isolated from the fruiting bodies of one commercially cultivated *G. lingzhi* (laboratory number strain P) in Jilin Province, China. The mating type was partial segregation. ISSR molecular markers proved that basidiospore monokaryons had rich genetic diversity. The clustering results can be used as the basis for the screening selfing of monokaryons. A similarity coefficient of 0.49 can distinguish monokaryons with different mating types. At the level of 0.76, the super parent rate of the selfing progenies of monokaryons randomly selected according to the grouping results reached 71.43%. This study proves that ISSR molecular marker-assisted self-breeding can effectively select new strains of *G. lingzhi* with high yield and high quality.

**Supplementary Materials:** The following supporting information can be downloaded at: https://www.mdpi.com/article/10.3390/agriculture14050745/s1, Figure S1: Morphology of monokaryotic strains mycelium; Figure S2: Morphology of hybrid mycelium; Figure S3: Fruiting body morphology of hybrids; Table S1: Antagonism between parents and hybrids.

**Author Contributions:** Conceptualization, M.Y. and J.L.; methodology, J.L.; validation, S.W., Q.F. and L.L.; formal analysis, J.L.; investigation, J.L. and S.W.; resources, Y.G.; writing—original draft preparation, J.L. and S.W.; writing—review and editing, M.Y. and J.L.; visualization, J.L.; supervision, M.Y. and C.S.; project administration, M.Y.; funding acquisition, M.Y. All authors have read and agreed to the published version of the manuscript.

**Funding:** This research was funded by the Agricultural Science and Technology Innovation Program (CAAS-ASTIP-2021-ISAPS) and Demonstration and Promotion of Key Agricultural Core Technologies in Jilin Province (Industrial Technology System 202400601).

**Institutional Review Board Statement:** Not applicable.

**Data Availability Statement:** The data presented in this study are available in the article.

**Conflicts of Interest:** The authors declare no conflicts of interest.

## References


1. Cao, Y.; Wu, S.H.; Dai, Y.C. Species clarification of the prize medicinal *Ganoderma* mushroom "Lingzhi". *Fungal Divers.* **2012**, *56*, 49–62. [CrossRef]
2. Xia, J.; He, X.Y.; Yang, W.; Song, H.Y.; Yang, J.H.; Zhang, G.L.; Yang, Z.Q.; Chen, H.M.; Liang, Z.S.; Kollie, L.; et al. Unveiling the distribution of chemical constituents at different body parts and maturity stages of *Ganoderma lingzhi* by combining metabolomics with desorption electrospray ionization mass spectrometry imaging (DESI). *Food Chem.* **2024**, *436*, 137737. [CrossRef]
3. Ren, L.; Zhang, J.; Zhang, T.H. Immunomodulatory activities of polysaccharides from *Ganoderma* on immune effector cells. *Food Chem.* **2021**, *340*, 127933. [CrossRef]
4. Zhao, Z.Z.; Ji, B.Y.; Wang, Z.Z.; Si, Y.Y.; Sun, Y.J.; Chen, H.; Feng, W.S.; Zheng, X.K.; Liu, J.K. Lanostane triterpenoids with anti-proliferative and anti-inflammatory activities from medicinal mushroom *Ganoderma lingzhi*. *Phytochemistry* **2023**, *213*, 113791. [CrossRef]
5. Ferreira, I.C.; Heleno, S.A.; Reis, F.S.; Stojkovic, D.; Queiroz, M.J.; Vasconcelos, M.H.; Sokovic, M. Chemical features of *Ganoderma* polysaccharides with antioxidant, antitumor and antimicrobial activities. *Phytochemistry* **2015**, *114*, 38–55. [CrossRef]
6. Ren, L. Protective effect of ganoderic acid against the streptozotocin induced diabetes, inflammation, hyperlipidemia and microbiota imbalance in diabetic rats. *Saudi J. Biol. Sci.* **2019**, *26*, 1961–1972. [CrossRef]
7. Wu, S.J.; Zhang, S.Y.; Peng, B.; Tan, D.C.; Wu, M.Y.; Wei, J.C.; Wang, Y.T.; Luo, H. *Ganoderma lucidum*: A comprehensive review of phytochemistry, efficacy, safety and clinical study. *Food Sci. Hum. Wellness* **2024**, *13*, 568–596. [CrossRef]
8. Zhou, X.W.; Su, K.Q.; Zhang, Y.M. Applied modern biotechnology for cultivation of *Ganoderma* and development of their products. *Appl. Microbiol. Biotechnol.* **2012**, *93*, 941–963. [CrossRef]
9. Liu, S.R.; Ke, B.R.; Zhang, W.R.; Liu, X.R.; Wu, X.P. Breeding of new *Ganoderma lucidum* strains simultaneously rich in polysaccharides and triterpenes by mating basidiospore-derived monokaryons of two commercial cultivars. *Sci. Hortic-Amst.* **2017**, *216*, 58–65. [CrossRef]
10. Tang, C.H.; Tan, Y.; Zhang, J.S.; Zhou, S.; Honda, Y.; Zhang, H.N. A novel strain breeding of *Ganoderma lucidum* UV119 (Agaricomycetes) with high spores yield and strong resistant ability to other microbes' invasions. *Foods* **2023**, *12*, 465. [CrossRef] [PubMed]
11. Li, J.T.; Liu, L.L.; Xu, L.; Wang, S.; Zhang, N.; Sun, C.W.; Yan, M.X. Interspecific hybridization between *Ganoderma lingzhi* and *G. resinaceum* by PEG-Induced double-inactivated protoplast fusion. *Horticulturae* **2023**, *9*, 1129. [CrossRef]
12. Liu, Y.N.; Wu, F.Y.; Tian, R.Y. Shi, Y.X.; Xu, Z.Q.; Liu, J.Y.; Huang, J.; Xue, F.F.; Liu, B.Y.; Liu, G.Q. The bHLH-zip transcription factor SREBP regulates triterpenoid and lipid metabolisms in the medicinal fungus *Ganoderma lingzhi*. *Commun. Biol.* **2023**, *6*, 1. [CrossRef] [PubMed]
13. Bao, D.P. Scientific problems in crossbreeding of edible fungi. *Acta Edulis Fungi* **2020**, *27*, 1–24. [CrossRef]
14. Pan, Y.J.; Chen, M.J.; Wang, Z.Y.; Ma, A.M.; Ling, X.F.; He, D.M.; Feng, Z.Y.; Bo, H.Y. Application of monokaryon and homokaryon protoplast technology in genetic and breeding of edible mushroom. *Acta Edulis Fungi* **1994**, *1*, 56–62. [CrossRef]
15. Cai, M.J.; Liang, X.W.; Liu, Y.C.; Hu, H.P.; Xie, Y.Z.; Chen, S.D.; Gao, X.; Li, X.M.; Xiao, C.; Chen, D.L.; et al. Transcriptional dynamics of genes purportedly involved in the control of meiosis, carbohydrate, and secondary metabolism during sporulation in *Ganoderma lucidum*. *Genes* **2021**, *12*, 504. [CrossRef] [PubMed]
16. Chai, H.M.; Zhou, H.M.; Zhao, J.; Chen, W.M.; Zhao, Y.C. Searching Development-deficient Genes in Edible Mushroom by Self-crossing. *J. Agr. Sci. Tech.-Iran.* **2012**, *13*, 2037–2043. [CrossRef]
17. Ha, B.S.; Kim, S.; Ro, H.S. Isolation and characterization of monokaryotic strains of *Lentinula edodes* showing higher fruiting rate and better fruiting body production. *Mycobiology* **2018**, *43*, 24–30. [CrossRef]
18. Paterson, A.H.; Tanksley, S.D.; Sorrells, M.E. DNA Markers in Plant Improvement. *Adv. Agron.* **1991**, *46*, 39–90. [CrossRef]
19. Liu, L.L.; Yang, Y.M.; Zhang, L.; Yan, M.X. Progresses on Application of Molecular Markers in *Ganoderma lucidum*. *Spec. Wild Econ. Anim. Plant Res.* **2024**, *46*, 163–167. [CrossRef]
20. Liu, X.B.; Feng, B.; Li, J.; Yan, C.; Yang, Z.L. Genetic diversity and breeding history of Winter Mushroom (*Flammulina velutipes*) in China uncovered by genomic SSR markers. *Gene* **2016**, *591*, 227–235. [CrossRef]
21. Zhang, R.Y.; Huang, C.Y.; Zheng, S.Y.; Zhang, J.X.; Ng, T.B.; Jiang, R.B.; Zuo, X.M.; Wang, H.X. Strain-typing of *Lentinula edodes* in China with inter simple sequence repeat markers. *Appl. Microbiol. Biotechnol.* **2007**, *74*, 140–145. [CrossRef] [PubMed]
22. Fu, L.Z.; Zhang, H.Y.; Wu, X.Q.; Li, H.B.; Wei, H.L.; Wu, Q.Q.; Wang, L.A. Evaluation of genetic diversity in *Lentinula edodes* strains using RAPD, ISSR and SRAP markers. *World J. Microbiol. Biotechnol.* **2009**, *26*, 709–716. [CrossRef]
23. Sun, S.J.; Gao, W.; Lin, S.Q.; Zhu, J.; Xie, B.G.; Lin, Z.B. Analysis of genetic diversity in *Ganoderma* population with a novel molecular marker SRAP. *Appl. Microbiol. Biotechnol.* **2006**, *72*, 537–543. [CrossRef] [PubMed]
24. Pradeep Reddy, M.; Sarla, N.; Siddiq, E.A. Inter simple sequence repeat (ISSR) polymorphism and its application in plant breeding. *Euphytica* **2002**, *128*, 9–17. [CrossRef]
25. Du, P.; Cui, B.K.; Zhang, C.F. Dai, Y.C. Genetic diversity of wild *Auricularia auricula-judae* revealed by ISSR analysis. *Biochem. Syst. Ecol.* **2013**, *48*, 199–205. [CrossRef]
26. Song, X.Y.; Xiao, Y.; Bian, Y.B. Application of ISSR marker in the genetics analysis of monokaryons from *Auricularia auricula*. *Mycosystema* **2007**, *26*, 528–533.
27. Xiong, D.K.; Wang, H.; Chen, M.J.; Xue, C.Q.; Li, Z.P.; Bian, Y.B.; Bao, D.P. Application of mating type genes in molecular marker-assisted breeding of the edible straw mushroom *Volvariella volvacea*. *Sci. Hortic.* **2014**, *180*, 59–62. [CrossRef]

28. Kües, U.; Nelson, D.R.; Liu, C.; Yu, G.J.; Zhang, J.; Li, J.; Wang, X.C.; Sun, H. Genome analysis of medicinal *Ganoderma* spp. with plant-pathogenic and saprotrophic life-styles. *Phytochemistry* **2015**, *114*, 18–37. [CrossRef] [PubMed]
29. Ke, B.R.; Lu, Z.H.; Wu, X.P.; Guo, L.X.; Lan, Q.X. Determination and diversity analysis of mating types of *Ganoderma lucidum* basidispores. *Chin. J. Trop. Crops* **2018**, *39*, 145–150.
30. Wang, N.; Zhang, Y.H.; Li, J.A.; Wu, L.L.; Xu, J.; Xiong, L.; Leng, J.M.; Wang, Y.J. Screening test of Camellia oleifera sawdust substrate formula for cultivation of *Stropharia rugosoannulata* under the forest. *Non-Wood For. Res.* **2022**, *40*, 95–103. [CrossRef]
31. Gupta, B.; Niranjan Reddy, B.P.; Kotasthane, A.S. Molecular characterization and mating type analysis of oyster mushroom (*Pleurotus* spp.) using single basidiospores for strain improvement. *World J. Microbiol. Biotechnol.* **2010**, *27*, 1–9. [CrossRef]
32. Nieuwenhuis, B.P.S.; Billiard, S.; Vuilleumier, S.; Petit, E.; Hood, M.E.; Giraud, T. Evolution of uni- and bifactorial sexual compatibility systems in fungi. *Heredity* **2013**, *111*, 445–455. [CrossRef] [PubMed]
33. Cheng, S.M.; Lin, F.X. Genetic analysis of distorted segregation ratio of mating types among basidiospores in *Lentinula edodes*. *Sci. Agric. Sin.* **2007**, 2296–2302. [CrossRef]
34. Bao, D.P. Research progress and insights of the biological properties of edible basidiomycete dukaryon. *Mycosystema* **2024**, *43*, 14–29. [CrossRef]
35. Gong, W.B.; Liu, W.; Lu, Y.Y.; Bian, Y.B.; Zhou, Y.; Kwan, H.S.; Cheung, M.K.; Xiao, Y. Constructing a new integrated genetic linkage map and mapping quantitative trait loci for vegetative mycelium growth rate in *Lentinula edodes*. *Fungal Biol.* **2014**, *118*, 295–308. [CrossRef] [PubMed]
36. Gong, W.B.; Xie, C.L.; Zhou, Y.J.; Zhu, Z.H.; Wang, Y.H.; Peng, Y.D. A resequencing-based ultradense genetic map of *Hericium erinaceus* for anchoring genome sequences and identifying genetic loci associated with monokaryon growth. *Front. Microbiol.* **2020**, *10*, 3129. [CrossRef] [PubMed]
37. Gong, W.B.; Liu, K.F.; Li, X.R.; Zhang, L.; Shen, N.; Bian, Y.B.; Xiao, Y. QTL mapping reveals mating type gene *LeHD1* regulating mycelial growth in shiitake mushroom, *Lentinula edodes*. *Sci. Hortic.* **2022**, *305*, 111417. [CrossRef]
38. Mei, Z.Q.; Yang, L.Q.; Khan, M.A.; Yang, M.N.; Wei, C.L.; Yang, W.C.; Peng, X.N.; Tania, M.; Zhang, H.; Li, X.T.; et al. Genotyping of *Ganoderma* species by improved random amplified polymorphic DNA (RAPD) and inter-simple sequence repeat (ISSR) analysis. *Biochem. Syst. Ecol.* **2014**, *56*, 40–48. [CrossRef]
39. Liu, X.X.; Wang, Q.; Zhang, B.B.; Wang, C.X.; Guo, J.Y.; Zheng, S.Y. Analysis of genetic diversity of *Pleurotus ostreatus* monokaryon strains based on esterase and ISSR technology. *Jiangsu Agric. Sci.* **2022**, *50*, 27–32. [CrossRef]
40. Tan, Q.; Yang, J.M.; Chen, M.J.; He, D.M.; Pan, Y.J.; Huang, W.Y. Analysis on genetic character between sporulated monokaryon and protoplasted monokaryon of *Lentinula edodes*. *Edible Fungi China* **2001**, *20*, 3–5+25.
41. Liu, S.R.; Zhang, W.R.; Kuang, Y.B. Production of stalk spawn of an edible mushroom (*Pleurotus ostreatus*) in liquid culture as a suitable substitute for stick spawn in mushroom cultivation. *Sci. Hortic.* **2018**, *240*, 572–577. [CrossRef]
42. Lin, F.X.; Li, H.S.; Feng, L.; Liang, B.D.; Bao, D.P. The relationships between mating types and mycelial growth rates of monokaryons and dikaryons in *Lentinula edodes*. *Acta Agric. Shanghai* **2013**, *29*, 15–22.
43. Shen, N.; Xie, H.Y.; Liu, K.F.; Li, X.R.; Wang, L.; Deng, Y.J.; Chen, L.F.; Bian, Y.B.; Xiao, Y. Near-gapless genome and transcriptome analyses provide insights into fruiting body development in *Lentinula edodes*. *Int. J. Biol. Macromol.* **2024**, *263*, 130610. [CrossRef] [PubMed]
44. Liu, T.X.; Li, H.R.; Ding, Y.T.; Qi, Y.C.; Gao, Y.Q.; Song, A.D.; Shen, J.W.; Qiu, L.Y. Genome-wide gene expression patterns in dikaryon of the basidiomycete fungus *Pleurotus ostreatus*. *Braz. J. Microbiol.* **2017**, *48*, 380–390. [CrossRef] [PubMed]
45. Zhou, S.; Zhang, J.S.; Ma, F.Y.; Tang, C.H.; Tang, Q.J.; Zhang, X.Y. Investigation of lignocellulolytic enzymes during different growth phases of *Ganoderma lucidum* strain G0119 using genomic, transcriptomic and secretomic analyses. *PLoS ONE* **2018**, *13*, e0198404. [CrossRef] [PubMed]
46. Li, J.T.; Duan, Y.C.; Hu, Z.Y.; Yang, F.; Wu, X.L.; Zhang, R.Y. Physiological mechanisms by which gypsum increases the growth and yield of *Lentinula edodes*. *Appl. Microbiol. Biot.* **2022**, *106*, 2677–2688. [CrossRef]
47. Wei, J.T.; Ma, X.Y.; Li, X.W.; Zhang, Z.J.; Li, C.; Ma, F.W. Drought resistance evaluation of eight apple germplasm resources. *J. Fruit. Sci.* **2024**, *41*, 569–578. [CrossRef]

agriculture

Article

# High Resistance and Yield: A New Cultivar 'ZJLZS002' of *Lyophyllum decastes* Suitable for Industrial Cultivation

**Qimeng Liu [1], Shaoxiong Liu [1], Jianying Li [1], Junbo Zhang [1], Fan Zhou [1], Xi Luo [1], Jianxiong Ma [2], Rong Hua [1,*] and Dafeng Sun [1,3,*]**

1 Kunming Edible Fungi Institute of All China Federation of Supply and Marketing Cooperatives, Kunming 650221, China; liuqimeng1952@126.com (Q.L.); lsxcary@163.com (S.L.); jianying0818@163.com (J.L.); 15170211710@163.com (J.Z.); zhoufan8f@126.com (F.Z.); fungixil@163.com (X.L.)
2 Jiangsu Hongsheng Biotechnology Co., Ltd., Xuzhou 221000, China; 18052252708@163.com
3 Yunnan Academy of Edible Fungi Industry Development Center, Kunming 650021, China
* Correspondence: huarong15797958114@126.com (R.H.); sdafeng@163.com (D.S.)

**Abstract:** *Lyophyllum decastes*, commonly known as Luronggu, is extensively cultivated across China. It exhibits rich germplasm in China. However, the number of cultivars available for commercial production is limited, highlighting the importance of targeted breeding programs. In this study, we utilized selected breeding and SSR molecular markers to develop improved strains of *L. decastes* for the first time. The breeding process strictly adhered to China's national standard for 'Technical inspection for mushroom selecting and breeding'. It encompassed pure strain isolation, biological classification, primary screening, secondary screening, physiological performance determination, molecular characterization, intermediate test, and demonstration cultivation. As a result, strain ZJLZS002, known for its high yield (380 ± 3.6 g·bag$^{-1}$), shortened growth period (75.6 ± 1.3 d), and stable traits, is well suited for industrial cultivation. This new cultivar has achieved a significant milestone as the first variety in China to be officially recognized at the provincial level, under the name 'Zhongjunluronggu No. 1'. Its development signifies a crucial advancement in achieving seed source independence and promotes the replacement of imported varieties with domestic ones, contributing to the sustainable development of China's edible fungi industry.

**Keywords:** germplasm; SSR molecular marker; selected breeding; Zhongjunluronggu No. 1; new variety

## 1. Introduction

*Lyophyllum decastes* (Fr.) Singer, a rare and valuable edible–medicinal mushroom, possesses exceptional nutritional and pharmacological properties [1]. This mushroom, belonging to the phylum Basidiomycota, order Agaricales, family Lyophyllaceae, and the genus *Lyophyllum* [2,3], exhibits a natural distribution across temperate regions of the northern hemisphere, including China, Japan, Korea, and European and American countries [4,5]. Particularly in China, in regions such as Yunnan, Guizhou, Liaoning, Jilin, Heilongjiang, and the Inner Mongolia Autonomous Region, its genetic diversity is abundant [6]. It is particularly prized for its delicate texture, aromatic grass-like scent, and delightful taste [7]. Both the fruiting body and mycelium are abundant in proteins, amino acids, and various vitamins, as well as critical trace elements, including iron, zinc, and selenium [8–10]. Notably, it has garnered significant attention for its bioactive substances.

Its polysaccharides exhibit a variety of potential health-promoting properties, such as antioxidant, hypolipidemic, antitumor, antibacterial, and antidiabetic effects [11–14].

As an emerging culinary-medicinal mushroom, *L. decastes* has experienced rapid production growth following the development of industrial cultivation [7]. China currently dominates global production, with official data from the China Edible Fungi Association Public Service Platform reporting 193,131 tons of output in 2023 (https://bigdata.cefa.org.cn/output.html, accessed on 28 April 2025). Major production regions are concentrated in the provinces of Guizhou, Shandong, Jiangxi, Fujian, Sichuan, and Gansu [15]. However, our research team has identified concerning genetic uniformity among commercial strains (non-published material). Market analysis revealed that most cultivated strains in China can be traced back to a single progenitor—the KX-HA092 variety, originally introduced in Japan by Shanghai Fengke Biotechnology Co., Ltd. (Shanghai, China) [16].

There have been reports on the breeding of new strains and varieties of *L. decastes*. Using selected breeding, Woo et al. [17] evaluated nine strains from South Korea and Japan, identifying SPA202 and SPA205 as high-yield stains. Yoshihama et al. [18] achieved fruiting body formation on artificial media, and applied for a US patent in 1994. Wei et al. [19] successfully isolated wild strains from China's Qilian Mountains using potato dextrose agar (PDA) enrichment medium. Qiu et al. [20] developed the dark-brown cultivar 'Changli No. 1'. Utilizing monosporous hybridization, the commercially valuable KX-HA092 variety was developed through crossing Japanese strains 5-99 and 62-33(2) with Swiss strain IFO32185, and its growth period was 78 days [21]. Pan et al. [22] developed three improved hybrids (KL4, KL10, KL17) with superior fruiting body characteristics, among which KL17 demonstrated optimal morphology and the shortest production cycle. Furthermore, Liang et al. [23] successfully screened a high-yield laccase strain HY1022-01 through UV-induced protoplast mutagenesis and selected a new high-polysaccharide-yield strain ZY481-1 by protoplast mutagenesis; its polysaccharide yield was up to 643.1961 mg/g, which revealed a 31.05% increase compared with the original strain, ZY48-1 [24]. Despite the successful establishment of large-scale industrial cultivation for *L. decastes*, several critical challenges persist that hinder its full commercial potential. These include (1) the scarcity of potential strains with independent intellectual property rights, (2) slow mycelial growth rates, (3) suboptimal yield performance, (4) rapid spawn degeneration, and (5) poor resistance [1,25]. These limitations collectively constrain its commercial development and market competitiveness.

Simple sequence repeat (SSR) molecular markers offer a powerful solution for germplasm improvement. As codominant markers that detect genetic variation at the DNA level, SSRs exhibit several superior characteristics: (1) environmental stability, (2) high reproducibility, and (3) exceptional discriminatory power. Notably, SSR molecular markers, developed from whole genome sequences, demonstrate enhanced polymorphism and more comprehensive genome coverage compared to other marker systems [26,27]. These attributes make SSR analysis an indispensable tool for assessing genetic diversity and germplasm resources in edible fungi; however, there are no reports on *L. decastes* to date.

In developing the new variety 'ZJLZS002', our research team implemented a comprehensive breeding strategy for (1) precise strain identification, (2) intellectual property protection, and (3) varietal registration.

Considering China's current cultivation scenario, this strategy facilitates (1) systematic development of improved industrial strains, (2) the creation of novel cultivars, and (3) the sustainable advancement of the industry. The implementation of this strategy is particularly significant for (1) optimizing *L. decastes* germplasm resources, and (2) enhancing China's independent germplasm innovation capacity.

## 2. Materials and Methods

### 2.1. Materials

#### 2.1.1. Tested Strains

The nine tested strains utilized in the present study are illustrated in Table 1 and are stored at the Germplasm Resource Bank of Edible Fungi, Kunming Edible Fungi Institute of All China Federation of Supply and Marketing Cooperatives. LR-A, obtained from Jiangsu Hongsheng Biotechnology Co., Ltd. (Xuzhou, China), was used as the benchmark control variety. We executed the preservation of patented strains at the China Center for Type Culture Collection of Wuhan University, and filed patent applications for potential strains.

**Table 1.** Tested strains and their origins.

| Code | Strain No. | Taxon | Type | Origin of Resource |
|---|---|---|---|---|
| 1 | ZJLZS001 | *Lyophyllum decastes* | Wild | Dandong City, Liaoning Province |
| 2 | ZJLZS002 | *Lyophyllum decastes* | Wild | Dandong City, Liaoning Province |
| 3 | ZJLZS003 | *Lyophyllum decastes* | Wild | Hohhot City, Inner Mongolia Autonomous Region |
| 4 | ZJLZS004 | *Lyophyllum decastes* | Wild | Yongping County, Yunnan Province |
| 5 | ZJLZS005 | *Lyophyllum decastes* | Wild | Xinjiang Uygur Autonomous Region |
| 6 | ZJLZS006 | *Lyophyllum decastes* | Wild | Yuxi County, Yunnan Province |
| 7 | ZJLZS007 | *Lyophyllum decastes* | Wild | Lijiang City, Yunnan Province |
| 8 | ZJLZS008 | *Lyophyllum decastes* | Wild | Diqing Tibetan Autonomous Prefecture |
| 9 | LR-A | *Lyophyllum decastes* | Cultivation | Jiangsu Hongsheng Biotechnology Co., Ltd. |

#### 2.1.2. Culture Conditions

PDA enrichment medium was used for pure strain isolation and culture. It included 200 g of potatoes (immersed), 20 g of glucose, 16 g of agar, 3 g of yeast powder, 3 g of potassium dihydrogen phosphate, and 1.5 g of magnesium sulfate, all dissolved in 1 L of water, with the pH adjusted to 6. Pure strains were maintained in darkness at 25 °C. A liquid medium was used to produce liquid spawn. It included 200 g of potatoes (immersed), 20 g of glucose, 3 g of yeast powder, 3 g of potassium dihydrogen phosphate, and 1.5 g of magnesium sulfate, combined with 1 L of water, with the pH adjusted to 6. The liquid spawn were incubated in a shaking incubator (in darkness at 25 °C, shaken at 160 rpm). The culture medium for cultivation comprised 35% wood chips, 35% corn cob, 20% wheat bran, 4% corn meal, 4% soybean meal, and 2% calcium bicarbonate, with the pH adjusted to 6. Mycelium was cultured in darkness at 25 °C, with a relative humidity of less than 70%.

### 2.2. Methods

Adhering to the national standard for 'Technical inspection for mushroom selecting and breeding' [28], the breeding process is shown in Figure 1.

#### 2.2.1. Tissue Isolation and Purification

First, a small piece of tissue at the junction of the cap and stipe of the fruiting body, approximately 5 mm × 5 mm in size, was excised using a sterile scalpel, transferred into the PDA enrichment medium, and purified 2–3 times. Strains exhibiting superior mycelium growth were then selected. These selected strains were incubated in darkness at 25 °C for 12–15 days.

#### 2.2.2. Biological Classification

DNA was successfully extracted from the mycelium using an optimized CTAB procedureDNA [29,30]. Primer sets LR0R-LR5 [31], were applied to amplify the nuclear ribosomal large subunit (nrLSU) sequence. The PCR reaction conditions followed the protocol described by Tang et al. [3]. The PCR products were dispatched to Tianyi Huayu Gene Technology Co., Ltd. (Wuhan, China) for subsequent sequencing analysis.

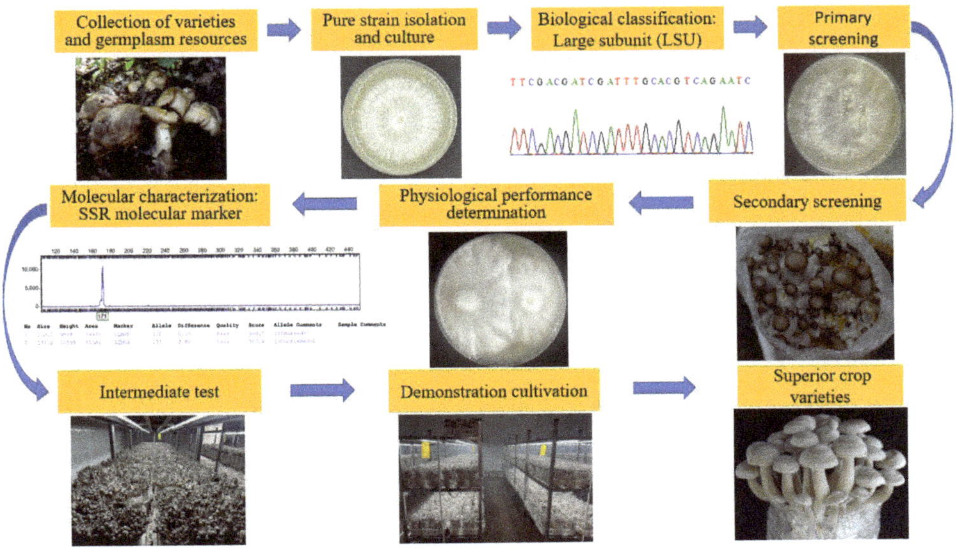


**Figure 1.** A flow chart of selected breeding.

The *L. decastes* sequences produced in this study were submitted to the NCBI database for Basic Local Alignment Search Tool (https://blast.ncbi.nlm.nih.gov/Blast.cgi?PROGRAM=blastn&PAGE_TYPE=BlastSearch&LINK_LOC=blasthome), and clusters with closely related sequences were downloaded. The nrLSU dataset was submitted to the MAFFT version 7 [32] online platform for alignment, and maximum likelihood (ML) analyses was conducted at IQTREE Web Server (http://iqtree.cibiv.univie.ac.at/, accessed on 28 April 2025). The optimal TNe+G4 model was obtained using 1000 replicates with default parameters. *Calocybe vinacea* was used as the outgroup. Only bootstrap support values of branches ≥70 were displayed on the phylogenetic tree.

2.2.3. Primary Screening

Strains of the same fungal age and with a diameter of 5 mm were connected to the center of the PDA enrichment medium and incubated at a constant temperature at 25 °C for 14 days. Utilizing colony characteristics, mycelial growth potential, and growth rate as criteria, each strain was tested three times to screen for the good-performing strains. Subsequently, the strains capable of fruiting on the PDA enrichment medium were selected for further secondary screening.

2.2.4. Secondary Screening

Polypropylene bags, each measuring 18 cm × 33 cm, were used to contain 400 g of dry cultivation substrate (comprising 35% wood chips, 35% corn cob, 20% wheat bran, 4% corn meal, 4% soybean meal, and 2% calcium bicarbonate). These bags were then sterilized at a temperature of 121 °C and pressure of 0.15 MPa for 120 min. A total of 60 bags were allocated for each strain. In a sterile environment, the liquid spawn of the dominant strains was introduced into the sterilized bags and maintained at a stable temperature of 25 °C in darkness for 50 days. Fruiting management was carried out according to the industrial cultivation method of *L. decastes* [33].

The potential strains, characterized by high yield and suitability for industrial production, were selected based on their agronomic characteristics, such as single fruit mass, cap diameter, stipe diameter, yield, biological efficiency, and growth period.

$$Z = \frac{X}{Y} \times 100\%$$

In the equation, $Z$ represents biological efficiency; $X$ denotes the yield weight per bag, measured in grams; $Y$ stands for the weight of the dry cultivation substrate per package, also measured in grams.

#### 2.2.5. Physiological Performance Determination of Potential Strains

##### Antagonistic Reaction Determination

So far, no varieties of *L. decastes* have been identified in China. The prevalent cultivar LR-A served as the control. Mycelium pieces of the same fungal age and dimensions from potential strains and LR-A were excised using a 5 mm punch and planted on PDA enrichment medium via the two-point inoculation technique [34]. Subsequent observation of the antagonistic response was conducted at a temperature of 25 °C.

##### Antibacterial TEST

Using a 5 mm hole punch, mycelium pieces of the same fungal age and dimensions were selected from the potential strains, *Penicillium brevicompactum* and *Trichoderma pleuroticola*, and were planted in the PDA enrichment medium via the two-point inoculation technique. The cultures were monitored for growth at a temperature of 25 °C for a designated duration, and the antibacterial efficacy was determined through calculation, with triple repetition. The antibacterial efficacy, denoted as $E$, was calculated using the following formula:

$$E = \frac{(R1 - R2)}{R1} \times 100\%$$

where $R1$ signifies the radius of mold growth under normal conditions, measured in centimeters, and $R2$ represents the radius of mold growth in the two-point confrontation experiment, also in centimeters.

#### 2.2.6. Molecular Characterization: SSR Molecular Markers

Based on the sequencing platform, SSR loci were identified in the sequence of the *L. decastes* genome (GCF_021015755.1) from the database for development and screening. Following DNA extraction [30] and PCR amplification [35], capillary electrophoresis (CE) was performed on an ABI 3730XL sequencer (Applied Biosystems, Carlsbad, CA, USA). Capillary electropherograms were analyzed with GeneMarker® version 2.7.0. Ten pairs of SSR primers with clear amplification bands, high reproducibility, and high specificity were obtained. These included LD045, LD063, LD089, LD093, LD094, LD095, LD098, LD129, LD149, and LD157. The details of these 10 pairs of SSR primers are presented in Table 2.

**Table 2.** Information of 10 pairs of SSR primers.

| Primer Name | Size of Primer Fragment |
|---|---|
| LD045 | F: CCGCACAACACCTCAACAAG; R: TCGTCGCACAAAGAGCAGTA |
| LD063 | F: TCATCGGGACAAAGCCGAAA; R: GTCAACCCAAGCGACAACAC |
| LD089 | F: CCAAACAGTGCCGTTGAGTG; R: CTCCGGCGTTGAGTGACTAG |
| LD093 | F: TGGGTGTCGTTTGGGTATGG; R: CACGAGGACAGGCACATTCT |
| LD094 | F: GGGACGGAAGGGAAGGAAAG; R: TCTTTAACGCAGCGGTCCAT |
| LD095 | F: TGATGATGAGGCTTCGACGG; R: GTCACGACAACGCACTGTTC |

**Table 2.** *Cont.*

| Primer Name | Size of Primer Fragment |
|---|---|
| LD098 | F: TCGGTGCGTAATCGTTTGGA; R: ATCGCCGCTTTCTGCAAATG |
| LD129 | F: CGTCGTTACATCGTTGACGC; R: TGGCACATCCATGAAGCAGT |
| LD149 | F: AAACTCAAGCTGTCCCGGTC; R: TCAAAGGAAGCTCCACCGAC |
| LD157 | F: ACGAGCTATTGGACGACCAC; R: TTTCGTTTCCACGCCGACTA |

F stands for forward primer; R stands for reverse primer.

2.2.7. Intermediate Test

Utilizing the industrial cultivation method of *L. decastes* [33], potential strains and LR-A were cultivated, with a total of 1500 bags set up for each strain across three replicate groups. The yield of each strain was measured with 100 bags randomly selected, and the agronomic traits of the fruiting bodies were documented to determine the yield and growth period.

2.2.8. Demonstration Cultivation

From April 2022 to April 2024, demonstration cultivation was carried out at Base 1: Jinning Baofeng Base, Kunming Institute of Edible Fungi, All China Supply and Marketing Cooperative Society; and Base 2: Jiangsu Hongsheng Biotechnology Co., Ltd., situated in Suining County, Xuzhou City, Jiangsu Province.

Following the provisions set forth in the national standard 'Technical inspection for mushroom selecting and breeding' [28], the experimental design for the demonstration cultivation was designed with 10,000 bags per group, across three groups. Utilizing the industrial cultivation method of *L. decastes* [33], potential strains and LR-A were cultivated. Metrics such as the fresh single fruit mass, first-tide yield, and biological efficiency were counted.

2.2.9. Determination of Nutrient Content in Fruiting Body

Dried fruiting body samples of the potential strains were sent to Yunnan Sanzheng Technical Testing Co., Ltd. (Kunming, China) for analysis, which included the detection of proteins [36], amino acids [37], polysaccharides [38], trace elements, and vitamins, among other components.

*2.3. Analysis of Data*

Statistical analyses were performed using SPSS version 22. Continuous data were presented as the mean ± standard deviation ($\overline{X} \pm s$). One-way ANOVA, followed by the LSD post hoc test, was used for multiple comparisons among the groups. A $p$-value < 0.05 was considered statistically significant.

## 3. Results

*3.1. Isolation of Strains*

A total of 22 wild fruiting bodies were collected from seven provinces across China. Following successful isolation and purification, eight pure strains were obtained.

*3.2. Phylogenetic Analyses*

Nine new nrLSU sequences were generated (Table 3). The phylogenetic tree based on nrLSU sequences indicates that the nine strains, along with two other known *L. decastes* strains, are clustered in the large clade (Figure 2).

**Table 3.** The data emp oyed in this study.

| Species | Voucher | nrLSU | Reference |
|---|---|---|---|
| *Lyophyllum decastes* | ZJLZS001 | PV570115 | This study |
| *L. decastes* | ZJLZS002 | PV570116 | This study |
| *L. decastes* | ZJLZS003 | PV570117 | This study |
| *L. decastes* | ZJLZS004 | PV570118 | This study |
| *L. decastes* | ZJLZS005 | PV570119 | This study |
| *L. decastes* | ZJLZS006 | PV570120 | This study |
| *L. decastes* | ZJLZS007 | PV570121 | This study |
| *L. decastes* | ZJLZS008 | PV570122 | This study |
| *L. decastes* | LR-A | PV570114 | This study |
| *L. decastes* | Sundberg091007a | HM572548 | [39] |
| *L. decastes* | F-855 | PQ652409 | Genbank |
| *L. ambustum* | F-810 | PQ652355 | Genbank |
| *L. fumosu* | Lipovac090903 | HM572538 | [39] |
| *L. shimeji* | Eilertsen090908 | HM572531 | [39] |
| *L. shimeji* | CBS 451.87 | AF223215 | [40] |
| *L. shimeji* | haukebo1982 | HM572529 | [39] |
| *Calocybe vinacea* | HMJU 5135 | NG243053 | Genbank |

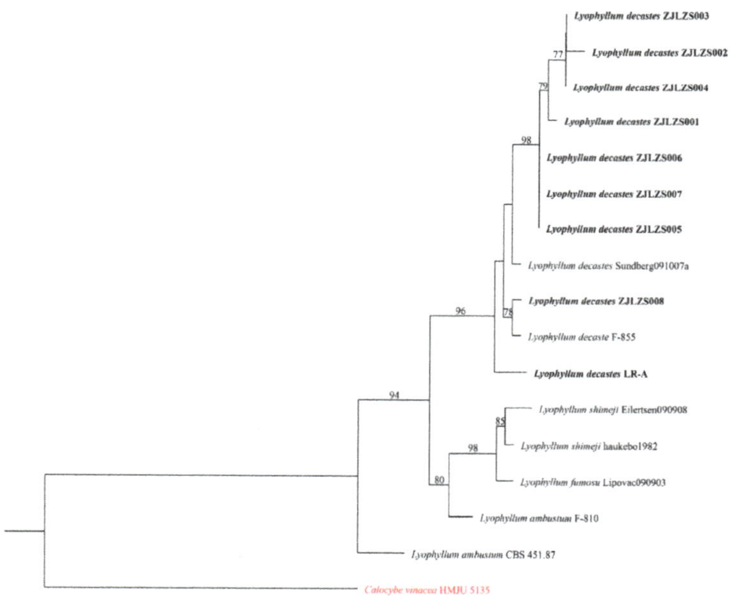


**Figure 2.** Phylogenetic analysis of Lyophyllum decastes and related species based on nuclear ribosomal large subunit. *Calocybe vinacea* was used as the outgroup.

### 3.3. Comparison of Preliminary Screening

Analysis of the mycelium growth rates revealed that, as detailed in Table 4, strains ZJLZS002 and ZJLZS003 exhibited thick and strong growth, with respective mycelium growth rates of 0.35 cm/d and 0.36 cm/d. ZJLZS002 and ZJLZS003 had significant differences compared with the other six strains ($p < 0.05$). Strains ZJLZS004, ZJLZS005, ZJLZZS006, and ZJLZS008 displayed thin and average growth, with mycelial growth rates of 0.31 cm/d, 0.30 cm/d, 0.30 cm/d, and 0.30 cm/d, respectively. ZJLZS001 and ZJLZS007, however, presented as thin and weak, with mycelial growth of 0.27 cm/d. As depicted in

Figure 3, ZJLZS001, ZJLZS002, ZJLZS003, and ZJLZS007 demonstrated their capability to form fruiting bodies at the later stage of mycelium culture in the PDA enrichment medium, indicating their fruiting potential. Consequently, strains ZJLZS002 and ZJLZS003, marked by their good growth and fruiting ability, were selected for the secondary screening.

**Table 4.** Preliminary screening results.

| Strain | Mycelium Growth Rate (cm/d) | Growth Vigor | Fruiting |
|---|---|---|---|
| ZJLZS001 | 0.27 ± 0.01 bc | + | T |
| ZJLZS002 | 0.35 ± 0.01 a | +++ | T |
| ZJLZS003 | 0.36 ± 0.04 a | +++ | T |
| ZJLZS004 | 0.31 ± 0.01 b | ++ | N |
| ZJLZS005 | 0.3 ± 0.00 b | ++ | N |
| ZJLZS006 | 0.3 ± 0.02 b | ++ | N |
| ZJLZS007 | 0.27 ± 0.02 bc | + | T |
| ZJLZS008 | 0.3 ± 0.02 b | ++ | N |

+ Mycelium is thin and weak. ++ Mycelium is thin and general. +++ Mycelium is thick and strong. Lowercase letters indicate significant difference at $p < 0.05$. T stands for fruiting; N stands for non-fruiting.

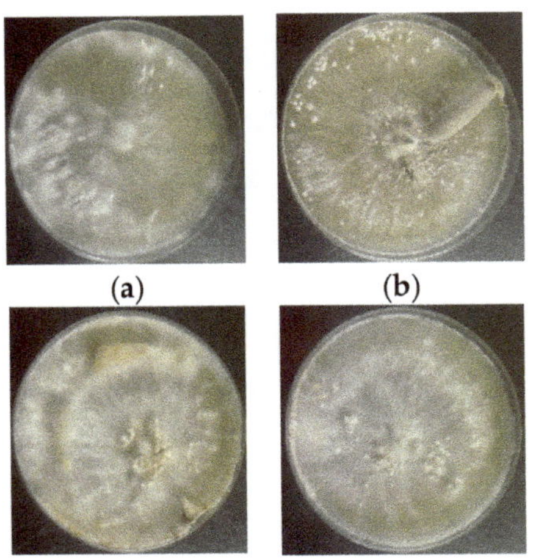

**Figure 3.** Strains forming fruited bodies on the PDA enrichment medium. (**a**) Strain ZJLZS001; (**b**) strain ZJLZS002; (**c**) strain ZJLZS003; (**d**) strain ZJLZS007.

### 3.4. Comparison of Secondary Screening

Compared to strain ZJLZS003, ZJLZS002 demonstrated superior fruiting body development, characterized by greater thickness and higher quality, as indicated in Figure 4 and Table 5. Its yield reached an impressive 374.67 ± 8.5 g per bag, significantly higher than that of ZJLZS003, with 337 ± 21 g ($p < 0.05$). The cap was of moderate size, with a diameter of 33.89 ± 0.57 mm. The length of the stipe was 9.8 ± 0.36 cm, its upper diameter was 10.92 ± 0.33 mm ($p < 0.05$), and its bottom diameter was 16.49 ± 1.00 mm ($p < 0.05$). The single fruit mass was 16 ± 0.26 g, and the growth period was 75.7 ± 1.5 d ($p < 0.05$). Consequently, ZJLZS002 was selected as a potential strain for physiological performance detection, distinguishing identification, intermediate testing, and demonstration cultivation.

**Figure 4.** Photos of fruiting. (**a**) Strain ZJLZS002; (**b**) strain ZJLZS003.

**Table 5.** Rescreening results.

| Strain | Growth Period /Day | Stipe Characteristics | | | Cap Diameter /mm | Single Fruit Mass/g | Yield /(g·bag$^{-1}$) | Biological Efficiency/% |
|---|---|---|---|---|---|---|---|---|
| | | Length /cm | Upper Diameter /mm | Bottom Diameter /mm | | | | |
| ZJLZS002 | 75.7 ± 1.5 b | 9.80 ± 0.36 | 10.92 ± 0.33 a | 16.49 ± 1.00 a | 33.89 ± 0.57 | 16.00 ± 0.26 | 374.67 ± 8.50 a | 93.67 ± 2.13 a |
| ZJLZS003 | 80.7 ± 1.5 a | 10.73 ± 0.92 | 8.53 ± 0.72 b | 10.87 ± 0.11 b | 32.61 ± 1.13 | 14.03 ± 1.76 | 337.00 ± 21.0 b | 84.25 ± 6.00 b |

Lowercase letters indicate significant difference at $p < 0.05$.

### 3.5. Results of Antagonistic Reaction

The results of the antagonism tests reveal an isolated antagonistic interaction between ZJLZS002 and LR-A, as depicted in Figure 5. Additionally, distinct differences were observed in the mycelial morphology.

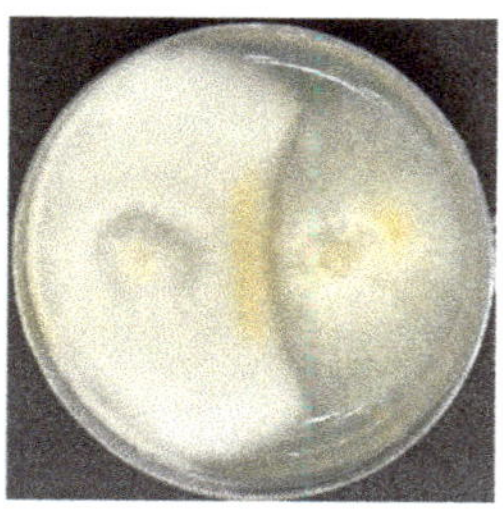

**Figure 5.** Antagonistic reaction between ZJLZS002 (**left**) and LR-A (**right**).

### 3.6. Results of Antibacterial Test

As depicted in Figure 6 and Table 6, during the antibacterial evaluation against *Penicillium brevicompactum* and *Trichoderma pleuroticola*, the potential strain ZJLZS002 demonstrated impressive antibacterial efficacy, with rates of 54.2% and 48.8%, respectively, outperforming LR-A, which achieved rates of only 33.3% and 21.9%. This indicates that ZJLZS002 possessed higher resistance.

**Table 6.** Antibacterial rates of ZJLZS002 and LR-A.

| Strain | *P. brevicompactum* | | | | *T. pleuroticola* | | | |
|---|---|---|---|---|---|---|---|---|
| | R1/cm | R2/cm | R1 − R2/cm | E/% | R1/cm | R2/cm | R1 − R2/cm | E/% |
| ZJLZS002 | 2.4 | 1.1 | 1.3 | 54.2 | 2.1 | 4.1 | 2 | 48.8 |
| LR-A | 2.4 | 1.6 | 0.8 | 33.3 | 3.2 | 4.1 | 0.9 | 21.9 |

R1 signifies the radius of mold growth under normal conditions, measured in centimeters; R2 represents the radius of mold growth in the two-point confrontation experiment, also in centimeters; *E* is the antibacterial rate.

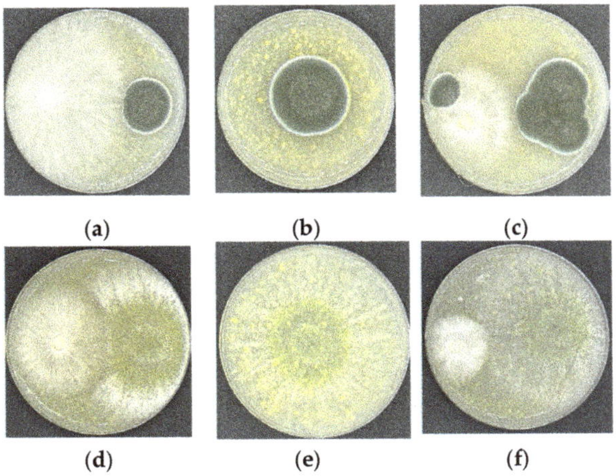

**Figure 6.** Antibacterial test. (**a**) ZJLZS002 (**left**) and *P. brevicompactum* (**right**); (**b**) *P. brevicompactum*; (**c**) LR-A (**left**) and *P. brevicompactum* (**right**); (**d**) ZJLZS002 (**left**) and *T. pleuroticola* (**right**); (**e**) *T. pleuroticola*; (**f**) LR-A (**left**) and *T. pleuroticola* (**right**).

### 3.7. Amplification Results of SSR

Following amplification with ten SSR primer pairs (LD045, LD063, LD089, LD093, LD094, LD095, LD098, LD129, LD149, and LD157), capillary electropherograms revealed distinct fragment size profiles between ZJLZS002 and LR-A,, as displayed in Figure 7 and Table 7. The amplified fragments of ZJLZS002 measured (167, 169), 281, 192, 162, 158, 200, (284, 299), (187, 190), 346, and (142, 145) bp, respectively. In contrast, LR-A produced fragments of 171, (288, 290), 189, 169, 181, 217, 286, (172, 187), 206, and 135 bp. These results demonstrate that the ten SSR markers effectively differentiate LR-A from ZJLZS002 based on their polymorphic fragment sizes.

**Table 7.** Comparison of amplified fragment sizes of ten primer pairs in ZJLZS002 and LR-A.

| Primer Name | LR-A | ZJLZS002 |
|---|---|---|
| LD045 | 171 | (167, 169) |
| LD063 | (288, 290) | 281 |
| LD089 | 189 | 192 |
| LD093 | 169 | 162 |
| LD094 | 181 | 158 |
| LD095 | 217 | 200 |
| LD098 | 286 | (284, 299) |
| LD129 | (172, 187) | (187, 190) |
| LD149 | 206 | 346 |
| LD157 | 135 | (142, 145) |

### 3.8. Results of Intermediate Test

In contrast to LR-A, ZJLZS002 exhibited a higher yield and biological efficiency. There was a significant disparity in the bottom diameter and single fruit mass ($p < 0.05$), with ZJLZS002 having a bottom diameter of 16.47 ± 0.53 mm, compared to LR-A's 10.93 ± 1.18 mm. Similarly, the single fruit mass of ZJLZS002 was 15.77 ± 0.95 g, whereas LR-A's was 11.5 ± 0.78 g. Notably, significant differences ($p < 0.05$) were also observed in the stipe length, upper diameter, and yield. Specifically, the stipe length of ZJLZS002 was 9.63 ± 0.49 cm, compared to LR-A's 10.9 ± 0.4 cm; the upper diameter was 10.38 ± 0.48 mm, while LR-A's was 7.87 ± 1.27 mm; and the yield was 380 ± 3.6 g, compared to LR-A's 345.7 ± 13.9 g. Although the growth period of ZJLZS002 was 2 days shorter than that of LR-A, this discrepancy was not statistically significant, nor was the discrepancy in cap size (Table 8). Intermediate testing revealed that ZJLZS002 was characterized by its stability, shortened growth period, and the advantages of high yield and biological efficiency.

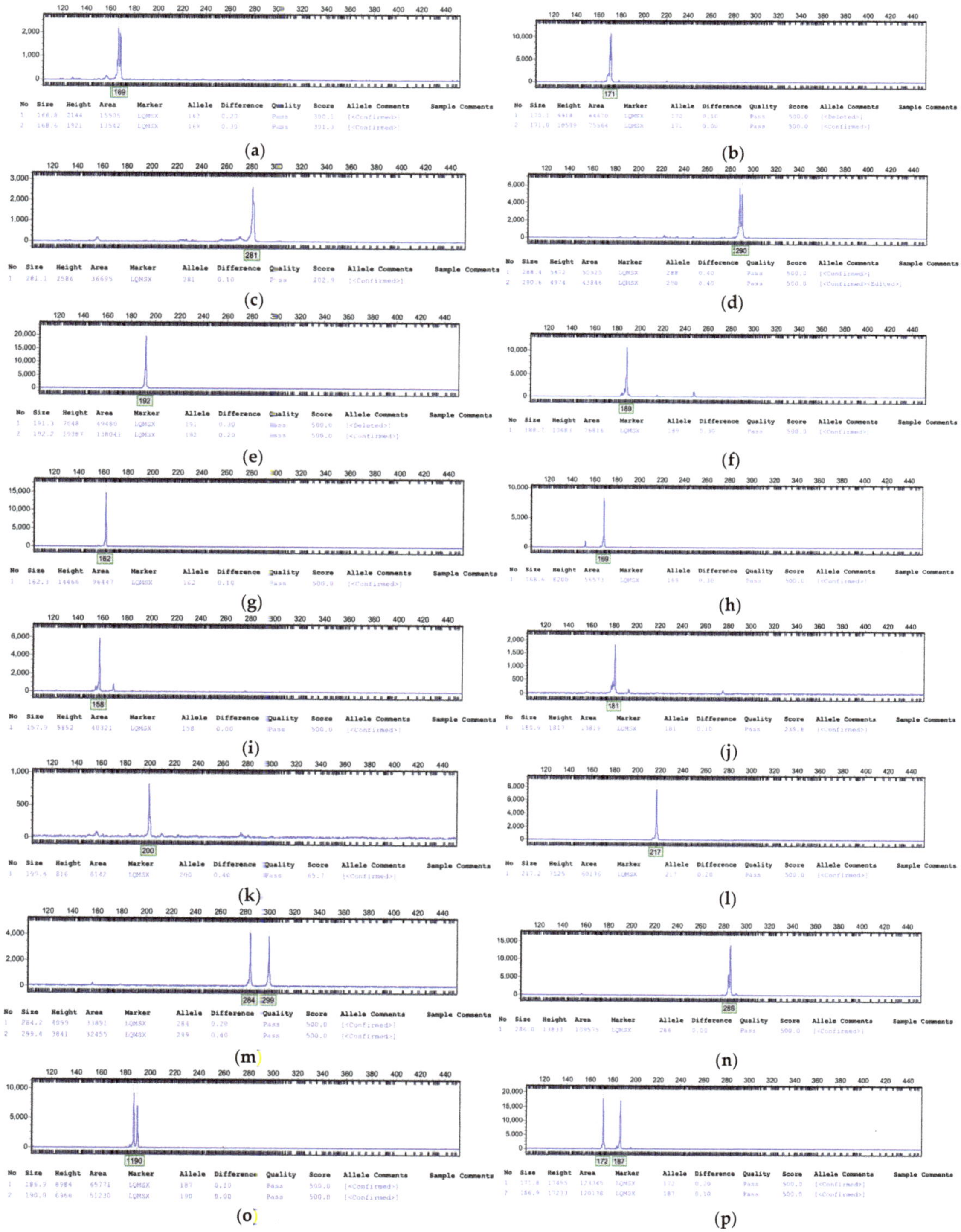

**Figure 7.** *Cont.*

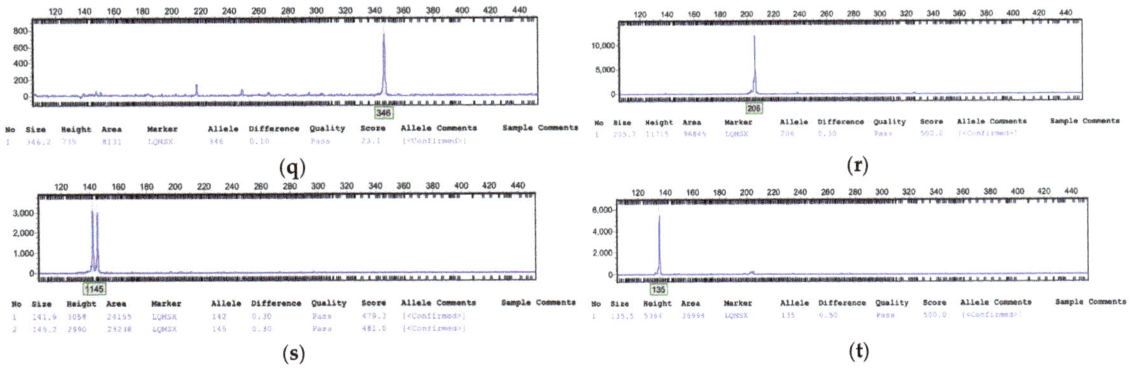

**Figure 7.** Capillary electropherogram of ZJLZS002 and LR-A. (**a**) Amplification of ZJLZS002 with LD045; (**b**) amplification of LR-A with LD045; (**c**) amplification of ZJLZS002 with LD063; (**d**) amplification of LR-A with LD063; (**e**) amplification of ZJLZS002 with LD089; (**f**) amplification of LR-A with LD089; (**g**) amplification of ZJLZS002 with LD093; (**h**) amplification of LR-A with LD093; (**i**) amplification of ZJLZS002 with LD094; (**j**) amplification of LR-A with LD094; (**k**) amplification of ZJLZS002 with LD095; (**l**) amplification of LR-A with LD095; (**m**) amplification of ZJLZS002 with LD098; (**n**) amplification of LR-A with LD098; (**o**) amplification of ZJLZS002 with LD129; (**p**) amplification of LR-A with LD129; (**q**) amplification of ZJLZS002 with LD149; (**r**) amplification of LR-A with LD149; (**s**) amplification of ZJLZS002 with LD157; (**t**) amplification of LR-A with LD157.

**Table 8.** Biological efficiencies of ZJLZS002 and LR-A in intermediate test.

| Strain | Growth Period /Day | Stipe Characters | | | Cap Diameter/mm | Single Fruit Mass/g | Yield /(g·bag$^{-1}$) | Biological Efficiency/% |
|---|---|---|---|---|---|---|---|---|
| | | Length /cm | Upper Diameter /mm | Bottom Diameter /mm | | | | |
| ZJLZS002 | 75.6 ± 1.3 | 9.63 ± 0.49 b | 10.38 ± 0.48 a | 16.47 ± 0.53 a | 33.89 ± 0.67 | 15.77 ± 0.95 a | 380 ± 3.6 a | 95 a |
| LR-A | 77.7 ± 1.5 | 10.9 ± 0.4 a | 7.87 ± 1.27 b | 10.93 ± 1.18 b | 33.87 ± 1.72 | 11.5 ± 0.78 b | 345.7 ± 13.9 b | 86.4 b |

Lowercase letters indicate significant difference at $p < 0.05$.

### 3.9. Results of Demonstration Cultivation

Compared to LR-A, as shown in Table 9, the biological efficiency of the ZJLZS002 groups 1, 2, and 3 were significantly higher ($p < 0.05$). Across different locations and multiple demonstration cultivations, the yield of ZJLZS002 increased by 8.31%.

**Table 9.** Biological efficiencies of ZJLZS002 and LR-A in demonstration cultivation.

| Strain | Biological Efficiency of the First Group/% | | Biological Efficiency of the Second Group/% | | Biological Efficiency of the Third Group/% | |
|---|---|---|---|---|---|---|
| | 1 | 2 | 1 | 2 | 1 | 2 |
| ZJLZS002 | 95.17 ± 0.38 a | 94.5 ± 1.64 a | 94.75 ± 0.66 a | 94.92 ± 0.63 a | 94.67 ± 0.58 a | 95.92 ± 1.26 a |
| LR-A | 86.5 ± 3.19 b | 86.5 ± 2.54 b | 86.67 ± 3.15 b | 86.58 ± 3.26 b | 86.42 ± 2.98 b | 87.42 ± 3.36 b |

Lowercase letters indicate significant difference at $p < 0.05$.

The results of these trials for ZJLZS002 satisfy the criteria for new cultivar identification, as shown in Figures 8 and 9. On 20 August 2024, ZJLZS002 passed the identification of non-major crop varieties in Yunnan Province by Yunnan Seed Management Station. The new cultivar was named 'Zhongjunlurongggu No. 1', which successfully bred the first new variety of *L. decastes* suitable for industrial cultivation in China.

(**a**) (**b**)

**Figure 8.** Industrial cultivation of ZJLZS002. (**a**) Fruiting for 15 days; (**b**) fruiting for 24 days.

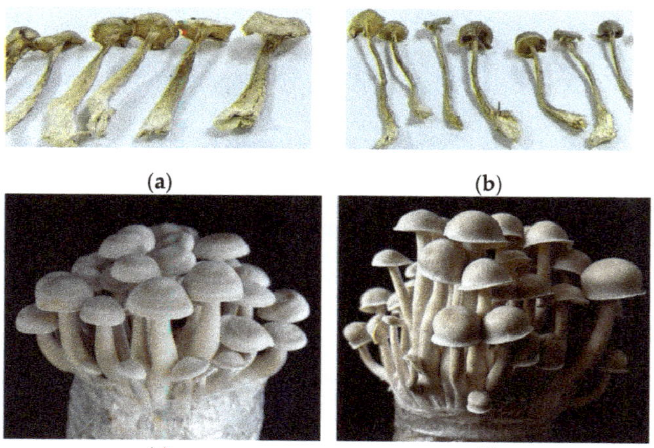

(**a**) (**b**) (**c**) (**d**)

**Figure 9.** Dried and fresh fruiting bodies. (**a**) Dried fruiting bodies of ZJLZS002; (**b**) dried fruiting bodies of LR-A; (**c**) fresh fruiting bodies of ZJLZS002; (**d**) fresh fruiting bodies of LR-A.

### *3.10. Nutrient Contents of ZJLZS002 Fruiting Bodies*

The nutrient contents of ZJLZS002 are shown in Table 10.

**Table 10.** Nutrient contents of ZJLZS002.

| Item | ZJLZS002 | Units | Item | ZJLZS002 | Units |
|---|---|---|---|---|---|
| Asp | 1.27 | g/100 g | Protein | 16.9 | g/100 g |
| Thr | 0.59 | g/100 g | Crude fiber | 24.1 | % |
| Ser | 0.63 | g/100 g | Fat | 1.6 | g/100 g |
| Glu | 1.65 | g/100 g | Total reducing sugar | 55.6 | % |
| Pro | 0.45 | g/100 g | Polysaccharide | 5.59 | g/100 g |
| Gly | 0.57 | g/100 g | Mg | 1217 | mg/kg |
| Ala | 0.64 | g/100 g | Se | 0.08 | mg/kg |
| Val | 0.58 | g/100 g | Mn | 9.09 | mg/kg |
| Met | 0.13 | g/100 g | Zn | 40.3 | mg/kg |
| Ile | 0.55 | g/100 g | Fe | 93.2 | mg/kg |
| Leu | 0.77 | g/100 g | Ca | 28.7 | mg/kg |
| Tyr | 0.43 | g/100 g | K | 4209 | mg/100 g |
| Phe | 0.54 | g/100 g | Na | 21.8 | mg/100 g |
| His | 0.33 | g/100g | Cu | 6.7 | mg/kg |
| Lys | 1.46 | g/100 g | P | 471 | mg/100 g |
| Arg | 0.72 | g/100 g | $VB_2$ | 1.37 | mg/100 g |
| Total amino acids | 11.31 | g/100 g | $VB_6$ | 4.57 | mg/kg |
| Moisture | 7.25 | g/100 g | $VB_3$ | 21 | mg/100 g |
| Ash content | 6.0 | g/100 g | - | - | - |

-: None.

## 4. Discussion

*Lyophyllum decastes* is known as Luronggu in China because of its striking resemblance to an antler [8]. It is also referred to as the fried chicken mushroom in Europe [41]. Unlike well-established cultivated mushrooms, such as *Lentinula edodes* [42] and *Agaricus bisporus* [43], which have centuries of domestication history, *L. decastes* has a relatively short cultivation timeline of approximately 50 years. It was first successfully cultivated outdoors in 1973 in Zhaotong city, China [44], and its industrial production was initially achieved in Japan in 1998 [7]. Subsequently, it has been widely cultivated in China. China's successful year-round facility cultivation of this mushroom began in 2006. The establishment of its mechanized and industrial cultivation followed in 2013. The production went from less than 1 ton in 2013 to 193,131 tons in 2023 [7].

Germplasm resources constitute the foundation of the edible fungi industry. However, the seed industry of edible fungi remains an underrecognized sector in China's agriculture [45]. Thus far, China's edible fungi industry has been reliant on imported spawn, particularly for industrial varieties. Production heavily depends on foreign-sourced spawns: (1) wood-rotting fungi (e.g., *L. decastes, Flammulina filiformis, L. edodes*) predominantly originate from Japan and South Korea; (2) straw-rotting fungi (e.g., *A. bisporus*) are mainly sourced from Europe and North America [46]. Moreover, the domestic capacity for strain development remains limited, with varietal improvement programs operating below the optimal efficiency levels [40]. This has created a critical 'jam neck' in China's edible fungi industry, affecting not only major varieties, such as *L. edodes, A. bisporus,* and *F. velutipes* [41,42], but also *L. decastes.*

Many studies on *L. decastes* breeding have been conducted globally, including selected breeding, monosporous hybridization, mutagenesis, and protoplasmic fusion. As the most conventional approach, selected breeding, also called natural selection, utilizes naturally occurring genetic variations within existing strains to develop new cultivars [28]. Although methodologically straightforward, this process is often time-intensive. Monosporous hybridization, a dominant breeding method, offers strong directional selection, operational feasibility, and hybrid vigor, albeit with significant time and labor requirements. While mutagenesis and protoplast fusion represent efficient breeding methodologies, they present significant challenges, including uncontrollable DNA mutation and genetic instability in the resulting strains. These limitations substantially complicate breeding efforts and strain stabilization [47].

Recently, SSR molecular markers have become indispensable tools for cultivar identification, population genetics studies, genetic mapping, and marker-assisted breeding. Their successful applications in edible fungi include *Lentinula edodes* [27], *Auricularia heimuer* [26], *Agaricus bisporus* [48], and *Flammulina velutipes* [49], but not *L. decastes.* Our investigation identified concerning homogeneity in the Chinese production of this mushroom. Most commercial strains are derived from the Japanese cultivar KX-HA092, and we confirm Tan Qi's [17] findings of severe uniformity.

Analysis of the Japanese Plant Variety Registration System revealed 28 registered cultivars of *L. decastes* from 1998 to 2012. However, 21 cultivars are currently inactive (https://www.hinshu2.maff.go.jp/vips/cmm/apCMM110.aspx?MOSS=1, accessed on 28 April 2025). No novel cultivars have been reported in other countries, particularly in China, which exhibits a rich germplasm. In order to bridge the domestic research gap concerning new cultivars of *L. decaste* and meet industry demands, it is crucial to breed new strains and cultivars. We have successfully developed a novel cultivar utilizing selected breeding and SSR molecular markers for the first time. The cultivar demonstrates superior agronomic characteristics, including high resistance and yield ($380 \pm 3.6$ g·bag$^{-1}$), a shortened growth period ($75.6 \pm 1.3$ d), and stable traits. Concurrently, we have provided a feasible strategy: (1) precise strain identification, (2) intellectual property protection, and (3) varietal registration. These advancements will be critical for breaking through current production bottlenecks, ensuring seed source security and enhancing the global competitiveness of China's edible fungi industry.

## 5. Conclusions

Utilizing selected breeding and SSR molecular markers, this study successfully developed an elite cultivar with significant commercial potential. The key findings include the following: (1) We collected 22 wild specimens from natural habitats and isolated and purified 8 strains with good mycelial growth. (2) after preliminary screening, we selected two stains (ZJLZS002 and ZJLZS003) based on their growth characteristics and fruiting potential. (3) We selected ZJLZS002 as the potential strain, which exhibited a shorter growth period (75.7 $\pm$ 1.5 d) and a higher yield (374.67 $\pm$ 8.50 g$\cdot$bag$^{-1}$) compared to ZJLZS003 based on secondary screening. (4) The results of the antagonism reaction revealed an isolated antagonistic interaction between ZJLZS002 and LR-A. (5) Compared to RL-A, ZJLZS002 had higher resistance based on antibacterial testing. (6) The results of the intermediate test and demonstration cultivation indicate that ZJLZS002 was characterized by its stability, shortened growth period (75.6 $\pm$ 1.3 d), high yield (380 $\pm$ 3.6 g$\cdot$bag$^{-1}$), and high biological efficiency (95%) compared to RL-A. (7) LR-A and ZJLZS002 were differentiated utilizing 10 pairs of SSR primers. (8) ZJLZS002 was officially certified as 'Zhongjunluronggu No.1' by the Yunnan Seed Management Station on 20 August 2024. We conclude that ZJLZS002 is well suited for industrial cultivation because of its high yield, shortened growth period, and stable traits.

## 6. Patents

The strain ZJLZS002 of *Lyophyllum decastes*, along with its application and SSR molecular marker primer, has the application number 202410492240.1. Application date: 23 April 2024.

**Author Contributions:** Conceptualization, Q.L., S.L., R.H. and D.S.; resources, Q.L., J.Z., F.Z. and J.M.; methodology, Q.L., S.L., J.Z., J.L., X.L. and J.M.; formal analysis, Q.L., S.L., J.Z. and J.L.; validation, Q.L., F.Z. and X.L.; writing—original draft preparation, Q.L.; writing—review and editing, R.H. and D.S.; supervision, R.H. and D.S.; project administration, R.H. and D.S.; funding acquisition, R.H. and D.S. All authors have read and agreed to the published version of the manuscript.

**Funding:** This research was funded by the National Key Research and Development Program of China (2023YFD2201800), the Yunnan Province Key Laboratory of Edible Fungi Germplasm Innovation and Functional Components (202402AN360003), and the Yunnan Province Science and Technology Talent and Platform Plan Project (202305AD160051).

**Institutional Review Board Statement:** Not applicable.

**Data Availability Statement:** The original contributions presented in this study are included in the article. Further inquiries can be directed to the corresponding author.

**Conflicts of Interest:** Author Ma Jianxiong was employed by the company Jiangsu Hongsheng Biotechnology Co., Ltd. He declares that he has no known competing financial interests or personal relationships that could have appeared to influence the work reported in this paper, and the other authors declare no conflicts of interest.

## References

1. Chen, J.H.; Zheng, H.F.; Ben, W.D.; Ma, S.Z. Industrial Cultivation of *Lyophyllum decastes*. *Acta Edulis Fungi* **2008**, *15*, 23–25. [CrossRef]
2. Li, S.H.; Tang, S.M.; He, J.; Zhou, D.Q. Two New *Lyophyllum* Species from Yunnan, China. *Diversity* **2024**, *16*, 210. [CrossRef]
3. Tang, S.M.; Yu, F.M.; Karunarathna, S.; Luo, Z.L.; Niu, K.Y.; Li, R.Y.; Li, L.; Su, X.J.; Li, S.H. Morpho-phylogenetic analyses of two novel edible mushrooms from China and a mini review of *Lyophyllum* (Agaricales, Lyophyllaceae) cultivation and bioactivities. *MycoKeys* **2025**, *112*, 307–334. [CrossRef] [PubMed]
4. Li, Y.; Li, T.H.; Liang, Y.Z.; Er, T.L.G.; Chen, D.Y. *Atlas of Chinese Macrofungal Resources*; Central Plains Farmers: Zhengzhou, China, 2015.

5. Chen, X.L.; Liu, Y.H.; Guo, W.Z.; Wang, M.F.; Zhao, J.X.; Zhang, X.; Zheng, W.G. The development and nutritional quality of *Lyophyllum decastes* affected by monochromatic or mixed light provided by light-emitting diode. *Front. Nutr.* **2024**, *11*, 1–13. [CrossRef]
6. Huang, J.Q.; Chen, H.; Lin, Y.; Ye, J.; Wang, Y.X.; Guo, Y.X.; Ma, L.Y. Genetic Differences and Mycelial Morphology of *Lyophyllum decastes* Strains. *Fujian J. Agric. Sci.* **2023**, *38*, 312–321.
7. Chen, J.H. Cultivation status and nutritional value of *Lyophyllum decastes*. *Edible Med. Mushroom* **2021**, *29*, 12–15.
8. Li, X.Y.; Jiang, W.Z.; Qin, W.X.; Long, W.S. Comparative Analysis of Nutrients in Fruit Body, Mycelia and Fermentation Broth of *Lyophyllum decastes*. *Food Sci.* **2010**, *31*, 155–157. [CrossRef]
9. Xi, Y.L.; Li, M.A.; Qin, W.X.; Hua, G.J.; Jiang, W.Z.; Long, W.S. Assessment for protein nutrition of fruit bodies, mycelia and fermentation broth of *Lyophyllum decastes*. *Mycosystema* **2010**, *29*, 603–607. [CrossRef]
10. Zhang, H.Y.; Zhang, F.Q.; Ming, W.X.; Long, W.S. Nutrition ingredient analysis and evaluation of *Lyophyllum decastes* fruit body. *Mycosystema* **2008**, *27*, 696–700.
11. Zhang, G.P.; Wang, Y.N.; Qin, C.Q.; Ye, S.M.; Zhang, F.M.; Linhardt, R. Structural characterization of an antioxidant polysaccharide isolated from the fruiting bodies of *Lyophyllum decastes*. *J. Mol. Struct.* **2023**, *1285*, 135507. [CrossRef]
12. Miura, T.; Kubo, M.; Itoh, Y.; Iwamoto, N.; Kato, M.; Park, S.; Ukawa, Y.; Kita, Y.; Suzuki, I. Antidiabetic Activity of *Lyophyllum decastes* in Genetically Type 2 Diabetic Mice. *Biol. Pharm. Bull.* **2002**, *25*, 1234–1237. [CrossRef] [PubMed]
13. Wei, Y.T.; Feng, N.; Zhan, J.S.; Han, W. Qualitative Chemical Analysis and Antitumor Activity of *Lyophyllum decastes* Fruit Body Extracts. *J-GLOBAL.* **2016**, *23*, 70–74.
14. Ukawa, Y.; Furuichi, Y.; Kokean, Y.; Nishii, T.; Hisamatsu, M. Effect of Hatakeshimeji (*Lyophyllum decastes* Sing.) Mushroom on Serum Lipid Levels in Rats. *J. Nutr. Sci. Vitaminol.* **2002**, *48*, 73–76. [CrossRef]
15. Xu, L.L.; Yang, W.J.; Qiu, T.M.; Gao, X.; Zhang, H.Y.; Zhang, S.L.; Cui, H.; Guo, L.Z.; Yu, H.L.; Yu, H. Complete genome sequences and comparative secretomic analysis for the industrially cultivated edible mushroom *Lyophyllum decastes* reveals insights on evolution and lignocellulose degradation potential. *Front. Microbiol.* **2023**, *14*, 1–14. [CrossRef]
16. Tan, Q.; Song, C.Y. *China Edible Fungus Seed Industry Development Report 2023*; China Agriculture Press: Beijing, China, 2025; pp. 261–264.
17. Woo, S.M.; Park, Y.H.; Yoo, Y.B.; Shin, P.G.; Jang, K.Y.; Lee, K.H.; Sung, J.M. Development on Artificial Cultivation method of Hatakeshimeji (*Lyophyllum decastes*) using fermented sawdust substrate. *J. Mushroom Sci. Prod.* **2009**, *7*, 156–162.
18. Yoshihama, Y.; Kusakabe, K.; Matsui, S.; Morita, H. Biologically Pure Mushroom Culture and Method for Mushroom Cultivation. U.S. Patent 5349121, 13 December 1994.
19. Wei, S.L.; Wang, Z.J.; Yu, H.P.; Chen, Y. Studies on Isolation and domestication characteristics of *Lypphyllum decastes*. *Southwest China J. Agric. Sci.* **2005**, *14*, 128–131.
20. Qiu, C.L.; Chen, C.M.; Zhuo, Z.M.; Lu, X. A Strain of Black Brown Lyophyllum decastes and Its Application. China Patent 202310624496.9, 30 May 2023.
21. Kimura, E. Factory cultivation of *Lyophyllum decastes*. *Edible Med. Mushrooms* **2019**, *27*, 237–240.
22. Pan, H.Y.; Ning, T.Y.; Xu, Y.T.; Zhou, H.Y.; Xu, B. Breeding of high yield strains of *Lyophyllum decastes*. *Edible Fungi China* **2023**, *42*, 42–46.
23. Liang, Q.Q.; Wei, S.L.; Ding, L.Q.; Yan, H.X.; Liang, Y.P.; Xi, Y.L. Mutational screening of *Lyophyllum decastes* with high Yield of Laccase and Characteristics of Laccase from the Mutant. *Nat. Prod. Res. Dev.* **2014**, *26*, 1178–1181.
24. Liang, Q.Q.; Wei, S.L.; Wang, H.; Wang, X.Q.; Ding, L.Q.; Xi, Y.L. Breeding high polysaccharide producing of *Lyophyllum decastes* strains by protolplast mutagensis. *Sci. Technol. Food Ind.* **2013**, *34*, 178–181.
25. Zhou, H.M.; Zhang, Y.Z.; Xi, Y.L.; Wei, S.L. Screening of *Lyophyllum decastes* highly productive cultivable strains. *Mycosystema* **2014**, *33*, 208–217. [CrossRef]
26. Yin, L.L.; Yao, F.J.; Shi, C.Y.; Lu, L.X.; Wang, T.; Liu, W. Genetic diversity of wild *Auricularia heimuer* germplasm resources based on SSR markers. *Acta Edulis Fungi* **2022**, *29*, 1–9.
27. Zhang, D.; Wu, P.; Zhang, L.J.; Tang, L.H.; Song, C.Y.; Shang, X.D.; Bao, D.P.; Tan, Q. Polymorphism of SSR markers based on the whole genime sequence of *Lentinula edodes* and use in strain identification. *Acta Edulis Fungi* **2012**, *19*, 7–10.
28. *GB/T 21125-2007*; Technical Inspection for Mushroom Selecting and Breeding. National Standardization Administration: Beijing, China, 2007.
29. Huang, J.; Ge, X.; Sun, M. Modified CTAB protocol using a silica matrix for isolation of plant genomic DNA. *BioTechniques* **2000**, *28*, 432–434. [CrossRef]
30. Pahlich, E.; Gerlitz, C. A rapid DNA isolation procedure for small quantities of fresh leaf tissue. *Phytochemistry* **1980**, *19*, 11–13. [CrossRef]
31. Vilgalys, R.; González, D. Organization of ribosomal DNA in the basidiomycete *Thanatephorus praticola*. *Curr. Genet.* **1990**, *18*, 277–280. [CrossRef]

32. Katoh, K.; Standley, D. MAFFT Multiple Sequence Alignment Software Version 7: Improvements in Performance and Usability. *Mol. Biol. Evol.* **2013**, *30*, 772–780. [CrossRef] [PubMed]
33. Ma, J.X.; Ma, J.H.; Meng, H. The Strain ZJLZS002 of Lyophyllum decastes, and Its Application, SSR Molecular Marker Primer. China Patent 202410492240.1, 23 April 2024.
34. *NY/T 1845-2010*; Identification of Distinctness for Edible Mushroom Cultivar by Antagonism. Ministry of Agriculture and Rural Affairs of China: Beijing, China, 2010.
35. Liu, Q.M.; Hua, R.; Sun, D.F.; Zhang, J.B.; Li, J.Y.; Liu, C.L.; Luo, X.; Liu, S.X. Breeding of a New *Cyclocybe chaxingu* Cultivar 'Zhongjunbaicha 1'. *Acta Edulis Fungi* **2023**, *30*, 57–66.
36. *GB5009.5-2016*; Determination of Protein in Food of National Standard for Food Safety. China Food and Drug Administation: Beijing, China, 2016.
37. *GB5009.124-2016*; Determination of Amino Acids in Food of National Standard for Food Safety. China Food and Drug Administation: Beijing, China, 2016.
38. *SN/T 4260-2015*; Determination of Crude Polysaccharides in Plant Source Foods for Export-Phehol-Sulfuric Acid Colorimetry. General Administration of Quality Supervision, Inspection and Quarantine of the People's Republic of China: Beijing, China, 2015.
39. Ellen, L.; Henrik, S. *Lyophyllum shimeji*, a species associated with lichen pine forest in northern Fennoscandia. *Mycoscience* **2011**, *52*, 289–295. [CrossRef]
40. Moncalvo, J.M.; Vilgalys, R.; Redhead, S.A.; Johnson, J.E.; James, T.Y.; Catherine Aime, M.; Hofstetter, V.; Verduin, S.J.; Larsson, E.; Baroni, T.J.; et al. One hundred and seventeen clades of euagarics. *Mol. Phylogenet. Evol.* **2002**, *23*, 357–400. [CrossRef]
41. Wei, S.W.; Qi, B.; Zhang, X.Z.; Peng, Z.W.; Li, Y.; Wang, Q. Global diversity and biogeography of *Lyophyllum* inferred from amplicon datasets. *Front. Ecol. Evol.* **2024**, *12*, 1328569. [CrossRef]
42. Zhu, X.K.; Zhang, Q.Y. Protection and development of the old skill "bark hacking method" of *Lentinula edodes* cultivation. *Edible Med. Mushrooms* **2019**, *27*, 96–98.
43. Wang, Z.S.; Liao, J.H.; Chen, M.Y.; Cai, Z.X. Breeding and industrial development of *Agaricus bisporus*. *Acta Edulis Fungi* **2012**, *19*, 1–14. [CrossRef]
44. Chen, S.Y. *Mushroom Cultivation manual*; Scientific and Technical Documents Publishing House: Beijing, China, 2003.
45. Liu, Q. Fungus seed industry: A neglected new seed industry. *China Dev. Obs.* **2021**, *14*, 43–45.
46. Zhao, J.; Wang, Y.F.; Sheng, C.G.; Shi, L.; Pan, C.L.; Wang, J.H.; Liu, Z.T.; Zhang, P.; Yu, H.Y.; Meng, X.H. The production status and development trend of edible fungi factory in China. *For. By-Prod. Spec. China* **2021**, *174*, 68–74.
47. Zhou, X.; Zhang, Y.; Zhang, C.X.; Yu, J.J. Breeding of high polysaccharide yield strains of *Lyophyllum decastes* by UV induced protoplast mutagenesis. *Jiangsu Agric. Sci.* **2017**, *45*, 126–128.
48. Yuan, B.; Ke, L.N.; Chen, G.X.; Lian, Y.P.; Zhang, Z.H.; Ji, P.W.; Wu, Z.Q. *Agaricus bisporus* germplasms in southern Fujian identified by somatic incompatibility test and SSR. *Fujian J. Agric. Sci.* **2020**, *35*, 950–956.
49. Gao, L.H.; Bao, D.P.; Xu, Z.; Li, Y.; Lu, H.; Tan, Y.S.; Shang, X.D.; Chen, H.Y.; Wang, R.J.; Wu, Y.Y. Construction of core collection and DNA fingerprinting of *Flammulina filiformis* based on genetic diversity analysis. *Mycosystema* **2021**, *40*, 3214–3230.

agriculture

Review

# Exploring the Potential of *Russula griseocarnosa*: A Molecular Ecology Perspective

**Yuanchao Liu, Tianqiao Yong, Manjun Cai, Xiaoxian Wu, Huiyang Guo, Yizhen Xie, Huiping Hu and Qingping Wu ***

National Health Commission Science and Technology Innovation Platform for Nutrition and Safety of Microbial Food, Guangdong Provincial Key Laboratory of Microbial Safety and Health, State Key Laboratory of Applied Microbiology Southern China, Institute of Microbiology, Guangdong Academy of Sciences, Guangzhou 510070, China; liuyc1020@163.com (Y.L.); tianqiao@mail.ustc.edu.cn (T.Y.); caimanjun@gdim.cn (M.C.); wuxiaoxian@gdim.cn (X.W.); guohuiyang0309@163.com (H.G.); xieyizhen@126.com (Y.X.); huhp@gdim.cn (H.H.)

* Correspondence: wuqp203@163.com

**Abstract:** *Russula griseocarnosa*, an edible and medicinal mushroom abundant in nutrients and notable bioactivities, is predominantly grown in the broad-leaved forest with trees of the family Fagaceae in southern China. This species forms ectomycorrhizal associations with plant roots and cannot be artificially cultivated currently. Previous research indicates a strong correlation between the growth of *R. griseocarnosa* and factors such as the host plant, climate variables (specifically mean temperature and precipitation from June to October), and the rhizosphere microbiota of its habitat. However, comprehensive studies on the fundamental biology of this species are lacking. The interaction between *R. griseocarnosa* and its host plant, as well as the mechanisms underlying the microbial community dynamics within its habitat, remain ambiguous. The limited repertoire and diversity of carbohydrate-active enzymes (CAZymes) in *R. griseocarnosa* relative to saprophytic fungi may contribute to its recalcitrance to cultivation on synthetic media. The specific core enzyme and the substances provided by the host plant to facilitate growth are yet to be elucidated, posing a significant challenge in the artificial cultivation of *R. griseocarnosa*. The habitat of *R. griseocarnosa* harbours unique microbial communities, indicating the presence of potentially beneficial microorganisms that could be exploited for artificial propagation and conservation efforts. However, the lack of definitive functional verification experiments hinders the realization of this promising prospect. This review offers a comprehensive overview of the nutritional profile and health benefits of *R. griseocarnosa*, emphasizing recent developments in its isolation, molecular ecology, and artificial cultivation. Additionally, it explores prospective advancements in *R. griseocarnosa* research, aiming to enrich our foundational understanding for applied purposes and fostering progress in the realm of ectomycorrhizal edible mushrooms.

**Keywords:** *R. griseocarnosa*; nutritional composition and bioactivity; molecular ecology; symbiotic characteristics; artificial domestication; habitat microorganisms

## 1. Introduction

The genus *Russula* is widely distributed globally and has significant economic and research importance. It represents a group of ectomycorrhizal fungi utilised for culinary and medicinal purposes [1,2]. Worldwide, there are 2759 identified taxonomic units identified within *Russula* [3]. A total of 193 taxonomic units have been reported in China, with 158 confirmed through specimen examination [4,5]. Among these, 82 species are edible, while 22 possess medicinal properties due to their various nutrients and bioactive compounds [6–12]. *Russula* spp. have a long medicinal history in China, being used alone or combined with other herbs in traditional Chinese medicine, as documented in ancient pharmacopoeias such as "Compendium of Materia Medica" and "Comprehensive Records of the Eight Min Regions". Valued for their health advantages, *Russula* spp. have been favoured by generations [13,14]. *Russula griseocarnosa* X.H. Wang, Zhu L. Yang, and Knudsen 2009 was a new species of *Russula* found in Southern China and could also be grown in Northeast India [15] and northeast Vietnam [16]. It has a distinct greyish-dense context, brownish-red pileus, and elongated spines [17,18]. Recognised as one of China's top ten wild edible fungi at the 2016 China Nanhua Wild Mushroom Conference, *R. griseocarnosa* was an ectomycorrhizal mushroom with high nutrients and health care properties [15,16], including anticancer, antioxidation, antifatigue, hypolipidemic, immunoregulation, and hematopoietic properties [19–22]. Highly prized in southern Chinese regions as a postpartum health supplement, its dried form fetches up to 2000 yuan/kg (277.78 USD/kg) in Meizhou (Guangdong, China) due to limited and unstable natural yield. As it requires symbiosis with Fagaceae tree roots [23,24], artificial cultivation is currently unfeasible using the cultivation method for saprophytic mushrooms. Existing research efforts on *R. griseocarnosa* have predominantly focused on evaluating its nutritional profile and therapeutic potential, complemented by investigations into its genomic characteristics and the microbial ecology of its natural habitat. Nevertheless, compared to other widely studied edible mushrooms our current knowledge of *R. griseocarnosa* remains relatively limited, lacking in-depth and comprehensive analyses. We used literature research methods to search for relevant research on *R. griseocarnosa* from databases such as the China National Knowledge Infrastructure (CNKI), Web of Science, and Google Scholar and conducted a review of this edible mushroom.

This review synthesizes the current research on the nutritional composition and bioactivity of *R. griseocarnosa*. It provides a comprehensive elaboration on the progress made in understanding the molecular ecology and artificial domestication of this mushroom. By integrating the ongoing research by the authors, this review offers perspectives on potential breakthroughs for the future for *R. griseocarnosa* research to drive fundamental investigations to enable its applications. Advancements in elucidating the nutritional profile, bioactive compounds, molecular mechanisms underlying its bioactivities, and strategies for successful cultivation will pave the way for developing *R. griseocarnosa* as a valuable resource for food, pharmaceutical, and other industrial applications.

## 2. Methods

Literature research methods were employed for this review. The literature sources are from the CNKI (China National Knowledge Infrastructure), Web of Science, and Google Scholar. The main keywords used are Russula, *R. griseocarnosa*, Ectomycorrhizal fungi, domestication, cultivation, symbiotic, microorganism, environmental microbiome, and genome. The retrieved literature was screened and deduplicated, low-quality literature was removed, and research not related to *R. griseocarnosa* was eliminated. At the same time, considering that *R. griseocarnosa* is a well-known ectomycorrhizal edible fungus, some of the literature related to the research background is supplemented, and then the literature was classified and recapitulated. The current research results are summarised, and the authors' insights are presented. Finally, an outlook is provided based on the current research status and the development of new technologies.

## 3. Results

### 3.1. Nutritional Composition of *R. griseocarnosa*

The fruiting body of *R. griseocarnosa* is replete with essential nutrients such as amino acids, proteins, and efficacious components like polysaccharides. Research indicates that its protein and crude fat content surpasses that of similar edible fungi such as *R. virescens* and *Cantharellus cibarius* Fr. 1821, with the pileus of the mushroom exhibiting higher concentrations of these nutrients compared to the stipe. Mineral element analysis reveals that *R. griseocarnosa* is rich in phosphorus, potassium, iron, sodium, calcium, magnesium, and zinc, displaying a characteristic of high potassium and low sodium content, which holds potential health benefits for individuals with hypertension. Furthermore, this mushroom contains seven types of fatty acids, with oleic acid predominant in the pileus and linoleic acid abundant in the stipe, surpassing common foods like eggs and chicken in fatty acid content, overall; both the pileus (80.69%) and the stipe (69.13%) showed advantages over eggs (15.1 ± 0.79%) and chicken (23.34 ± 3.56%), presenting healthier characters (Table 1). Moreover, *R. griseocarnosa* serves as a high-quality protein source, containing a significant amount of essential amino acids such as valine, lysine, phenylalanine, tyrosine, and threonine, far exceeding the FAO/WHO standard protein model [25].

**Table 1.** Comparations of fatty acid content of *R. griseocarnosa* with eggs, chicken, and deep-sea fish oil.

| Fatty Acids [a] | Pileus | Stipe | Egg | Chicken | Deep-Sea Fish Oil |
|---|---|---|---|---|---|
| C16:0 | 19.30 ± 1.11 | 30.87 ± 4.42 | 25.56 ± 1.20 | 29.37 ± 1.58 | 19.10 |
| C16:1 | 0.42 ± 0.04 | 0.90 ± 0.11 | 3.58 ± 0.56 | 0.20 ± 0.05c | 5.68 |
| C16:3 | 0.38 ± 0.09 | 2.04 ± 0.23 | - | - | - |
| C18:1 | 47.92 ± 6.47 | 14.20 ± 2.06 | 42.81 ± 1.47 | 19.80 ± 2.37 | 17.18 |
| C18:2 | 28.40 ± 5.79 | 50.22 ± 8.07 | 18.79 ± 0.63 | 21.29 ± 1.05 | 11.88 |
| C18:3 | 1.02 ± 0.11a | 1.29 ± 0.16 | 0.32 ± 0.30 | 0.19 ± 0.10 | 4.28 |
| C20:4 | 2.55 ± 0.10 | 0.48 ± 0.07 | 1.87 ± 0.10 | - | 2.2 |
| total saturates | 19.30 | 30.87 | 34.34 ± 1.20 | 55.94 ± 3.85 | - |
| total unsaturates | 80.69 | 69.13 | 15.1 ± 0.79 | 23.34 ± 3.56 | - |

[a] Values are expressed as percentage of total fatty acids, references for chicken [26], and deep-sea fish oil [27].

Owing to the close association between free amino acids and the taste and nutritional value of mushrooms, researchers have employed LC-MS/MS methods to determine the levels of free amino acids across different developmental stages of *R. griseocarnosa*. The results indicate that among the three stages of fruiting body development, the total content of free amino acids exhibits a slight increase from 19.88 mg/g (dry weight) in the first stage (Young period, egg shape) to 20.57 mg/g in the second stage (cap is not fully open), then significantly decreases to 14.42 mg/g in the third stage (cap entirely open). Concurrently, during the three developmental stages, compounds resembling monosodium glutamate, including aspartic acid and glutamic acid, exhibit substantial variations in content, all showing a decreasing trend [28]. The flavour of *R. griseocarnosa* is also influenced by its volatile oil components. Researchers have utilised gas chromatography-mass spectrometry (GC-MS) to analyse volatile oil extracts of *R. griseocarnosa*. The results indicate that the volatile oil components from the fruiting bodies of *R. griseocarnosa* primarily consist of compounds such as ketones, aldehydes, acids, alcohols, esters, alkanes, and alkenes. It is suggested that the unique aroma of *R. griseocarnosa* is attributed to its abundant content of ketones and aldehydes [29]. Combined with our own test results, we compiled Appendix A Table A1 for the volatile oil components of *R. griseocarnosa*. Additionally, *R. griseocarnosa* is also rich in other bioactive substances such as ergosterol, phenolic compounds (e.g., quercetin), and crude polysaccharides, which can serve as potential raw materials for developing health products [25,30,31].

### *3.2. Health Benefits of R. griseocarnosa*

Over the past two decades, the efficacy of *Russula* mushroom extracts have been extensively studied in China, with reported benefits primarily encompassing antioxidation, free radical scavenging, antifatigue, antiaging, hepatoprotection, blood sugar and lipid reduction, tumour inhibition, and antibacterial properties. For instance, it has been reported that the extract from *Russula* mushroom fruit bodies can ameliorate oxidative damage induced by formaldehyde inhalation in mice [20], as well as increase the levels of glutathione and superoxide dismutase in mouse serum, enhancing their ability to adapt to exercise loads, resisting fatigue generation, and accelerating fatigue elimination [32–34]. Additionally, the extract exhibits antioxidative effects in aged mice [35]. Moreover, polysaccharides of *Russula* demonstrate significant scavenging effects on superoxide anions and hydroxyl radicals [36]. Feeding with powder and polysaccharides of *Russula* has been found to significantly reduce blood glucose, total cholesterol, triglycerides, and low-density lipoprotein levels in mice with hyperglycaemia and hyperlipidaemia models, showing a dose-dependent response and indicating notable hypoglycaemic and lipid-lowering effects of *Russula*. Similarly, injection of polysaccharides of *Russula* has been shown to reduce cholesterol levels in rats with hyperlipidaemia, resulting in a significant 45.2% reduction in total cholesterol compared to the control group [37,38]. The alcohol extracts from *Russula* fruit bodies have been tested and found to possess certain antibacterial properties against *Escherichia coli* and *Staphylococcus aureus* [39]. Additionally, lectins isolated from fresh fruit bodies of *Russula* exhibit potent inhibitory effects on the proliferation of HepG2 hepatoma and MCF7 breast cancer cells, although interspecies differences may exist in their inhibitory activities against ribonuclease and HIV-1 reverse transcriptase [40–42]. Furthermore, polysaccharides of *Russula* have been shown to promote lymphocyte activity, achieving an inhibitory effect on SiHa cancer cell proliferation [43].

It should be noted that most of the materials used in the above studies come from Rongxian County, Guangxi, China. Due to the difficulty of classifying the genus *Russula* [44], and the fact that there are many closely related species of *R. griseocarnosa* that have not yet been identified [45], although these materials in some studies may be *R. griseocarnosa*, it has not been clearly stated. Instead, some of the research indicates that the materials are *Russula vinosa* Lindblad 1901, *Russula delica* Fr. 1838, and *Russula lepida* Fr. 1836, which has caused confusion in accurately understanding the efficacy of *R. griseocarnosa*. Fortunately, to ensure the accuracy of this review, we have primarily incorporated research materials that were unambiguously identified as *R griseocarnosa*. The subsequent research progress discussed in this review largely stems from experiments where *R. griseocarnosa* was the main study material utilised.

Polysaccharides extracted from *R. griseocarnosa* fruit bodies demonstrate a certain clearance effect on hydroxyl radicals and superoxide radicals. In vitro cell experiments confirm the significant inhibitory effect of red mushroom polysaccharides on cancer cells Hela and SiHa, along with a noticeable enhancement in the phagocytic capability of peritoneal macrophages in mice. This enhancement promotes the secretion of NO and cytokine IL-6, demonstrating strong immunomodulatory activity [31,46]. Volatile oils and petroleum ether extracts from *R. griseocarnosa* fruit bodies exhibit strong antibacterial effects against Staphylococcus aureus and others, while the ethanol extract shows less significant antibacterial activity. Additionally, methanol extracts from fruit bodies demonstrate inhibition and scavenging effects on hydroxyl radicals generated by DPPH and Fenton reactions, confirming the natural antimicrobial, antioxidant, and free radical scavenging properties of *R. griseocarnosa*. [25]. Of these, the bioactive compounds may be phytochemicals such as phenolics, flavonoids, ergosterol, and β-carotene (Table 2). The major antioxidative component in *R. griseocarnosa* was quercetin, which was detected in levels up to 95.82 μg/g. On the other hand, the chemical constituents of *R. vinosa* were examined carefully, leading to the discovery of 15 compounds. Wherein four sesquiterpenes, eight triterpenes together, and three compounds containing N element were included, some of which showed inhibitory effects on nitric oxide (NO) production.

**Table 2.** Chemical composition *versus* biological activity of *Russula species*.

| Compound | Name | Structure | Biological Activity | Target | Species | Source |
|---|---|---|---|---|---|---|
| 1 | ergosterol | 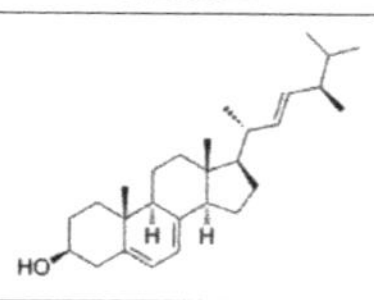 | antioxidant | DPPH | *R. griseocarnosa* | [30] |
| 2 | β-carotene | 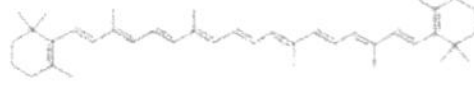 | antioxidant | DPPH | as above | [30] |

**Table 2.** *Cont.*

| Compound | Name | Structure | Biological Activity | Target | Species | Source |
|---|---|---|---|---|---|---|
| 3 | quercetin | 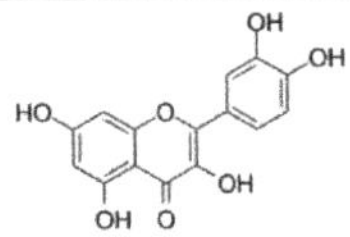 | antioxidant | DPPH | as above | [30] |
| 4 | caffeic acid | 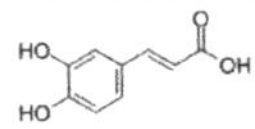 | antioxidant | DPPH | as above | [30] |
| 5 | protocatechuicacid | 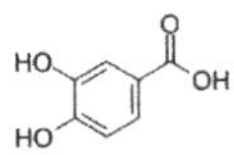 | antioxidative, antibacterial, and antimutagenic activities | DPPH | as above | [30] |
| 6 | vinosane | 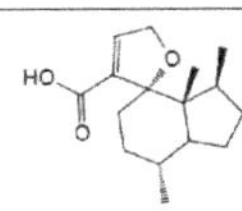 | inhibiting NO production | - | *R. vinosa* | [47] |
| 7 | rulepidadione C | 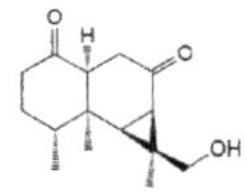 | - | - | as above | [47] |
| 8 | 7α,8α,13-trihydroxy-marasm-5-oic acid-lactone | 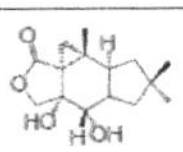 | inhibiting NO production | - | as above | [47] |
| 9 | aristolone | 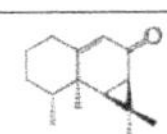 | inhibiting NO production | - | as above | [47] |
| 10 | (24E)-3,4-seco-cucurbita-4,24-diene-26,29-dioic acid-3-methyl ester | 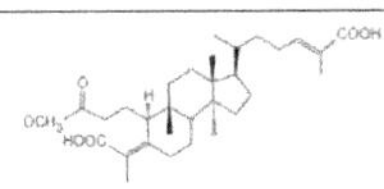 | inhibiting NO production | - | as above | [47] |
| 11 | (24E)-3,4-seco-cucurbita-4,24-diene-26-oic acid-3-ethyl ester | 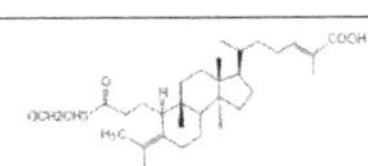 | inhibiting NO production | - | as above | [47] |
| 12 | (24E)-3β-hydroxycucurbita-5,24-diene-26,29-dioic acid | 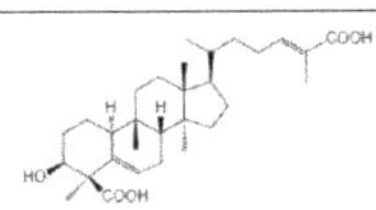 | - | - | as above | [47] |
| 13 | (24E)-3,4-secocucurbita-4,24-diene-3,26,29-trioic acid | 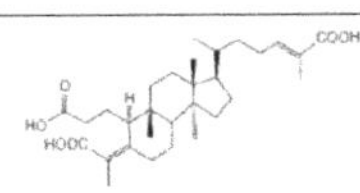 | inhibiting NO production | - | as above | [47] |
| 14 | (24E)-3,4-secocucurbita-4,24-diene-3,26-dioic acid | 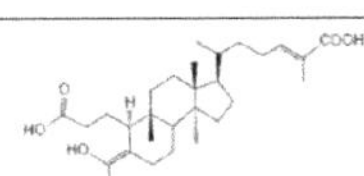 | - | - | as above | [47] |
| 15 | (24E)-3β-hydroxycucurbita-5,24-diene-26-oic acid | 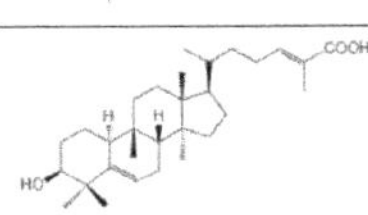 | - | - | as above | [47] |
| 16 | rosacea acid B | 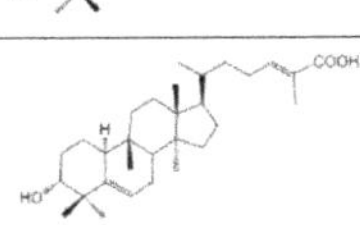 | - | - | as above | [47] |

**Table 2.** *Cont.*

| Compound | Name | Structure | Biological Activity | Target | Species | Source |
|---|---|---|---|---|---|---|
| 17 | rosacea acid A | 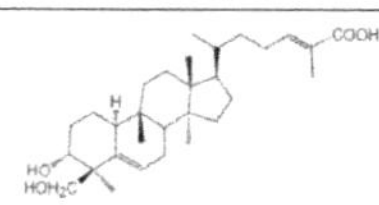 | - | - | as above | [47] |
| 18 | (2S,3S,4R,20R)-2-(20-hydroxydocosanoylamino) eicosane-1,3,4-triol | 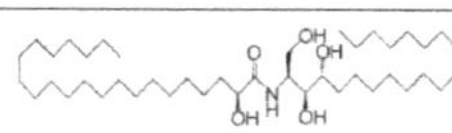 | - | - | as above | [47] |
| 19 | 7,8-dimethylalloxazine | 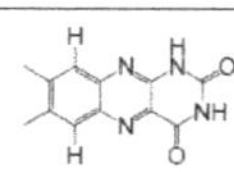 | - | - | as above | [47] |
| 20 | L-pyroglutamic acid | 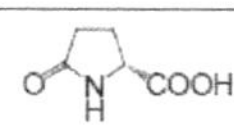 | - | - | as above | [47] |

Studies have also shown that fresh fruit bodies of *R. griseocarnosa* were observed to stimulate the activities of phenylalanine ammonia-lyase (PAL) and chalcone synthase, consequently leading to an enhanced accumulation of phenolic and flavonoid contents when treated with nitric oxide fumigation. This can enhance the bioactive compounds and improve antioxidant activities in the mushrooms [48]. This result was similar to the studies on other fungi that demonstrated the promotive effect of NO on the biosynthesis of fungal secondary metabolites [49].

The polysaccharide PRG1-1, isolated from the fruiting body of *R. griseocarnosa*, exhibited antiproliferative effects on HeLa and SiHa cervical carcinoma cells. It significantly reduced cell viability, increased lactate dehydrogenase (LDH) and reactive oxygen species (ROS) production, and enhanced the apoptotic rate. These findings highlight the promising potential of the bioactive PRG1-1 as a natural agent for inhibiting tumour cell proliferation in the treatment of cervical cancer [50]. Also, PRG1-1 has the capacity to activate macrophages via the NF-κB and MAPK pathways, which exhibit immunomodulatory potential [51].

The latest research confirms that polysaccharide (RGP1) derived from *R. griseocarnosa* can improve hematopoietic function in K562 cells. Mechanism studies show that RGP1 could alleviate hematopoietic dysfunction by promoting the activation of CD4+ T cells and the Janus kinase/signal transducer and activator of the transcription three pathway; this study provides compelling evidence for the application of *R. griseocarnosa* to improve hematopoietic dysfunction [52]. This is similar to the reported active polysaccharide function of edible mushrooms [53]. Still, it is the first edible fungus with a blood-tonifying effect confirmed via animal experiments, and it is of great significance for the development of blood-enhancing dietary supplements. The polysaccharide RGP2 isolated from the fruiting body of *R. griseocarnosa* modulated gut microbiota composition and serum metabolite expression, and regulated T cell differentiation to enhance immune function in cyclophosphamide (CTX)-induced immunosuppressed mice [54]. This discovery provides a potential drug lead compound for the development of new natural immune modulators and has important application prospects for improving immune-related diseases and enhancing the body's immunity. The above results showed that Polysaccharides from *R. griseocarnosa* exhibit antioxidant, antimicrobial, immunomodulatory and anticancer properties. Further research on these bioactive compounds from *R. griseocarnosa* holds great promise for developing novel therapeutic agents and nutraceuticals.

### *3.3. Isolation and Identification of 'Strains' from R. griseocarnosa*

The mycelium of *R. griseocarnosa* is considered the premise and basis for its research and application. Therefore, the focus of researchers has consistently been on obtaining effective culture conditions. The predecessors mainly adopted the method of tissue and basidiospore separation to isolate the strain. Research has primarily centred on identifying suitable nutrients and conditions for its growth, such as carbon and nitrogen sources [55–64], mineral elements [65], vitamins, plant growth regulators [66], rare earth elements [67], and extracts from tree roots and insect dung [68] in the medium, as well as pH, incubation temperature, time [69–71], and light [72]. However, identification of most of the strains has not been confirmed. A small number of strains have been identified as other fungi [73–75], such as *Monochaetia* Sacc [56] and *Nectria* sp. [76], by morphological or molecular biology methods, suggesting the existence of a variety of endophytes in the fruiting bodies and making it more challenging to isolate the pure culture of *R. griseocarnosa* compared to other saprophytic mushrooms [77].

The work of HINTIKKA et al. suggests that the slow growth of mycorrhizal fungal mycelium may be attributed to a deficiency in specific nutrients, such as vitamins. In contrast, it is proposed that the mycelium of *R.* spp. can achieve better growth in media lacking an additional carbon source, implying that the nitrogen source and trace elements may be more critical factors. However, the mycelium mentioned in the study has not been determined with molecular identification [78].

Several media listed in the <*Handbook of Microbiological Media (Fourth Edition)*> [79], such as MMN, Hagem's Modess Medium, and Oddoux Medium, which are suitable for isolating ectomycorrhizal fungi, were tested by the authors. However, attempts to obtain a pure culture of *R. griseocarnosa* were unsuccessful, and instead, multiple endophytes were isolated and identified. Additionally, efforts were made to obtain the *R. griseocarnosa* strain from ectomycorrhizas, referencing the work of Yamada et al. [80], as well as co-culturing basidiocarp tissue with the host plant's callus [81]. Yet, the mycelia obtained through these methods were identified as plant endophytes using molecular techniques.

Furthermore, analyses of genome-wide genes have revealed that the types and quantities of carbohydrate-active enzymes in *R. griseocarnosa* are significantly less than those found in saprophytic mushrooms, which provides a molecular-level interpretation of the difficulty in obtaining a pure culture of this species on artificial media [82,83]. Given these findings, it is concluded that a verified pure strain of *R. griseocarnosa* cannot be obtained under artificial conditions.

### 3.4. The Symbiotic Characteristics of R. griseocarnosa

Ectomycorrhizal fungi are commonly observed to exhibit preferential associations with specific host plants, and these host preferences are believed to be important evolutionary drivers of ectomycorrhizal fungal diversification [84]. Current studies have shown that Fagaceae trees are necessary and play a vital role in the formation of fruiting bodies of *R. griseocarnosa* [31]. However, the symbiotic mechanism between *R. griseocarnosa* and its host plant remains unclear. It is generally accepted that ectomycorrhizal fungi are indispensable mutual aid partners for many trees and shrubs in forest ecosystems. More than 80% of plant nitrogen and phosphorus is provided by these mycorrhizal fungi, and many plant species depend on these symbionts for their growth and survival [85–87]. Additionally, ectomycorrhizal fungi are thought to have acquired the ability to colonise plant root tip tissues like plant pathogens to obtain host-derived glucose [88,89].

Ectomycorrhizal fungi have been thought to possess the ability to decompose organic matter in the soil. However, it is now understood that they do not directly utilise the carbon sources that are released into the soil through decomposition processes [90]. Furthermore, the ability of ectomycorrhizal fungi to decompose lignocellulose and cellulose has been found to be much lower than that of wood-decay and soil-dwelling saprotrophic microbes [87,91].

Ectomycorrhizal fungi have evolved to become highly dependent on their host plants, primarily relying on the latter to provide simple carbohydrates. This evolutionary trajectory has been shaped by the convergent loss of gene families responsible for the degradation of organic matter. Notably, ectomycorrhizal fungi do not require the ability to degrade lignocellulose, as they can maintain the integrity of host plant cells by avoiding the release of plant cell wall-degrading enzymes (PCWDEs). In this process, a substantial number of PCWDE and lignin-oxidizing class II peroxidase genes have been lost, with peroxidases for lignin oxidation and cellulases for cellulose degradation being the most extensively lost. Furthermore, ectomycorrhizal fungi have almost entirely lost the GH6 and GH7 family genes encoding cellobiohydrolase, which are present in soil saprophytes and white-rot fungi [92–94]. These observations suggest that ectomycorrhizal fungi have evolved repeatedly and independently from ecologically diverse ancestors, such as brown-rot fungi, white-rot fungi, and other saprotrophs, and the loss of genes encoding PCWDEs is considered a primary process in their evolution from saprophytic ancestors. However, some essential genes with a potential role in mycelial growth and fruit body development, such as those encoding lytic polysaccharide monooxygenase (LPMO), laccase, decolourising oxidase, and heme thiol peroxidase, have been maintained in ectomycorrhizal fungi [94–96], indicating that these genes provide apparent advantages in adaptation.

The diversity of degradative enzymes retained by different ectomycorrhizal fungi is thought to reflect their polyphyletic origins and may also indicate changes in their decay abilities [94]. During the process of symbiosis with the host, the ectomycorrhizal fungus *Laccaria bicolor* (Maire) P.D. Orton 1960 can block the expression of defence genes related to the jasmonic acid signaling pathway by secreting effectors such as the MiSSP7 protein, which binds to the inhibitor of the plant jasmonic acid signal pathway. This allows the fungal mycelium to be successfully colonised in the cortical cytoplasmic exosome area, ultimately forming a Hartig net. Meanwhile, the host plant can also secrete small secreted proteins (SSPs) into the fungal mycelium, affecting the growth and morphology of the mycelium [97], to maintain the mutual benefit state of ectomycorrhizal symbiosis [98,99].

Research has shown that the development of symbiosis requires the active functioning of multiple gene networks [100]. While some progress has been made in the study of the symbiosis between *L. bicolor* and host plants, the molecular mechanisms underlying this symbiosis have not been fully elucidated. In particular, the specific genes responsible for the establishment and maintenance of the ectomycorrhizal symbiosis remain unclear [85].

Research on the symbiotic mechanisms of *R. griseocarnosa* is quite limited. In the present study, the genome of *R. griseocarnosa* was analysed, revealing that the type and number of carbohydrate-metabolizing enzymes are significantly reduced compared to saprophytic fungi. However, the cellulose-degrading enzyme LPMO and GH6 and GH7 family genes are still found, slightly differing from previous conclusions. Notably, no homologous genes of MiSSP7 were identified in the predicted coding genes of *R. griseocarnosa*, suggesting that there may be other mycorrhizal-induced small secreted proteins (MiSSPs) involved in the symbiotic recognition between *R. griseocarnosa* and its host plant. Phylogenetic tree analysis based on whole-genome construction has shown that *R. griseocarnosa* and other ectomycorrhizal fungi are grouped together, clearly distinguished from saprophytic fungi, indicating their significant symbiotic properties [83].

### 3.5. Habitat Molecular Ecology of *R. griseocarnosa*

Previous studies have suggested that ectomycorrhizal fungi, such as *Russula* spp., tend to appear primarily during the later stages of forest succession [78,101]. Environmental selection and dispersal limitation are considered the two main processes shaping the construction of biological communities in ecosystems [102]. The dispersal of basidiospores is an essential factor affecting the diversity of fungal communities [103]. Bioinformatic analyses of operational taxonomic units (OTUs) have shown that ectomycorrhizal fungal groups in forests at different successional stages (young, intermediate, and old) are diverse and can be affected by various types of factors [104]. Furthermore, there is competition between different ectomycorrhizal fungi in forests, which in turn impacts species richness [105], the species abundance of *Russula* has exhibited unstable characteristics, with both increases and decreases observed along the forest succession gradient. Based on our present investigation, the appearance of *R. griseocarnosa* is typically associated with forests older than 50 years and without large-scale deforestation. Although the altitudinal gradient has been identified as an important factor affecting the diversity of ectomycorrhizal fungi [106,107], the findings demonstrate that *R. griseocarnosa* can be widely distributed, ranging from 244 m to 2100 m in the Guangdong, Fujian, and Yunnan provinces of China.

Soil is an integral component of terrestrial ecology. Soil characteristics, such as pH, organic carbon, nitrogen, and phosphorus, have significantly influenced the structure of soil microbial communities [108–110]. Additionally, the vertical stratification of the soil profile has been observed to exert similar effects, with the abundance of ectomycorrhizal fungi decreasing with increasing depth of the soil layer [111]. It has been observed that *R. griseocarnosa* generally grows in the shallow soil layer, with only a small portion of the stipe penetrating the soil. The microbiome compositions of soil at different depths within the rhizosphere of *R. griseocarnosa* were determined through 16S rRNA gene sequencing, revealing a significant difference in the soil diversity index between the surface and deeper soil layers in the habitat of *R. griseocarnosa* (unpublished data).

Among the various factors that can affect the occurrence of *R. griseocarnosa*, biological interactions are considered to be of critical importance. Soil bacteria, for instance, have been shown to influence the abundance of ectomycorrhizal fungi by affecting the cycling of soil nutrients [112,113]. Previous studies on mycorrhiza helper bacteria (MHB) have suggested that there are physical and metabolic interactions between soil microorganisms, mycorrhizal fungi, and host plants.

*R. griseocarnosa* is widely distributed across southern China, a region characterized by diverse climatic conditions. Notably, ectomycorrhizal fungi have been observed to associate with different MHB under varying climatic regimes [114]. Considering the differential composition of mycorrhizal helper bacteria in response to climatic factors, more extensive analyses of habitat microbiomes are warranted to identify the common and unique MHB associated with *R. griseocarnosa* throughout this region. By examining the microbial communities across the diverse environments inhabited by this fungal species, researchers can gain a comprehensive understanding of the microbial partners involved in facilitating ectomycorrhizal symbioses, and how these partnerships adapt or diverge in response to the heterogeneous climatic conditions prevalent in southern China.

The other microorganisms present in the rhizosphere affect the establishment of mycorrhizal symbiosis on plant roots [115]. For example, MHB and host plants have been reported to release substances akin to vitamins, which can promote the formation of mycorrhizal symbiosis [116,117].

Studies on soil microorganisms in the habitat of *R. griseocarnosa* were conducted using Illumina high-throughput sequencing, which revealed that the main rhizosphere bacteria exhibited a similar community structure. However, the diversity of rhizobacteria was significantly lower than that of non-rhizosphere bacteria. Notably, the bacterial secretion system, tyrosine metabolism, biosynthesis of unsaturated fatty acids, and vitamin metabolism were much more abundant in the rhizosphere. The results suggest that soil pH and available nitrogen were the primary factors influencing the microbial community structure, and these rhizosphere bacteria play a vital role in the growth of *R. griseocarnosa* [118].

Similarly, research employing comparable methods has been conducted to study the dynamic composition of microbial communities associated with *Russula* in the *Russula-Fagaceae* nature areas of Fujian province, China. The findings revealed that the fungal diversity of the *Russula* habitat was negatively correlated with the occurrence of *Russula*. These potential indicator species associated with sporocarp production in *Russula* may provide a new strategy for improving *Russula* symbiosis and sporocarp yield [119]. The research highlighted above has focused on exploring the microbial community structure within the native habitat of *R. griseocarnosa*. However, it is essential to note that the wild environment contains not only *R. griseocarnosa* but also other microorganisms, particularly those similar to *R. griseocarnosa*. Are there any similarities in the habitat microorganisms associated with these related species? In addition, rare microorganisms with relatively low abundance in ectomycorrhizal fungal habitats are often overlooked. Wei Ge et al. [110]. conducted a study on the bacterial community in the fruiting body and mycosphere of *Cantharellus cibarius*. Similar to previous research findings, they observed a higher proportion of specialist bacteria compared to generalist bacteria in the fruiting body and mycosphere. Analysis of the metabolic functions and phenotypes of abundant and rare bacteria revealed that while abundant bacteria exhibited specific potential functions, rare bacteria may contribute supplementary or unique metabolic pathways (such as sulfite oxidizer and sulfur reducer) that enhance the ecological function of *C. cibarius*. This study elucidates the distribution and function of specific microorganisms from the perspective of the fruiting bodies of ectomycorrhizal fungi and rare bacteria in their mycosphere, providing a new perspective for deepening our understanding of the ecological functions of microbiota on ectomycorrhizal fungi.

The habitat microbial populations of *R. griseocarnosa* and its similar species were also studied. The results indicate that the evenness of soil fungi in the habitats of various species' fruiting bodies is not significantly different. However, *R. griseocarnosa* was found to have higher species richness, and there was no notable difference in the abundance of the top 10 soil microorganisms corresponding to the relative abundance of each species. Interestingly, the soil fungus *Aspergillus citocrescens*, which has a lower relative abundance, was significantly higher in the habitat of *R. griseocarnosa*. So, *A. citocrescens* was inferred to be a key species in the soil microhabitat under the fruiting bodies of *R. griseocarnosa* and may play an essential role in the formation of the rhizosphere microecology [120]. The study has also revealed that different ectomycorrhizal fungi may be able to reshape the soil microecology within their respective habitats. An analogous investigation for mycosphere soil microbiomes associated with *R. griseocarnosa* and its similar species *R. rosea* by Yu Fei et al. [121] revealed findings that were not entirely congruent; although significant differences were observed in the dominant microbial flora between the two ectomycorrhizal rhizospheres, certain bacterial and fungal taxa emerged as dominant ectomycorrhizal helper microorganisms shared by both *Russula* species. The bacterial taxa identified as prominent in this mutualistic association included *Variibacter*, *Candidatus_Solibacter*, *Sorangium*, *Mycobacterium*, *Singulisphaera*, *Isosphaera*, *Bdellovibrio*, and *Paenibacillus*. Additionally, the fungal genera *Trichoderma*, *Penicillium*, and *Hypomyces* were found to be prevalent ectomycorrhizal helper microorganisms. Remarkably, these bacterial groups possess diverse functional capabilities that facilitate the establishment of symbiotic relationships between the mushrooms and plant roots. Notably, these bacteria exhibit traits conducive to plant cell wall degradation, atmospheric nitrogen fixation, and solubilization of phosphorus, thereby enhancing nutrient acquisition and promoting the formation of ectomycorrhizal association. The study highlights the significance of these microbial consortia in mediating the intricate symbiotic interactions between ectomycorrhizal fungi and their plant hosts.

The research findings described above have indeed indicated the potential presence of indicator microorganisms or growth-promoting microbes within the habitat of *R. griseocarnosa*. However, field investigations revealed that, despite significant variations in soil physical and chemical properties, such as soil particle size and soil humus content, across different growth sites—which can lead to alterations in the microbial community composition [109]—*R. griseocarnosa* was still capable of thriving normally as long as the host plant was present within the forest ecosystem. This suggests that the specific microbial community structure observed within the habitat of *R. griseocarnosa* may not be solely attributable to the presence of the fungus itself. Rather, the root exudates of the host plant, which can potentially influence the physical and chemical properties of the habitat soil, may be an essential factor affecting the growth and development of *R. griseocarnosa*.

It also further indicates that the complex interactions between *R. griseocarnosa*, its host plant, and the soil environment, including the physicochemical characteristics and microbial community composition, collectively contribute to the successful establishment and thriving of this ectomycorrhizal fungus. The influence of host plant-derived metabolites on the soil habitat appears to be a crucial component in understanding the ecology and niche requirements of *R. griseocarnosa*. Further research is needed to elucidate the specific mechanisms by which the host plant and its root exudates shape the soil environment and microbiome, thereby supporting the growth and fruiting of *R. griseocarnosa*. This holistic perspective, considering the tripartite relationship between the fungus, host plant, and soil, is essential for gaining a comprehensive understanding of the ecology and habitat preferences of this economically and ecologically critical ectomycorrhizal species.

### 3.6. Artificial Domestication

As an ectomycorrhizal fungus, effective isolation of pure cultures of *R. griseocarnosa* has not been achieved to date [122,123]. Existing studies have attempted to increase yield or obtain fruiting bodies by improving the forest environment or inoculating the wild habitat. Related reports [124–127] have indicated that spraying spore suspensions of *R. griseocarnosa* into the forest or applying solid inoculum formulations containing the isolated 'strain' under seedbeds or seedlings can be beneficial for the formation of mycorrhizal seedlings or increasing the diversity of ectomycorrhizal fungi in the forest. These approaches are expected to improve the production of *R. griseocarnosa* eventually. Transforming the woodland environment has also been considered as a means to increase the yield [128,129].

Current research on the soil microbiome has shown that mycorrhiza helper bacteria (MHB), such as *Mycobacterium* spp. and *Acidophilus* spp., may be present when comparing the bacterial communities in the rhizosphere and non-rhizosphere soil of *R. griseocarnosa* [118]. Another analysis of the network structure and interactions within the sporocarp soil of *R. griseocarnosa* revealed that the genera *Bacillus*, *Burkholderia*, and *Streptomyces* exhibited the highest abundance. Furthermore, numerous effective partnerships and close associations were observed between the genera *Bacillus* and *Burkholderia*, suggesting a potential role in promoting the growth of *R. griseocarnosa*. However, the high-throughput sequencing results indicated a relatively higher abundance only for the genus *Burkholderia*, while *Bacillus* did not exhibit a similarly high abundance [130]. Therefore, a combinatorial approach employing traditional pure culture methods and high-throughput sequencing technology can provide a more comprehensive understanding of the rhizosphere microbial diversity associated with *R. griseocarnosa*. This integrated strategy could unravel the intricate microbial interactions and elucidate their contributions to the growth and development of this economically and ecologically significant mushroom species.

Additionally, applying nitrogen fertilisers and microbial fertilisers containing MHB may promote the protection, reproduction, and sustainable use of *R. griseocarnosa*. The existing literature in the field has put forward the notion that various interventions, including increasing surface water content, humidity control, raising soil temperature, and winter pruning of mycorrhizal trees, could be implemented in the woodland habitat where *R. griseocarnosa* grows and could significantly augment mushroom yield [131].

Artificially manipulating the number of basidiospores and the living environment of *R. griseocarnosa* through the methods described in the above studies may have a positive effect on increasing production. However, it has been difficult to conclusively prove the effectiveness of the process of inoculating 'cultures' that have not been accurately identified, in terms of increasing yield. It may also be that the inoculation of MHB into the habitat of *R. griseocarnosa* could increase production [132], but the mechanisms are not yet clear.

## 4. Discussion

With its delicious flavour, significant nutrition, and efficacy, *R. griseocarnosa* is highly sought after by consumers and researchers. However, the fundamental research related to hyphal physiology was relatively backward. Currently, there are still difficulties in the separation and identification of *R. griseocarnosa* fruiting bodies and strains. The challenge of distinguishing species similar to *R. griseocarnosa* and the existence of chaotic molecular sequences have led to confusion about the *Russula* species produced in regions such as Fujian, Guangxi, and Yunnan. The use of molecular identification techniques to verify the identity of the isolated cultures has not been adequately demonstrated in the existing research. While numerous attempts have been made to separate and cultivate strains, the reliability of the so-called "cultures" mentioned in the literature has not been conclusively established through the application of rigorous molecular biological methods. It is widely accepted that achieving a sequence alignment similarity of over 97% is requisite for categorizing specimens within the same species [133–135]. However, there is no documented instance where the sequence similarity between the isolated mycelium and the fruiting body of *R. griseocarnosa* has exceeded 97%. Research efforts aimed at augmenting *R. griseocarnosa* production through artificial field expansion and other methodologies confront a multitude of variables, rendering them challenging to substantiate through data or effective technical methodologies scientifically and objectively. Leveraging our existing research and integrating recent advancements in ectomycorrhizal fungi studies, plant root-microbe communication, and the application of cutting-edge technology, we have synthesised and speculated upon the molecular ecological interplay among *R. griseocarnosa*, associated microorganisms, and host plants (see Figure 1), in this process, the application of multi-omics technology should be valued and attempted to make up for the disadvantage of difficulty in isolating the mycelium and/or habitat microorganisms of *R. griseocarnosa* [136]. We contend that the following vital issues warrant firm attention:

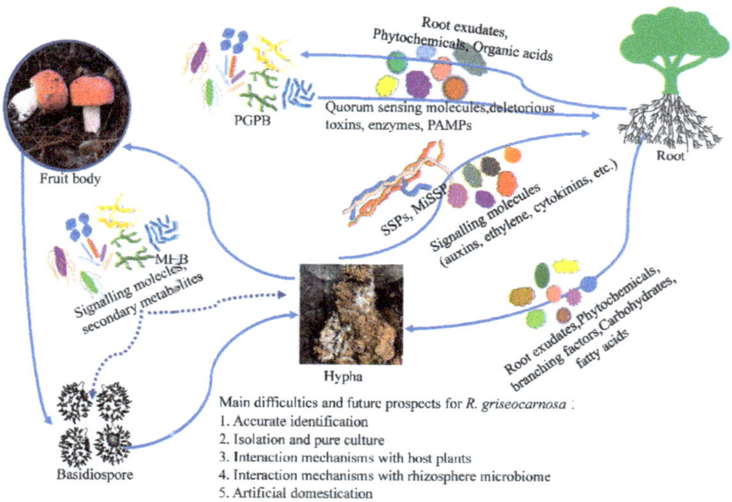


**Figure 1.** The challenges and pertinent issues encountered by *R. griseocarnosa*.

1. Enhancing the accurate identification of *R. griseocarnosa*

Traditional mushroom species identification predominantly relies on holotype characteristics, encompassing macroscopic and microscopic features alongside molecular sequence data [137]. The taxonomy of the genus *Russula* presents significant challenges, particularly in the classification of closely related species to *R. griseocarnosa* [44,138]. Notably, the taxonomic status of many closely related species remains contentious [45] although phylogenetic trees constructed based on ITS sequences were able to distinguish a few *Russula* species [139], and this is compounded by the extensive genetic diversity and differentiation within *R. griseocarnosa* populations [140,141]. In our investigation, we utilised the ITS sequence [17] of the *R. griseocarnosa* holotype to compare it with recently published genomic data [142,143]. Our analysis revealed a maximum similarity of 94.1%, contrasting with up to 99% similarity observed with our own genomic data of *R. griseocarnosa* [83]. Given the advancing precision of contemporary sequencing technologies, ensuring accurate identification of sequencing materials is fundamental to research integrity. During our preliminary resource survey, numerous specimens exhibited ITS sequences closely resembling those of *R. griseocarnosa*. However, discernible disparities in macroscopic morphology persisted, particularly evident in variations in pileus colour and thickness, dried lamellae colour, and fresh fruit body hardness. Consequently, there is an urgent imperative to establish standardised gene barcodes to facilitate accurate and expedient identification of *R. griseocarnosa*.

2. Navigating challenges for isolation of *R. griseocarnosa* culture

Culture isolation of *R. griseocarnosa* presents significant challenges. As previously noted, achieving pure cultures of this fungus proves arduous. Given the success in obtaining pure cultures of ectomycorrhizal fungi through mycorrhizal separation [144], it merits investigation whether this approach is applicable to *R. griseocarnosa*. Various media were employed in attempts to isolate *R. griseocarnosa* strains at different growth stages; however, these efforts yielded no pure cultures. Instead, numerous endophytic fungi were isolated. Further investigation is warranted to ascertain whether these endophytic fungi play a role in the growth and development of *R. griseocarnosa*. Additionally, whole-genome analysis revealed significant disparities in carbohydrate-related enzymes compared to saprophytic fungi. Nonetheless, it remains inconclusive whether the absence of certain enzyme genes directly hinders growth on artificial media. Transcriptome research may provide complementary insights for verification.

3. The Symbiotic Interaction between *R. griseocarnosa* and its host plant

Investigation into the interaction between *R. griseocarnosa* and its host plant reveals *Castanopsis hystrix* as the primary host, fostering a mutually beneficial relationship. Analogous to the pivotal role of other ectomycorrhizal fungi within forest ecosystems [145] the formation of mycorrhiza prompts *R. griseocarnosa* mycelium to release various molecular secretions into the soil. These secretions facilitate the growth of the host plant's root system and enhance nutrient uptake. Concurrently, they bolster the host plant's resilience to stress and its survival rate during transplantation [146]. However, the specific substance provided by the host plant to *R. griseocarnosa* remains undetermined. Whether it pertains to sugars or fatty acids, reminiscent of the metabolic processes of arbuscular mycorrhizal fungi [147], necessitates elucidation through further interaction research.

4. Clarify the functional interaction of *R. griseocarnosa* with microorganisms in habitat soil and endophytic fungi in the fruiting body

The study focuses on examining the interaction between *R. griseocarnosa* and microorganisms present in soil and fruiting bodies. Soil microorganisms, integral to terrestrial ecosystems, play pivotal roles in biogeochemical cycles and exert significant influence across diverse environmental conditions [108]. They also impact the abundance of ectomycorrhizal fungi [113]. Furthermore, eukaryotic organisms, including mushrooms, engage in intricate interactions with microbial communities [148], with specific microbial helper bacteria (MHB) identified in *R. griseocarnosa* habitats [118]. Nonetheless, the precise mechanism by which these microorganisms promote growth remains unclear. Similarly, akin to the beneficial effects of ectomycorrhizal fungi on plant growth and development, endophytic microorganisms are presumed to positively influence the growth of *R. griseocarnosa* [149]. However, whether endophytic microorganisms associated with the host of *R. griseocarnosa* exert similar effects on *R. griseocarnosa* itself remains uncertain. Furthermore, the function of endophytic microorganisms from *R. griseocarnosa* remains unexplored. Additionally, investigating potential variations in microbial structures associated with different growth stages of *R. griseocarnosa*, akin to other ectomycorrhizal edible mushrooms [150], is imperative.

5. Challenges in the artificial domestication of *R. griseocarnosa*

Research on the artificial domestication of *R. griseocarnosa* remains the ultimate goal of many researchers. At present, the cultivation and development of ectomycorrhizal edible mushrooms mainly rely on the introduction of mycorrhizal seedlings to simulate the natural ecological conditions of the mushroom and establish artificially cultivated plantations (forests) to achieve semi-artificial or wild-like cultivation [151,152]. Using this method, various ectomycorrhizal mushrooms have been successfully cultivated. However, there is still controversy over the artificially cultivated *Suillus luteus* [153] and *Phlebopus portentosus* [154,155]; after the first report of the former, no follow-up research and progress have been made, while the latter was considered not to be an ectomycorrhizal fungus, with some researchers suggesting it might be a facultative ectomycorrhizal fungus [156]. Controversy also surrounds *Morchella* spp., as studies have shown that it can form mycorrhiza with plants [157–161], while some cultivated *Morchella* spp. were considered to be saprophytes [162], a finding supported by the latest research results [163]. Whether there is a saprophytic species in the genus *Russula* is currently unclear and requires further research through large-scale isolate tests or comparative genomics analysis based on enough *Russula* species. In addition, research on mycorrhizal seedlings requires continuous efforts, as under the current situation, only by obtaining mycorrhizal seedlings is it possible to induce or promote the formation of fruiting bodies by improving soil biological and abiotic factors and to achieve semi-artificial domestication.

## 5. Conclusions

*R. griseocarnosa* was regarded as a rare and precious edible ectomycorrhizal mushroom in southern China, possessing outstanding nutritional value and health benefits due to its rich composition of nutrients and bioactive compounds. It has garnered substantial market demand and has become a significant source of income for local mountain communities. However, the lack of systematic and in-depth research on its symbiotic relationships with host plants and microorganisms in its habitat has hindered successful isolation and artificial cultivation of its mycelium to date. To overcome this challenge, future research should leverage multi-omics technologies, particularly proteomics and metabolomics, to conduct comprehensive investigations into the material basis provided by host plants and microorganisms in the *Russula*'s habitat. Specifically, studies should focus on elucidating the intricate exchange of carbon sources, signalling molecules, and other key factors that govern the symbiotic associations. By unravelling these intricate mechanisms, researchers could potentially develop interventional strategies to promote and facilitate the artificial cultivation and domestication of this valuable ectomycorrhizal mushroom.

The successful implementation of such interventional approaches would not only enable sustainable development and application of *R. griseocarnosa* as an ectomycorrhizal edible mushroom resource but also contribute to a deeper understanding of the complex ecological interactions within its natural ecosystem. Moreover, the insights gained from this research could potentially inform sustainable cultivation practices for other ectomycorrhizal mushroom, thereby unlocking their vast potential for diverse applications in agriculture, food production, and biotechnology. In conclusion, by synergistically integrating multi-omics technologies and garnering comprehensive insights into the intricate symbiotic relationships of *R. griseocarnosa*, researchers are poised to harness the full potential of this valuable ectomycorrhizal mushroom, paving the path for its sustainable cultivation and utilization, while concomitantly advancing the overarching field of mycorrhizal biology and ecology.

**Author Contributions:** Conceptualization, Methodology, Writing—original draft, Y.L.; Review and editing, T.Y., M.C., X.W., H.G. and Y.X.; Conceptualization, Methodology, Funding acquisition, Review, H.H. and Q.W. All authors have read and agreed to the published version of the manuscript.

**Funding:** This research was financially supported by the National Key R&D Program of China (2023YFF1000801), GDAS'Project of Science and Technology Development (2022GDASZH-2022010101), Guangdong Edible Mushroom (Shaoguan) Seed Industry Innovation Park (2022-440000-43010404-9497).

**Institutional Review Board Statement:** Not applicable.

**Data Availability Statement:** Data are contained within the article.

**Conflicts of Interest:** The authors declare no conflicts of interest.

## Appendix A

**Table A1.** Volatile oil components of *R. griseocarnosa*.

| Volatile Compouds | Secondary (CAS) | Name | Formula |
|---|---|---|---|
| 1 | 57-10-3 | Palmitic acid | $C_{16}H_{32}O_2$ |
| 2 | 60-12-8 | 2-Phenylethanol | $C_8H_{10}O$ |
| 3 | 1002-43-3 | 3-Methylundecane | $C_{12}H_{26}$ |
| 4 | 1004-29-1 | 2-butyl tetrahydrofuran | $C_8H_{16}O$ |
| 5 | 100-52-7 | Benzaldehyde | $C_7H_6O$ |
| 6 | 1014-60-4 | Benzene,1,3-bis(1,1-dimethylethyl)- | $C_{14}H_{22}$ |
| 7 | 104-46-1 | cis-Anethol | $C_{10}H_{12}O$ |
| 8 | 10482-56-1 | (-)-α-Terpineol | $C_{10}H_{18}O$ |
| 9 | 107-50-6 | Tetradecamethyl Cycloheptasiloxane | $C_{14}H_{42}O_7Si_7$ |
| 10 | 109-08-0 | methylpyrazine | $C_5H_6N_2$ |
| 11 | 110-43-0 | 2-Heptanone | $C_7H_{14}O$ |
| 12 | 111150-30-2 | Pyrazine, 3,5-dimethyl-2-(3-methylbutyl)-(9CI) | $C_{11}H_{18}N_2$ |
| 13 | 111-71-7 | Heptanal | $C_7H_{14}O$ |
| 14 | 1120-21-4 | Undecane | $C_{11}H_{24}$ |
| 15 | 112-12-9 | 2-Undecanone | $C_{11}H_{22}O$ |
| 16 | 112-31-2 | Decanal | $C_{10}H_{20}O$ |
| 17 | 112-40-3 | Dodecane | $C_{12}H_{26}$ |
| 18 | 1124-11-4 | Tetramethylpyrazine | $C_8H_{12}N_2$ |
| 19 | 112-41-4 | dodecene | $C_{12}H_{24}$ |
| 20 | 112-44-7 | Undecanal | $C_{11}H_{22}O$ |
| 21 | 1125-21-9 | 2,6,6-Trimethyl-2-cyclohexene-1,4-dione | $C_9H_{12}O_2$ |
| 22 | 116-53-0 | 2-Methylbutyric acid | $C_5H_{10}O_2$ |
| 23 | 118-65-0 | isocaryophyllene | $C_{15}H_{24}$ |
| 24 | 120-94-5 | 1-Methylpyrrolidine | $C_5H_{11}N$ |
| 25 | 122-78-1 | Phenylacetaldehyde | $C_8H_8O$ |
| 26 | 123-32-0 | 2,5-Dimethylpyrazine | $C_6H_8N_2$ |
| 27 | 124-13-0 | Octanal | $C_8H_{16}O$ |
| 28 | 124-19-6 | Nonanal | $C_9H_{18}O$ |
| 29 | 13019-16-4 | 2-Butyl-2-octenal | $C_{12}H_{22}O$ |
| 30 | 13152-44-8 | Butylcyclobutane | $C_8H_{16}$ |
| 31 | 13286-73-2 | 3-Ethyltridecane | $C_{15}H_{32}$ |
| 32 | 13360-65-1 | 2-Ethyl-3,5-dimethylpyrazine | $C_8H_{12}N_2$ |
| 33 | 140-67-0 | Estragole | $C_{10}H_{12}O$ |
| 34 | 142-50-7 | Nerolidol, cis-(+) | $C_{15}H_{26}O$ |
| 35 | 14309-57-0 | 3-Nonen-2-one | $C_9H_{16}O$ |
| 36 | 1472-09-9 | octylcyclopropane | $C_{11}H_{22}$ |
| 37 | 1502-38-1 | methylcyclooctane | $C_9H_{18}$ |
| 38 | 15870-10-7 | 2-Methyl-1-heptene | $C_8H_{16}$ |
| 39 | 17301-25-6 | 2,8-Dimethylundecane | $C_{13}H_{28}$ |
| 40 | 17301-28-9 | Undecane,3,6-dimethyl- | $C_{13}H_{28}$ |
| 41 | 17301-29-0 | Undecane,3,7-dimethyl- | $C_{13}H_{28}$ |
| 42 | 17301-30-3 | 3,8-Dimethylundecane | $C_{13}H_{28}$ |
| 43 | 17301-32-5 | Undecane,4,7-dimethyl- | $C_{13}H_{28}$ |
| 44 | 17302-28-2 | NONANE,2,6-DIMETHYL- | $C_{11}H_{24}$ |
| 45 | 17312-68-4 | 4,4-Dimethylundecane | $C_{13}H_{28}$ |
| 46 | 17312-80-0 | 2,4-Dimethyl-undecane | $C_{13}H_{28}$ |
| 47 | 17453-93-9 | 5-methyldodecane | $C_{13}H_{28}$ |
| 48 | 17615-91-7 | Undecane,5,6-dimethyl- | $C_{13}H_{28}$ |
| 49 | 19132-06-0 | (+)-2,3-Butanediol | $C_4H_{10}O_2$ |

**Table A1.** *Cont.*

| Volatile Compouds | Secondary (CAS) | Name | Formula |
|---|---|---|---|
| 50 | 192823-15-7 | 2,3,5,8-tetramethyldecane | $C_{14}H_{30}$ |
| 51 | 19780-34-8 | Tridecane, 3-methylene- | $C_{14}H_{28}$ |
| 52 | 19780-74-6 | 5-Ethyl-1-nonene | $C_{11}H_{22}$ |
| 53 | 2027-47-6 | octadec-9-enoic acid | $C_{18}H_{34}O_2$ |
| 54 | 2471-84-3 | 1H-Indene,1-methilene- | $C_{10}H_8$ |
| 55 | 25117-31-1 | 5-Methyltridecane | $C_{14}H_{30}$ |
| 56 | 25117-33-3 | 5-methylpentadecane | $C_{16}H_{34}$ |
| 57 | 2801-84-5 | 2,4-dimethyldecane | $C_{12}H_{26}$ |
| 58 | 2882-96-4 | 3-Methylpentadecane | $C_{16}H_{34}$ |
| 59 | 295-17-0 | cyclotetradecane | $C_{14}H_{28}$ |
| 60 | 31295-56-4 | 2,6,11-Trimethyldodecane | $C_{15}H_{32}$ |
| 61 | 3391-86-4 | Oct-1-en-3-ol | $C_8H_{16}O$ |
| 62 | 3393-45-1 | 5,6-DIHYDRO-2H-PYRAN-2-ONE | $C_5H_6O_2$ |
| 63 | 3777-69-3 | 2-Amylfuran | $C_9H_{14}O$ |
| 64 | 3879-26-3 | neryl acetone | $C_{13}H_{22}O$ |
| 65 | 4126-78-7 | Methylcycloheptane | $C_8H_{16}$ |
| 66 | 41446-67-7 | (Z)-tetradec-3-ene | $C_{14}H_{28}$ |
| 67 | 4292-19-7 | 1-Iodododecane | $C_{12}H_{25}I$ |
| 68 | 4411-89-6 | 2-phenyl-2-butenal | $C_{10}H_{10}O$ |
| 69 | 4457-00-5 | hexylcyclopentane | $C_{11}H_{22}$ |
| 70 | 503-74-2 | 3-Methylbutanoic acid | $C_5H_{10}O_2$ |
| 71 | 50656-61-6 | (3aR,8aS)-2,2,8-trimethyl-3,3a,6,8a-tetrahydro-1H-azulene-5,6-dicarbal dehyde | $C_{15}H_{20}O_2$ |
| 72 | 51756-29-7 | 3-Butyl-3-methylcyclohexanone | $C_{11}H_{20}O$ |
| 73 | 51945-98-3 | 1,5-Heptadiene-3,4-diol | $C_7H_{12}O_2$ |
| 74 | 540-97-6 | Dodecamethylcyclohexasiloxane | $C_{12}H_{36}O_6Si_6$ |
| 75 | 541-02-6 | Decamethylcyclopentasiloxane | $C_{10}H_{30}O_5Si_5$ |
| 76 | 541-05-9 | hexamethylcyclotrisiloxane | $C_6H_{18}O_3Si_3$ |
| 77 | 544-76-3 | Hexadecane | $C_{16}H_{34}$ |
| 78 | 556-67-2 | Octamethylcyclotetrasiloxane | $C_8H_{24}O_4Si_4$ |
| 79 | 556-68-3 | hexadecamethylcyclooctasiloxane | $C_{16}H_{48}O_8Si_8$ |
| 80 | 563-16-6 | 3,3-Dimethylhexane | $C_8H_{18}$ |
| 81 | 5876-87-9 | 1,11-Dodecadiene | $C_{12}H_{22}$ |
| 82 | 590-86-3 | Isovaleraldehyde | $C_5H_{10}O$ |
| 83 | 61141-72-8 | dodecane,4,6-dimethyl | $C_{14}H_{30}$ |
| 84 | 62016-37-9 | 2,4,6-trimethyl octane | $C_{11}H_{24}$ |
| 85 | 62108-21-8 | 6-ethyl-2-methyl-decane | $C_{13}H_{28}$ |
| 86 | 62108-22-9 | 2,5,9-trimethyldecane | $C_{13}H_{28}$ |
| 87 | 62108-23-0 | Trimethyldecane, 2,5,6- | $C_{13}H_{28}$ |
| 88 | 622-39-9 | 2-Propylpyridine | $C_8H_{11}N$ |
| 89 | 62338-50-5 | (E)-8-Methyl-4-decene | $C_{11}H_{22}$ |
| 90 | 629-50-5 | Tridecane | $C_{13}H_{28}$ |
| 91 | 629-59-4 | Tetradecane | $C_{14}H_{30}$ |
| 92 | 6418-41-3 | 3-methyltridecane | $C_{14}H_{30}$ |
| 93 | 6418-43-5 | 3-methylhexadecane | $C_{17}H_{36}$ |
| 94 | 64-19-7 | acetic acid | $C_2H_4O_2$ |
| 95 | 66-25-1 | Hexanal | $C_6H_{12}O$ |
| 96 | 67-64-1 | Acetone | $C_3H_6O$ |
| 97 | 6831-17-0 | aristolone | $C_9H_{11}N_3$ |
| 98 | 693-54-9 | 2-Decanone | $C_{10}H_{20}O$ |
| 99 | 69460-62-4 | (4aS,8R)-4,4a,5,6,7,8-Hexahydro-4a,8-dimethyl-2(3H)-naphthalenone | $C_{12}H_{18}O$ |
| 100 | 71138-64-2 | Undecane, 3-methylene- | $C_{12}H_{24}$ |
| 101 | 7154-79-2 | 2,2,3,3-Tetramethylpentane | $C_9H_{20}$ |
| 102 | 74630-39-0 | 4-Methyl-1-undecene | $C_{12}H_{24}$ |
| 103 | 74645-98-0 | DODECANE,2,7,10-TRIMETHYL- | $C_{15}H_{32}$ |
| 104 | 74663-85-7 | Nonylcyclopropane | $C_{12}H_{24}$ |
| 105 | 74685-36-2 | Oxacyclotetradecane-2,11-dione, 13 methyl- | $C_{14}H_{24}O_3$ |

**Table A1.** *Cont.*

| Volatile Compouds | Secondary (CAS) | Name | Formula |
|---|---|---|---|
| 106 | 7473-98-5 | 2-Hydroxy-2-methyl propiophenone | $C_{10}H_{12}O_2$ |
| 107 | 75-50-3 | Trimethylamine | $C_3H_9N$ |
| 108 | 78-84-2 | Isobutyraldehyde | $C_4H_8O$ |
| 109 | 79-31-2 | Isobutyric acid | $C_4H_8O_2$ |
| 110 | 79-50-5 | DL-Pantolactone | $C_6H_{10}O_3$ |
| 111 | 84-69-5 | Diisobutyl phthalate | $C_{16}H_{22}O_4$ |
| 112 | 91010-41-2 | 2-methyl-6-(3-methyl-butyl)-pyrazine | $C_{10}H_{16}N_2$ |
| 113 | 91-20-3 | Naphthalene | $C_{10}H_8$ |
| 114 | 96-17-3 | 2-Methylbutanal | $C_5H_{10}O$ |
| 115 | 96-76-4 | 2,4-Di-t-butylphenol | $C_{14}H_{22}O$ |
| 116 | 98-55-5 | alpha-Terpineol | $C_{10}H_{18}O$ |

## References

1. Tang, C.; Chen, Y.; Liu, R. Advances in studies of edible mycorrhizal fungi. *Mycosystema* **2011**, *30*, 367–378. (In Chinese) [CrossRef]
2. Cao, B.; Li, G.; Zhao, R. Species diversity and geographic components of *Russula* from the Greater and Lesser Khinggan Mountains. *Biodivers. Sci.* **2019**, *27*, 854. [CrossRef]
3. Algaebase. *Index Fungorum*; CABI Bioscience: Wallingford, UK; CBS: New York, NY, USA; Landcare Research: Lincoln, New Zealand, 2018. Available online: https://www.indexfungorum.org/names/names.asp (accessed on 12 March 2024).
4. Kirk, P.M.; Cannon, P.F.; Stalpers, J.A. *Dictionary of the Fungi*, 10th ed.; CABI Europe: Wallingford, UK, 2008.
5. Li, G.J. The Taxonomy of *Russula* in China. Ph.D. Thesis, University of Chinese Academy of Sciences, Beijing, China, 2014. (In Chinese)
6. Bo, L. *Chinese Medicinal Fungi*; Shanxi People's Publishing House: Taiyuan, China, 1984. (In Chinese)
7. Dai, Y.C.; Yang, Z.L. A revised checklist of medicinal fungi in China. *Mycosystema* **2008**, *27*, 801–824. (In Chinese) [CrossRef]
8. Dai, Y.C.; Zhou, L.W.; Yang, Z.L.; Wen, H.A.; Bau, T.; Li, T.H. A revised checklist of edible fungi in China. *Mycosystema* **2010**, *29*, 1–21. (In Chinese) [CrossRef]
9. Li, G.J.; Li, S.F.; Wen, H.A. Economic value of *Russula* species resources in China. In Proceedings of the 9th National Edible Fungi Academic Symposium, Shanghai, China, 15 October 2010; pp. 155–160. (In Chinese)
10. Dai, Y.C.; Bau, T.; Cui, B.K.; Qin, G.F. *Illustration of Chinese Medicinal Fungi*; Northeast Forestry University Press: Harbin, China, 2012. (In Chinese)
11. Li, C.H.; Qu, M.Q.; Cao, H.; Deng, W.Q.; Shang, X.D.; Song, B.; Tan, Q. Checklist of Common Names of Mushrooms in China. *Acta Edulis Fungi* **2013**, *20*, 50–72. (In Chinese) [CrossRef]
12. Cheng, Y.; Gan, J.; Yan, B.; Wang, P.; Wu, H.; Huang, C. Polysaccharides from *Russula*: A review on extraction, purification, and bioactivities. *Front. Nutr.* **2024**, *11*, 1406817. [CrossRef]
13. Zhou, X.P.; Lu, Q.; Wang, X.P.; Wang, J. Research progress on *Russula* mushrooms. *Edible Fungi* **2010**, *32*, 1–2. (In Chinese)
14. Huang, N.; Lin, Z.; Chen, G. *Medicinal and Edible Fungi in China*; Shanghai Science and Technology Literature Publishing House: Shanghai, China, 2010; p. 1414. (In Chinese)
15. Das, K.; Van de Putte, K.; Buyck, B. New or interesting Russula from Sikkim Himalaya (India). *Cryptogam. Mycol.* **2010**, *31*, 373–387.
16. Anh, C.N.; Chi, N.M.; Kiet, T.T.; Long, P.D.; Thuy, P.T.T.; Van Loi, V.; Dell, B. Morphological and molecular identification of an edible *Russula* mushroom in northeast Vietnam. *J. For. Sci. Technol.* **2023**, 50–59. [CrossRef]
17. Wang, X.; Yang, Z.; Li, Y.; Knudsen, H.; Liu, P. *Russula griseocarnosa* sp. nov. (Russulaceae, Russulales), a commercially important edible mushroom in tropical China: Mycorrhiza, phylogenetic position, and taxonomy. *Nova Hedwig.* **2009**, *88*, 269–282. [CrossRef]
18. Liu, H.; Chen, X. Morphological and structural characteristics of *Russula griseocarnosa* sp. nov. in Guangdong. *Guangdong Agric. Sci.* **2011**, *38*, 140–143. (In Chinese) [CrossRef]
19. Li, Z.; Zhong, Y.Y.; Chen, Y.X. Investigation on the resources of wild *Russula griseocarnosa* in Meizhou, Guangdong. *Edible Fungi* **2012**, *34*, 8–9. (In Chinese)
20. Lou, X.-H.; Gan, Y.-K.; Wang, L.-M.; Yan, L.; Yang, X. Protective effect of extract from *Russula* sp. on oxidative damage caused by formaldehyde. *J. Toxicol.* **2007**, *21*, 225–226. (In Chinese)
21. Chen, X.J.; Zhang, Y.Q. Polysaccharide Extract from *Russula* and Its Role of Lowering Blood Glucose and Lipid. *Food Sci.* **2010**, *31*, 255–258. (In Chinese)
22. Liu, X. Study on the Structural Characterization of *Russula griseocarnosa* Polysaccharide and Its Improvement on Hematopoietic Function Based on Immunoregulation. Ph.D. Thesis, Jilin University, Changchun, China, 2023. (In Chinese)
23. Smith, S.E.; Read, D.J. *Mycorrhizal Symbiosis*, 3rd ed.; Section 2 Ectomycorrhizas; Elsevier: Amsterdam, The Netherlands, 2008.
24. Hall, I.R.; Zambonelli, A. Laying the Foundations. In *Edible Ectomycorrhizal Mushrooms: Current Knowledge and Future Prospects*; Zambonelli, A., Bonito, G.M., Eds.; Springer: Berlin/Heidelberg, Germany, 2012; pp. 3–16.

25. Chen, X. Studies on Morphological and Molecular Identifications, Nutrient Components Analysis and Biological Properties of Commercial *Russula* sp. Produce n Guangdong. Ph.D. Thesis, Central South university, Changsha, China, 2010. (In Chinese)
26. Liang, F. Effects of Summer and Winter on Slaughter Performance, Meat Quality and Fat Metabolism of Pheasants. Master's Thesis, Jilin University, Changchun, China, 2023. (In Chinese)
27. Li, J. Fatty Acid Enrichment and Metabolomics of Egg Yolk in Diets Supplemented with Fish Oil. Master's Thesis, Yunnan Agricultural University, Kunming, China, 2023. (In Chinese)
28. Ming, T.; Li, J.; Huo, P.; Wei, Y.; Chen, X. Analysis of Free Amino Acids in *Russula griseocarnosa* Harvested at Different Stages of Maturity Using iTRAQ®-LC-MS/MS. *Food Anal. Methods* **2014**, *7*, 1816–1823. [CrossRef]
29. Wei, Y.; Li, J.; Xu, S.Y.; Zhao, L.Y.; Chen, X.H. Study on the chemical constituents of the essential oil from *Russula griseocarnosa* by steam distillation and petrol ether extract. *Sci. Technol. Food Ind.* **2013**, *34*, 91–93+117. (In Chinese) [CrossRef]
30. Chen, X.H.; Xia, L.X.; Zhou, H.B.; Qiu, G.Z. Chemical Composition and Antioxidant Activities of *Russula griseocarnosa* sp. nov. *J. Agric. Food Chem.* **2010**, *58*, 6966–6971. [CrossRef]
31. He, T. Study on Structure and Bioactivity of Polysaccarides from *Russula griseocarnosa*. Master's Thesis, South China University of Technology, Guangzhou, China, 2015. (In Chinese)
32. Gan, Y.K.; Lou, X.H.; Li, Y.; Wang, L.M.; Yang, X. Preliminary study on the antioxidant properties of *Russula* fruiting bodies. *Edible Fungi* **2007**, *29*, 57–59. (In Chinese)
33. Gan, Y.K.; Chen, X.B.; Zeng, S.Y.; Wei, X.J.; Xiong, X.L. Study on the anti-exercise fatigue effect of *Russula* extract. *Edible Fungi* **2010**, *32*, 66–67. (In Chinese)
34. Gan, Y.K.; Liu, Y.Z.; Liang, J.S.; Zeng, S.Y.; Yu, J.R.; Zhou, Z.Y. Preliminary Exploration about the Anti-aging Effect of Liquid Extraction from the Russula Fruitbodyon. *J. Yulin Norm. Univ.* **2011**, *32*, 81–85+82. (In Chinese) [CrossRef]
35. Zeng, S.Y.; Gang, Y.K.; Ye, C.F. Study on Effect of *Russula* Extracting Solution on Anti-oxidation of the Elder Mice. *J. Anhui Agric. Sci.* **2009**, *37*, 7464–7466. [CrossRef]
36. Zhao, F.L.; Zhang, Y.G.; Ning, L.D. Extraction, isolation and antioxidant activity of *Russula*-polysaccharide. *China Brew.* **2009**, *28*, 98–101. (In Chinese)
37. Cai, X.; Zhang, P.; He, Y.; Kuang, X.; Yin, L. The Role of Reducing Blood Cholesterin of *Auricularia* Polysaccharide and *Russula* Polysaccharide. *Shenzhen J. Integr Tradit. Chin. West. Med.* **2002**, *12*, 137–139. (In Chinese) [CrossRef]
38. Cai, X.L.; Hu, C.Y.; Liu, M.P.; Xue, X.W. Effects of *Russula* and its Polysaccharides on Hemorrhagic Anemia in Mice. *Edible Fungi* **2002**, *24*, 40–41. (In Chinese)
39. Liu, Y.Z.; Gan, Y.K.; Chen, X.J.; Ming, T.H.; Zeng, X.Y. Comparison of Bacteriostatic Effects between Extracts from *Russula* and *Cyclobalanopsis glauca*. *Food Sci.* **2011**, *32*, 36–38. (In Chinese)
40. Zhao, S.; Zhao, Y.; Li, S.; Zhao, J.; Zhang, G.; Wang, H.; Ng, T.B. A novel lectin with highly potent antiproliferative and HIV-1 reverse transcriptase inhibitory activities from the edible wild mushroom *Russula delica*. *Glycoconj. J.* **2010**, *27*, 259–265. [CrossRef] [PubMed]
41. Zhang, G.; Sun, J.; Wang, H.; Ng, T.B. First isolation and characterization of a novel lectin with potent antitumor activity from a *Russula* mushroom. *Phytomedicine* **2010**, *17*, 775–781. [CrossRef] [PubMed]
42. Zhang, G.; Chen, Q.; Zhao, S.; Wang, S.; Wang, H. Purification and comparison of two lectins from the genus *Russula*. *Mycosystema* **2012**, *31*, 110–118. (In Chinese) [CrossRef]
43. Chen, J.; Shen, C.; He, T.; Yan, S.E. Anticancer and Immunoregulation Activities of a Polysaccharide from *Russula vinosa*. *Mod. Food Sci. Technol.* **2016**, *32*, 16–21. (In Chinese) [CrossRef]
44. Wang, X.H. Taxonomic comments on edible species of Russulaceae. *Mycosystema* **2020**, *39*, 1617–1639. [CrossRef]
45. Li, M.; Liang, J.; Li, Y.; Feng, B.; Yang, Z.L.; James, T.Y.; Xu, J. Genetic Diversity of Dahongjun, the Commercially Important "Big Red Mushroom" from Southern China. *PLoS ONE* **2010**, *5*, e10684. [CrossRef] [PubMed]
46. Yuan, Y.; Liu, Y.; Liu, M.; Chen, Q.; Jiao, Y.; Liu, Y.; Meng, Z. Optimization extraction and bioactivities of polysaccharide from wild *Russula griseocarnosa*. *Saudi Pharm. J.* **2017**, *25*, 523–530. [CrossRef] [PubMed]
47. Zhang, G.; Geng, H.; Zhao, C.; Li, F.; Li, Z.F.; Lun, B.; Wang, C.; Yu, H.; Bie, S.; Li, Z. Chemical Constituents with Inhibitory Activity of NO Production from a Wild Edible Mushroom, *Russula vinosa* Lindbl, May Be Its Nutritional Ingredients. *Molecules* **2019**, *24*, 1305. [CrossRef] [PubMed]
48. Dong, J.; Zhang, M.; Lu, L.; Sun, L.; Xu, M. Nitric oxide fumigation stimulates flavonoid and phenolic accumulation and enhances antioxidant activity of mushroom. *Food Chem.* **2012**, *135*, 1220–1225. [CrossRef] [PubMed]
49. Zhao, Y.; Lim, J.; Xu, J.; Yu, J.H.; Zheng, W. Nitric oxide as a developmental and metabolic signal in filamentous fungi. *Mol. Microbiol.* **2020**, *113*, 872–882. [CrossRef] [PubMed]
50. Liu, Y.; Zhang, J.; Meng, Z. Purification, characterization and anti-tumor activities of polysaccharides extracted from wild *Russula griseocarnosa*. *Int. J. Biol. Macromol.* **2018**, *109*, 1054–1060. [CrossRef] [PubMed]
51. Chen, Q.; Qi, C.; Peng, G.; Liu, Y.; Zhang, X.; Meng, Z. Immune-enhancing effects of a polysaccharide PRG1-1 from *Russula griseocarnosa* on RAW264. 7 macrophage cells via the MAPK and NF-κB signalling pathways. *Food Agric. Immunol.* **2018**, *29*, 833–844. [CrossRef]
52. Liu, X.; Dong, M.; Li, Y.; Li, L.; Zhang, Y.; Zhou, A.; Wang, D. Structural characterization of *Russula griseocarnosa* polysaccharide and its improvement on hematopoietic function. *Int. J. Biol. Macromol.* **2024**, *263*, 130355. [CrossRef] [PubMed]

53. Martinez-Medina, G.A.; Chávez-González, M.L.; Verma, D.K.; Prado-Barragán, L.A.; Martínez-Hernández, J.L.; Flores-Gallegos, A.C.; Thakur, M.; Srivastav, P.P.; Aguilar, C.N. Bio-funcional components in mushrooms, a health opportunity: Ergothionine and huitlacohe as recent trends. *J. Funct. Food.* **2021**, *77*, 104326. [CrossRef]
54. Liu, X.; Dong, M.; Li, Y.; Li, L.; Zhang, Y.; Wang, C.; Wang, N.; Wang, D. Structural properties of glucan from *Russula griseocarnosa* and its immunomodulatory activities mediated via T cell differentiation. *Carbohydr. Polym.* **2024**, *339*, 122214. [CrossRef]
55. Liu, B.; Mo, T.Y. Separation of *Russula* and Observation of Its Culture Characteristics. *J. Guangxi Agric. Univ.* **1994**, *13*, 345–347. (In Chinese)
56. Cai, X.-L.; Ge, G.; Guo, Y.; He, Z.-Z. Isolated and purified cultrue of *Russula lepida* Fr. *Guangzhou Food Sci. Technol.* **2003**, *19*, 30–31. (In Chinese) [CrossRef]
57. Chen, S.-Z.; Huang, S.-L.; Bi, Z.-Q.; Cai, B.-H.; Chen, L.-X.; Xia, L.; Su, G.-X. Isolation of Mycellia from Wild *Russula* sp. *Chin. Agric. Sci. Bull.* **2004**, *20*, 5–7. (In Chinese)
58. Chen, H.-T.; Zhang, Y.-H.; Zhai, D.-P.; Luo, Z.-W. Selection of the Optimum Medium for Mother Species of *Russula vinosa*. *Edible Fungi* **2012**, 26–27. (In Chinese)
59. Zhang, B.; Zhou, Z.Z.; Chen, Y.; Liang, K.N.; Yu, X.B. Screening of Nutrient Elements for Culturing Mycelia of *Russula* sp. *Chin. J. Trop. Crops* **2009**, *30*, 827–831. (In Chinese)
60. Zhang, B.; Zhou, Z.Z.; Chen, Y.; Liang, K.N.; Yu, X.B. Research on the Optimum Combination of Nutritional Factors of Liquid Culture of *Russual* sp. *J. Anhui Agric. Sci.* **2009**, *37*, 11871–11873. (In Chinese)
61. Chen, X.-J.; Gan, Y.-K. Preliminary Study on Nutritional Factors of *Russula* in Deep Culture. *Edible Fungi* **2007**, 21–22. (In Chinese)
62. Zhang, Y.X.; Zhang, K.; Li, M.M. Research about Submerged Fermentation Culture Conditions of *Russula lepida*. *North. Hortic.* **2010**, 220–221. (In Chinese)
63. Chai, D.-D.; Guo, S.-J.; Sun, X.-B.; Qin, T.-T. Optimization of culture conditions for 7 chestnut ectomycorrhizal fungi in Yanshan Area. *J. Northwest AF Univ. (Nat. Sci. Ed.)* **2014**, *42*, 109–116. (In Chinese) [CrossRef]
64. Liu, D.-M.; Yuan, D.-Y.; Zou, F.; Zhang, X.-H.; Zhu, Z.-J.; Tan, L.-M. Optimization of Culture Conditions for 3 *Castanea henryi* Ectomycorrhizal fungi. *J. Northwest For. Univ.* **2016**, *31*, 195–200. (In Chinese) [CrossRef]
65. Huang, F.-C.; Mo, T.-Y.; Liu, B. A Study on the Physiological Character of the Pure Culture of *Russula vinosa*. *J. Guangxi Agric. Univ.* **1998**, *17*, 33–39. (In Chinese)
66. Tang, L.-H.; Liu, Z.-H.; Shi, Q.-Q.; Wu, S.-G. Effect of Plant Growth Regulators and Vitamins on *Russula vinosa* Growth. *J. Fujian Teach. Univ. (Nat. Sci.)* **2001**, *17*, 84–87. (In Chinese)
67. Li, Z.-L. Effects of Rare Earth Elements of Praseodymium, Neodymium and Erbium on Production of Polysaccharide by Submerged Fermentation of *Russula* sp. *Food Sci.* **2007**, *28*, 312–316. (In Chinese)
68. Chen, X.-J.; Gan, Y.-K.; Wang, M.-G.; Chen, B.-Y. The Initial Research in Separating Culture of *Russular*. *J. Yulin Teach. Coll. (Nat. Sci.)* **2005**, *26*, 64–65+77. (In Chinese)
69. Gan, Y.-K.; Chen, X.-J.; Wei, Q.-C.; Peng, S.-N.; Chen, L. The Influence of Temperature upon the Growth of the Natural Red Mushroom Hypha. *J. Yulin Teach. Coll. (Nat. Sci.)* **2007**, *28*, 58–60+64. (In Chinese) [CrossRef]
70. Gan, Y.-K.; Chen, X.-J.; Su, L.; He, Y. Effects of Aquatic Extraction Substance from Tenebrio molitor Linnaeus Feces on Mycelium Growth of Five Edible Fungi. *J. Anhui Agric. Sci.* **2008**, *36*, 11295–11296+11320. (In Chinese)
71. Li, Z.-L. Effect of initial pH and culture time on production of polysaccharide by submerged fermentation of *Russula* sp. *Food Sci. Technol.* **2006**, *32*, 21–25. (In Chinese) [CrossRef]
72. Mo, T.-Y.; Huang, F.-C. Studies on the Physiological Characterestics of *Russufa vinosa*. *J. Jilin Agric. Univ.* **1998**, *20*, 113. (In Chinese)
73. Wang, G.-W.; Sun, W.B. Nucleotide Sequence Analysis on ITS rDNA of Fruitbodies and Isolates of *Russula* in Guangxi. *Guangxi Sci.* **2004**, *11*, 261–265. (In Chinese)
74. Wu, X.; Li, H.; Fu, L.; Wei, H.; Wu, Q. Pure culture of ectomycorrhizal fungi with RAPD and ITS molecular markers for identification. In Proceedings of the First National Academic Exchange Conference of Young and Middle-Aged Edible Fungi Experts, Wuhan, China, 1 June 2006; pp. 47–54. (In Chinese)
75. Li, H.; Wu, X.; Wei, H.; Fu, L.; Wu, Q. Identification of Pure Culture for Macrofungi with RAPD and ITS Molecular Markers. *Sci. Silvae Sin.* **2007**, *43*, 94–100. (In Chinese)
76. Zeng, X.L.; Liu, D.L. Isolation and Mycelial Growth Characteristics of *Nectria* sp. Isolated from *Russula griseocarnosa*. *Edible Fungi China* **2013**, *32*, 32–34. (In Chinese)
77. Luan, Q.-S.; Jin, R.-Z.; Yun, L.-L. Study on Isolation and Cultivation of Wild Macro Fungi in Forest. In Proceedings of the 11th Chapter of the 2004 China Association for Science and Technology Annual Conference, Hainan, China, 20–21 November 2004; pp. 489–491. (In Chinese)
78. Hintikka, V.; Niemi, K. Aseptic culture of slowly growing mycorrhizal *Russula* and *Cortinarius* species. *Karstenia* **1999**, *39*, 39–41. [CrossRef]
79. Atlas, R.M. *Handbook of Microbiological Media*, 4th ed.; CRC Press (Taylor & Francis Group): Washington, DC, USA, 2010; pp. 1–2043.
80. Yamada, A.; Katsuya, K. Mycorrhizal association of isolates from sporocarps and ectomycorrhizas with *Pinus densiflora* seedlings. *Mycoscience* **1995**, *36*, 315–323. [CrossRef]

81. Huang, G.W.; Zhang, P. Symbiosis Cultivation of Ectomycorrhizal Mycelium. *Edible Fungi China* **2005**, *24*, 16–17. (In Chinese) [CrossRef]
82. Yang, R.-H.; Li, Y.; Tang, L.-H.; Li, C.-H.; Bao, D.-P. Genome-wide comparison of lignocellulose degradation enzymes in Agaricales. *Mycosystema* **2017**, *36*, 705–717. (In Chinese) [CrossRef]
83. Liu, Y.; Hu, H.; Cai, M.; Liang, X.; Wu, X.; Wang, A.; Chen, X.; Li, X.; Xiao, C.; Huang, L.; et al. Whole genome sequencing of an edible and medicinal mushroom, *Russula griseocarnosa*, and its association with mycorrhizal characteristics. *Gene* **2022**, *808*, 145996. [CrossRef] [PubMed]
84. van Galen, L.G.; Orlovich, D.A.; Lord, J.M.; Nilsen, A.R.; Dutoit, L.; Larcombe, M.J. Correlated evolution in an ectomycorrhizal host-symbiont system. *New Phytol.* **2023**, *238*, 1215–1229. [CrossRef] [PubMed]
85. van Der Heijden, M.G.; Martin, F.M.; Selosse, M.A.; Sanders, I.R. Mycorrhizal ecology and evolution: The past, the present, and the future. *New Phytol.* **2015**, *205*, 1406. [CrossRef] [PubMed]
86. Baskaran, P.; Hyvonen, R.; Berglund, S.L.; Clemmensen, K.E.; Agren, G.I.; Lindahl, B.D.; Manzoni, S. Modelling the influence of ectomycorrhizal decomposition on plant nutrition and soil carbon sequestration in boreal forest ecosystems. *New Phytol.* **2017**, *213*, 1452–1465. [CrossRef]
87. Treseder, K.K.; Torn, M.S.; Masiello, C.A. An ecosystem-scale radiocarbon tracer to test use of litter carbon by ectomycorrhizal fungi. *Soil Biol. Biochem.* **2006**, *38*, 1077–1082. [CrossRef]
88. Martin, F.; Duplessis, S.; Ditengou, F.; Lagrange, H.; Voiblet, C.; Lapeyrie, F. Developmental cross talking in the ectomycorrhizal symbiosis: Signals and communication genes. *New Phytol.* **2001**, *151*, 145–154. [CrossRef] [PubMed]
89. Spanu, P.D. The Genomics of Obligate (and Nonobligate) Biotrophs. *Annu. Rev. Phytopathol.* **2012**, *50*, 91–109. [CrossRef] [PubMed]
90. Lindahl, B.D.; Tunlid, A. Ectomycorrhizal fungi—Potential organic matter decomposers, yet not saprotrophs. *New Phytol.* **2015**, *205*, 1443–1447. [CrossRef] [PubMed]
91. Read, D.; Leake, J.R.; Perezmoreno, J. Mycorrhizal fungi as drivers of ecosystem processes in heathland and boreal forest biomes. *Botany* **2004**, *82*, 1243–1263. [CrossRef]
92. Martin, F.; Aerts, A.; Ahren, D.; Brun, A.; Danchin, E.G.J.; Duchaussoy, F.; Gibon, J.; Kohler, A.; Lindquist, E.; Pereda, V. The genome of *Laccaria bicolor* provides insights into mycorrhizal symbiosis. *Nature* **2008**, *452*, 88–92. [CrossRef] [PubMed]
93. Martin, F.; Kohler, A.; Murat, C.; Balestrini, R.; Coutinho, P.M.; Jaillon, O.; Montanini, B.; Morin, E.; Noel, B.; Percudani, R. Périgord black truffle genome uncovers evolutionary origins and mechanisms of symbiosis. *Nature* **2010**, *464*, 1033–1038. [CrossRef] [PubMed]
94. Kohler, A.; Kuo, A.; Nagy, L.; Morin, E.; Barry, K.; Buscot, F.; Canback, B.; Choi, C.; Cichocki, N.; Clum, A. Convergent losses of decay mechanisms and rapid turnover of symbiosis genes in mycorrhizal mutualists. *Nat. Genet.* **2015**, *47*, 410–415. [CrossRef] [PubMed]
95. Lundell, T.K.; Mäkelä, M.R.; de Vries, R.P.; Hildén, K.S. Chapter Eleven—Genomics, Lifestyles and Future Prospects of Wood-Decay and Litter-Decomposing Basidiomycota. *Adv. Bot. Res.* **2014**, *70*, 329–370. [CrossRef]
96. Peter, M.; Kohler, A.; Ohm, R.A.; Kuo, A.; Krutzmann, J.; Morin, E.; Arend, M.; Barry, K.; Binder, M.; Choi, C. Ectomycorrhizal ecology is imprinted in the genome of the dominant symbiotic fungus *Cenococcum geophilum*. *Nat. Commun.* **2016**, *7*, 12662. [CrossRef]
97. Plett, J.M.; Yin, H.; Mewalal, R.; Hu, R.; Li, T.; Ranjan, P.; Jawdy, S.S.; De Paoli, H.C.; Butler, G.A.; Burchsmith, T.M. *Populus trichocarpa* encodes small, effector-like secreted proteins that are highly induced during mutualistic symbiosis. *Sci. Rep.* **2017**, *7*, 382. [CrossRef]
98. Plett, J.M.; Kemppainen, M.; Kale, S.D.; Kohler, A.; Legué, V.; Brun, A.; Tyler, B.M.; Pardo, A.G.; Martin, F. A Secreted Effector Protein of *Laccaria bicolor* Is Required for Symbiosis Development. *Curr. Biol.* **2011**, *21*, 1197–1203. [CrossRef] [PubMed]
99. Plett, J.M.; Daguerre, Y.; Wittulsky, S.; Vayssières, A.; Deveau, A.; Melton, S.J.; Kohler, A.; Morrell-Falvey, J.L.; Brun, A.; Veneault-Fourrey, C.; et al. Effector MiSSP7 of the mutualistic fungus *Laccaria bicolor* stabilizes the *Populus* JAZ6 protein and represses jasmonic acid (JA) responsive genes. *Proc. Natl. Acad. Sci. USA* **2014**, *111*, 8299–8304. [CrossRef] [PubMed]
100. Larsen, P.E.; Sreedasyam, A.; Trivedi, G.; Desai, S.; Dai, Y.; Cseke, L.J.; Collart, F. Multi-Omics Approach Identifies Molecular Mechanisms of Plant-Fungus Mycorrhizal Interaction. *Front. Plant Sci.* **2016**, *6*, 1061. [CrossRef] [PubMed]
101. Twieg, B.D.; Durall, D.M.; Simard, S.W. Ectomycorrhizal fungal succession in mixed temperate forests. *New Phytol.* **2007**, *176*, 437–447. [CrossRef] [PubMed]
102. Cottenie, K. Integrating environmental and spatial processes in ecological community dynamics. *Ecol. Lett.* **2010**, *8*, 1175–1182. [CrossRef] [PubMed]
103. Peay, K.G.; Bruns, T.D. Spore dispersal of basidiomycete fungi at the landscape scale is driven by stochastic and deterministic processes and generates variability in plant-fungal interactions. *New Phytol.* **2014**, *204*, 180–191. [CrossRef] [PubMed]
104. Gao, C.; Zhang, Y.U.; Shi, N.N.; Zheng, Y.; Chen, L.; Wubet, T.; Bruelheide, H.; Both, S.; Buscot, F.; Ding, Q.; et al. Community assembly of ectomycorrhizal fungi along a subtropical secondary forest succession. *New Phytol.* **2015**, *205*, 771–785. [CrossRef]
105. Kennedy, P.G.; Hortal, S.; Bergemann, S.E.; Bruns, T.D. Competitive interactions among three ectomycorrhizal fungi and their relation to host plant performance. *J. Ecol.* **2007**, *95*, 1338–1345. [CrossRef]

106. Bahram, M.; Polme, S.; Koljalg, U.; Zarre, S.; Tedersoo, L. Regional and local patterns of ectomycorrhizal fungal diversity and community structure along an altitudinal gradient in the Hyrcanian forests of northern Iran. *New Phytol.* **2012**, *193*, 465–473. [CrossRef]
107. Geml, J.; Pastor, N.; Fernandez, L.; Pacheco, S.; Semenova, T.A.; Becerra, A.G.; Wicaksono, C.Y.; Nouhra, E. Large-scale fungal diversity assessment in the Andean Yungas forests reveals strong community turnover among forest types along an altitudinal gradient. *Mol. Ecol.* **2014**, *23*, 2452–2472. [CrossRef]
108. Cao, H.; Chen, R.; Wang, L.; Jiang, L.; Yang, F.; Zheng, S.; Wang, G.; Lin, X. Soil pH, total phosphorus, climate and distance are the major factors influencing microbial activity at a regional spatial scale. *Sci. Rep.* **2016**, *6*, 25815. [CrossRef] [PubMed]
109. Baeza-Guzmán, Y.; Camargo-Ricalde, S.L.; Trejo Aguilar, D.; Montaño, N.M. Fungal and bacterial communities in a forest relict of *Pinus pseudostrobus* var. coatepecensis. *Iforest-Biogeosci. For.* **2023**, *16*, 299–306. [CrossRef]
110. Ge, W.; Ren, Y.; Dong, C.; Shao, Q.; Bai, Y.; He, Z.; Yao, T.; Zhang, Y.; Zhu, G.; Deshmukh, S.K. New perspective: Symbiotic pattern and assembly mechanism of *Cantharellus cibarius*-associated bacteria. *Front. Microbiol.* **2023**, *14*, 1074468. [CrossRef] [PubMed]
111. Gittel, A.; Barta, J.; Kohoutova, I.; Mikutta, R.; Owens, S.M.; Gilbert, J.A.; Schnecker, J.; Wild, B.; Hannisdal, B.; Maerz, J. Distinct microbial communities associated with buried soils in the *Siberian tundra*. *ISME J.* **2014**, *8*, 841–853. [CrossRef] [PubMed]
112. Llado, S.; Lopezmondejar, R.; Baldrian, P. Forest Soil Bacteria: Diversity, Involvement in Ecosystem Processes, and Response to Global Change. *Microbiol. Mol. Biol. Rev.* **2017**, *81*, e00063-00016. [CrossRef] [PubMed]
113. Mediavilla, O.; Geml, J.; Olaizola, J.; Oria-de-Rueda, J.A.; Baldrian, P.; Martín-Pinto, P. Effect of forest fire prevention treatments on bacterial communities associated with productive *Boletus edulis* sites. *Microb. Biotechnol.* **2019**, *12*, 1188–1198. [CrossRef] [PubMed]
114. Bera, I.; Nidhin, I.K.; Hembrom, M.E.; Das, K.; Chattopadhyay, I. Metagenomics offers insights into the rhizospheric bacterial diversity of mushrooms from a tropical forest and temperate forest of India. *Ecol. Genet. Genom.* **2023**, *29*, 100203. [CrossRef]
115. Garbaye, J. Tansley Review No. 76 Helper bacteria: A new dimension to the mycorrhizal symbiosis. *New Phytol.* **1994**, *128*, 197–210. [CrossRef] [PubMed]
116. Freyklett, P.; Garbaye, J.; Tarkka, M.T. The mycorrhiza helper bacteria revisited. *New Phytol.* **2007**, *176*, 22–36. [CrossRef]
117. Aspray, T.J.; Freyklett, P.; Jones, J.; Whipps, J.M.; Garbaye, J.; Bending, G.D. Mycorrhization helper bacteria: A case of specificity for altering ectomycorrhiza architecture but not ectomycorrhiza formation. *Mycorrhiza* **2006**, *16*, 533–541. [CrossRef]
118. Yu, F.; Liang, J.F.; Song, J.; Wang, S.K.; Lu, J.K. Bacterial Community Selection of *Russula griseocarnosa* Mycosphere Soil. *Front. Microbiol.* **2020**, *11*, 347. [CrossRef] [PubMed]
119. Yu, W.Y.; Peng, M.H.; Wang, J.J.; Ye, W.Y.; Li, Y.L.; Zhang, T.; Wang, A.R.; Zhang, D.M.; Wang, Z.H.; Lu, G.D.; et al. Microbial community associated with ectomycorrhizal *Russula* symbiosis and dominated nature areas in southern China. *FEMS Microbiol. Lett.* **2021**, *368*, fnab028. [CrossRef] [PubMed]
120. Qi, L.L.; Wu, X.-J.; Li, L.-Y.; Mo, C.-M.; Lang, N.; Chen, Z.-N. Diversity Research of Soil Fungi under Fruiting Bodies of *Russula griseocarnosa* and Its Related Species. *Chin. J. Trop. Crops* **2022**, *43*, 430–437. (In Chinese) [CrossRef]
121. Fei, Y.; Liang, J.-F. The Effect of *Russula rosea* and *Russula griseocarnosa* on Microorganism Structure of Mycosphere Soil. *For. Res.* **2022**, *35*, 52–63. [CrossRef]
122. Wand, R.; Liu, C.-G.; Xu, J. The Separation and Cultivation of Ectomycorrhizal Fungi. *Edible Fungi China* **2013**, *32*, 4–7. (In Chinese)
123. Looney, B.P.; Meidl, P.; Piatek, M.J.; Miettinen, O.; Martin, F.M.; Matheny, P.B.; Labbé, J.L. Russulaceae: A new genomic dataset to study ecosystem function and evolutionary diversification of ectomycorrhizal fungi with their tree associates. *New Phytol.* **2018**, *218*, 54–65. [CrossRef] [PubMed]
124. Chen, Y.H. The theories and application of the cultivation of *Russula vinosa*. In Proceedings of the First Cross-Strait Symposium on Edible (Medicinal) Fungi, Fuzhou, China, 1 November 2005; pp. 97–100. (In Chinese)
125. Chen, Y.H.; Chen, Z.M. Mycorrbizal biodiversity's influence on the fine root mass of Cyclobalanopsis chungii and basidiocarp collection of *Russula vinosa*. *Mycosystema* **2005**, *24*, 94–96. (In Chinese)
126. Chen, Y.-H.; Chen, Z.-M. Effect of mycorrhizal biodiversity on fine root mass of *Cyclobalanopsis chungii* and basidiocarp collection of *Russula vinosa*. *Chin. J. Eco-Agric.* **2007**, *15*, 171–173. (In Chinese)
127. Chen, Y.; Liang, J.F.; Zhou, Z.Z.; Zhong, C.L.; Chen, Z. The Effect of Seedling Inoculation of Three Local Tree Species Inoculated by *Russula lepida* and *R. vinosa*. *Guangdong For. Sci. Technol.* **2010**, *26*, 22–28. (In Chinese)
128. Li, K.X.; Liang, W.H.; Zeng, G.Y.; Li, Y.J.; Pang, X.; Chen, Q.S. Technique of Promoting Propagation of *Russula lepida* by Land Transformation in Guangxi. *Agric. Res. Appl.* **2016**, 30–33. (In Chinese)
129. Zeng, G.; Li, K.; Liang, W. Effect of Different Reforms of Forest Environments on *Russula lepida* Fruit Body Yields. *Acta Edulis Fungi* **2016**, *23*, 23–26. (In Chinese)
130. Wang, Y.; Man, J.; Lian, C. Diversity and Network Structure Analysis of Culturable Bacteria in Sporophore Site Soil of *Russula griseocarnosa*. *Fujian J. Agric. Sci.* **2022**, *37*, 529–537. (In Chinese) [CrossRef]
131. Zeng, G.-Y.; Zhao, Z.-H.; Gao, Z.-H.; Li, B.-C. Effect of woodland environmental regulation on yield of *Russula griseocarnosa*. *Acta Edulis Fungi* **2021**, *28*, 108–112. (In Chinese) [CrossRef]
132. Xu, M.; Fu, W.-Q.; Dai, C.-C.; Jia, Y. Ecological function of promoting microorganisms associated with ectomycorrhizal fungi. *Chin. J. Ecol.* **2018**, *37*, 1246–1256. (In Chinese)

133. Muhlmann, O.; Bacher, M.; Peintner, U. Polygonum viviparum mycobionts on an alpine primary successional glacier forefront. *Mycorrhiza* **2008**, *18*, 87–95. [CrossRef] [PubMed]
134. Muehlmann, O.; Peintner, U. Mycobionts of Salix herbacea on a glacier forefront in the Austrian Alps. *Mycorrhiza* **2008**, *18*, 171–180. [CrossRef]
135. Muehlmann, O.; Peintner, U. Ectomycorrhiza of Kobresia myosuroides at a primary successional glacier forefront. *Mycorrhiza* **2008**, *18*, 355–362. [CrossRef] [PubMed]
136. Cao, L.; Zhang, Q.; Miao, R.; Lin, J.; Feng, R.; Ni, Y.; Li, W.; Yang, D.; Zhao, X. Application of omics technology in the research on edible fungi. *Curr. Res. Food Sci.* **2023**, *6*, 100430. [CrossRef] [PubMed]
137. Adamčík, S.; Looney, B.; Caboň, M.; Jančovičová, S.; Adamčíková, K.; Avis, P.G.; Barajas, M.; Bhatt, R.P.; Corrales, A.; Das, K. The quest for a globally comprehensible *Russula* language. *Fungal Divers.* **2019**, *99*, 369–449. [CrossRef]
138. Buyck, B.; Zoller, S.; Hofstetter, V. Walking the thin line… ten years later: The dilemma of above-versus below-ground features to support phylogenies in the Russulaceae (Basidiomycota). *Fungal Divers.* **2018**, *89*, 267–292. [CrossRef]
139. Xiao, D.-L.; Chen, Y.-H.; Yang, J. Phylogenetic Diversity of *Russula griseocarnosa* from Fujian. *Fujian J. Agric. Sci.* **2013**, *28*, 902–905. (In Chinese) [CrossRef]
140. Feng, Y.L.; Fang, Y.; Yu, J.F.; Ma, M.; Lui, C.L.; Yang, Z.F.; Zhang, J.B.; Guo, X. Genetic Diversity and Population Genetic Differentiation of *Russula Griseocarnosa*. *Acta Edulis Fungi* **2020**, *27*, 7–15. (In Chinese) [CrossRef]
141. Chen, Y.H.; Chen, P.D.; Chen, L.Y.; Ma, L.Z. Phylogeographic Messages Encoded in the rDNA of the Commercial Mushroom Zhenghonggu® From Fujian, China. *Korean Soc. Mycol.* **2014**, *S5*, 45.
142. Fei, Y.; Liang, J. Data on the genome analysis of the wild edible mushroom, *Russula griseocarnosa*. *Data Brief* **2019**, *25*, 104295.
143. Yu, F.; Song, J.; Liang, J.; Wang, S.; Lu, J. Whole genome sequencing and genome annotation of the wild edible mushroom, *Russula griseocarnosa*. *Genomics* **2020**, *112*, 603–614. [CrossRef] [PubMed]
144. Ogawa, W.; Endo, N.; Takeda, Y.; Kodaira, M.; Fukuda, M.; Yamada, A. Efficient establishment of pure cultures of yellow chanterelle *Cantharellus anzutake* from ectomycorrhizal root tips, and morphological characteristics of ectomycorrhizae and cultured mycelium. *Mycoscience* **2019**, *60*, 45–53. [CrossRef]
145. Yang, Z.; Wang, H.-W.; Sha, T. Advances Research on Ectomycorrhizal Fungi. *Edible Fungi China* **2016**, *35*, 1–7. (In Chinese)
146. Jiang, Q.B.; Zhong, C.L.; Chen, Y.; Zhang, Y.; Chen, Z. Study on inoculating *Russula* fungi with *Pinus massoniana* seedling. *J. Cent. South Univ. For. Technol.* **2016**, *36*, 6–9+38. (In Chinese)
147. Jiang, Y.; Wang, W.; Xie, Q.; Liu, N.A.; Liu, L.; Wang, D.; Zhang, X.; Yang, C.; Chen, X.; Tang, D.; et al. Plants transfer lipids to sustain colonization by mutualistic mycorrhizal and parasitic fungi. *Science* **2017**, *356*, 1172–1175. [CrossRef]
148. Lareen, A.; Burton, F.; Schäfer, P. Plant root-microbe communication in shaping root microbiomes. *Plant Mol. Biol.* **2016**, *90*, 575–587. [CrossRef]
149. Salme, T.; Lawrence, B.; Julia, M.; Anthony, M.; Anne-Charlotte, A. Perspectives and Challenges of Microbial Application for Crop Improvement. *Front. Plant Sci.* **2017**, *8*, 49. [CrossRef] [PubMed]
150. Li, Q.; Li, X.-L.; Huang, W.-L.; Xiong, C.; Yang, Y.; Yang, Z.-R.; Zheng, L.-Y. Community structure and diversity of entophytic bacteria in *Tricholoma matsutake* in Sichuan Province, Southwest China. *Chin. J. Appl. Ecol.* **2014**, *25*, 3316–3322. (In Chinese) [CrossRef]
151. Miguel, A.M.D.; Águeda, B.; Sánchez, S.; Parladé, J. Ectomycorrhizal fungus diversity and community structure with natural and cultivated truffle hosts: Applying lessons learned to future truffle culture. *Mycorrhiza* **2014**, *24*, S5–S18. [CrossRef] [PubMed]
152. Lian, C.L. Research Progress on Cultivating Russula in Harvested Forest Lands. *Fujian For.* **2023**, 22–23. (In Chinese)
153. Ying, G.-H.; Lv, M.-L.; Li, L.-L.; Wang, Y.; Xue, Z.-W. Study on the Artificial Cultivation of *Suillus luteus*. *Edible Fungi China* **2009**, *28*, 14–15. (In Chinese)
154. Cao, Y.; Ji, K.; Liu, J.; Zhang, C.; He, M.; Wang, W. Effect of Casing on Fruiting of *Phlebopus portentosus* in Bottle Culture. *Acta Edulis Fungi* **2010**, *17*, 29–32+91. (In Chinese)
155. Cao, Y.; Ji, K.; Liu, J.; Zhang, C.; He, M.; Wang, W. Effect of Different Casing Soils on the Fruiting of *Phlebopus portentosus*. *Acta Edulis Fungi* **2011**, *18*, 25–27. (In Chinese)
156. Sanmee, R.; Lumyong, P.; Dell, B.; Lumyong, S. In vitro cultivation and fruit body formation of the black bolete, *Phlebopus portentosus*, a popular edible ectomycorrhizal fungus in Thailand. *Mycoscience* **2010**, *51*, 15–22. [CrossRef]
157. Buscot, F.; Kottke, I. The association of *Morchella rotunda* (Pers.) Boudier with roots of *Picea abies* (L.) Parst. *New Phytol.* **1990**, *116*, 425–430. [CrossRef]
158. Buscot, F. Synthesis of two types of association between *Morchella esculenta* and *Picea abies* under controlled culture conditions. *J. Plant Physiol.* **1993**, *141*, 12–17. [CrossRef]
159. Dahlstrom, J.L.; Smith, J.E.; Weber, N.S. Mycorrhiza-like interaction by Morchella with species of the Pinaceae in pure culture synthesis. *Mycorrhiza* **2000**, *9*, 279–285. [CrossRef]
160. Stark, C.; Babik, W.; Durka, W. Fungi from the roots of the common terrestrial orchid *Gymnadenia conopsea*. *Mycol. Res.* **2009**, *113*, 952–959. [CrossRef] [PubMed]
161. Li, Q.L.; Ding, C.; Fan, L. Trophic manner of morels analyzed by using stable carbon isotopes. *Mycosystema* **2013**, *32*, 213–223. (In Chinese)

162. Hobbie, E.A.; Weber, N.S.; Trappe, J.M. Mycorrhizal vs. saprotrophic status of fungi: The isotopic evidence. *New Phytol.* **2001**, *150*, 601–610. [CrossRef]
163. Hao, H.; Zhang, J.; Wang, H.; Wang, Q.; Chen, M.; Juan, J.; Feng, Z.; Chen, H. Comparative transcriptome analysis reveals potential fruiting body formation mechanisms in *Morchella importuna*. *AMB Express* **2019**, *9*, 103. [CrossRef]

MDPI AG
Grosspeteranlage 5
4052 Basel
Switzerland
Tel.: +41 61 683 77 34

*Agriculture* Editorial Office
E-mail: agriculture@mdpi.com
www.mdpi.com/journal/agriculture

www.ingramcontent.com/pod-product-compliance
Lightning Source LLC
Chambersburg PA
CBHW061557210926
51850CB00009B/2134